ADVANCES IN
HEAT TRANSFER

Volume 18

Advances in

HEAT TRANSFER

Edited by

James P. Hartnett

Energy Resources Center
University of Illinois
Chicago, Illinois

Thomas F. Irvine, Jr.

Department of Mechanical Engineering
State University of New York at Stony Brook
Stony Brook, New York

Volume 18

 1987

ACADEMIC PRESS, INC.
Harcourt Brace Jovanovich, Publishers
Orlando   San Diego   New York   Austin
Boston   London   Sydney   Tokyo   Toronto

ACADEMIC PRESS, INC.
Orlando, Florida 32887

United Kingdom Edition published by
ACADEMIC PRESS INC. (LONDON) LTD.
24–28 Oval Road, London NW1 7DX

LIBRARY OF CONGRESS CATALOG CARD NUMBER: 63-22329

ISBN 0–12–020018–X

PRINTED IN THE UNITED STATES OF AMERICA

87 88 89 90   9 8 7 6 5 4 3 2 1

CONTENTS

Natural Convection in Active and Passive Solar Thermal Systems

REN ANDERSON AND FRANK KREITH

Heat Transfer from Tubes in Crossflow

A. ŽUKAUSKAS

Thermal Conductivity of Structured Liquids

A. DUTTA AND R. A. MASHELKAR

Transition Boiling Heat Transfer

E. K. KALININ, I. I. BERLIN, AND V. V. KOSTIOUK

Thermodynamic and Transport Properties of Pure and Saline Water

DAVID J. KUKULKA, BENJAMIN GEBHART, AND JOSEPH C. MOLLENDORF

Natural Convection in Active and Passive Solar Thermal Systems

REN ANDERSON AND FRANK KREITH

Solar Energy Research Institute, Golden, Colorado 80401

I. Introduction

Recently, architects have successfully built many types of solar-heated buildings, and the solar industry has produced many thousands of reliable and efficient flat-plate collectors (FPCs). In addition, several new types of collectors have been introduced such as the compound parabolic collector (CPC) and the single-axis tracking parabolic trough collector (TPT). Furthermore, central solar receiver power plants with external as well as cavity-type receivers have been placed into operation and their performance has been monitored. In all of these solar thermal systems, natural convection heat transfer occurs. As a result, a large amount of natural convection heat transfer research has been motivated by solar-related applications. The objective of this monograph is to provide a review of those aspects of natural convection research that are applicable to solar system design and analysis. Solar ponds are not included because their convection phenomena are double diffusive and require a different kind of analysis. Double diffusive phenomena have recently been reviewed by Chen and Johnson [1] and Viskanta, Bergman, and Incropera [2].

A. COLLECTOR TYPES AND OPERATING TEMPERATURES

Solar collectors are generally categorized as "active" or "passive." Active collectors depend upon the use of external pumps or fans in order to

1

function properly. Passive collectors depend entirely upon natural convection to transport energy from the point of collection to the point of use.

There are three basic types of thermal collectors [3]:

1. Nonconcentrating and stationary (solar buildings, active and passive flat-plate collectors, and solar ponds) for low temperatures.

2. Slightly concentrating, with or without periodic adjustments (CPCs and V-troughs) for intermediate temperatures.

3. Concentrating, with either one- or two-axis tracking for intermediate and high temperatures.

Figure 1 shows schematic diagrams of the most common types of active solar-thermal collectors. Figure 2 shows the approximate operating tem-

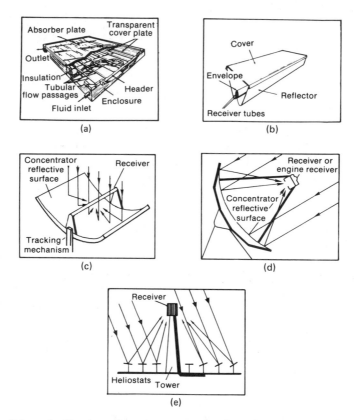

FIG. 1. Schematic diagrams of common types of active solar thermal collectors: (a) flat plate-statinary (b) compound parabolic collector (CPC), (c) single-axis-tracking parabolic trough, (d) dual-axis-tracking, paraboloid dish, (e) central receiver system with dual-axis-tracking heliostats.

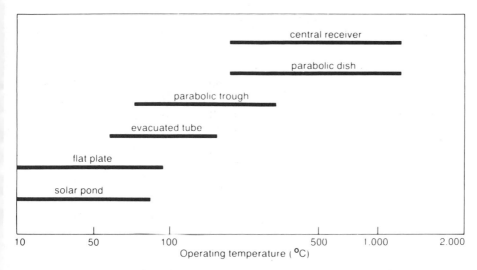

Fig. 2. Approximate operating temperatures of active solar collectors.

perature ranges for each type of active collector [4]. Flat-plate collectors are suitable for supplying hot water or hot air at temperatures up to 80°C with relatively good efficiency. They require no moving parts, have good durability, and can collect both direct and diffuse radiation. The key elements for a flat-plate collector are a frame, one or two transparent covers, a receiver or "absorber" plate with flow channels connected to inlet and outlet headers through which a working fluid passes, and some back-side insulation. The cover, usually of glass or a plastic, is transparent to solar radiation but opaque to infrared radiation from the receiver plate. Thus, the receiver plate absorbs solar radiation, heats up, and transfers heat to the working fluid, usually water or air. One of the most cost-effective applications of flat-plate collectors is domestic hot water heating.

To heat a fluid to temperatures above 80°C with good efficiency, solar energy must be concentrated on the receiver. The concentration ratio C is defined as [5]

$$C = A_a/A_r \tag{1}$$

where A_a is the aperture area and A_r is the receiver area.

Concentration reduces the size and surface area of the solar receiver and, therefore, reduces the heat losses, which are proportional to the surface area. The reduction in heat losses contributes to higher temperature capability. Concentration can be achieved by refraction or reflection, but reflection is used in most solar applications. Compound parabolic concentrators (CPC) can concentrate without tracking and utilize the beam as well as the diffuse radiation within the acceptance angle of the collector. They can achieve a concentration ratio of about two without adjustment and up to five with periodic tilt adjustment [5].

Tracking increases the complexity of the collector system and limits collection to the beam part of solar radiation. Line-focus collectors that track the sun in one direction can achieve concentration ratios of the order of 50 and deliver temperatures up to about 350°C. To heat a fluid to temperatures above 350°C with good efficiency requires a concentration ratio of 200 or more. Such a concentration ratio can only be achieved by means of dual-axis (azimuth and altitude) tracking of the sun by point-focus receiver systems. Basically two designs are available for high solar concentrations:

1. Central receiver systems (CRS) in which radiation is reflected from tracking mirrors (heliostats) onto a stationary receiver that can have the configuration of a cylinder with vertical flow tubes (external type) or a cavity lined with flow tubes (cavity type) as shown in Fig. 3.

2. Dual-axis-tracking paraboloid dish systems (PDS) with point-focus receivers in which the reflector as well as the receiver move to track the sun;

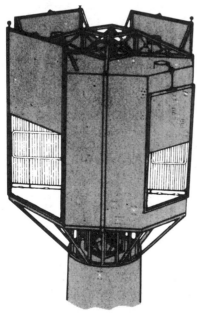

cavity receiver

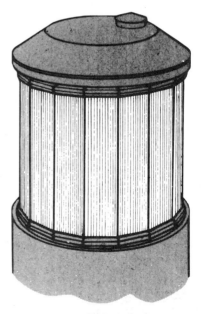

external receiver

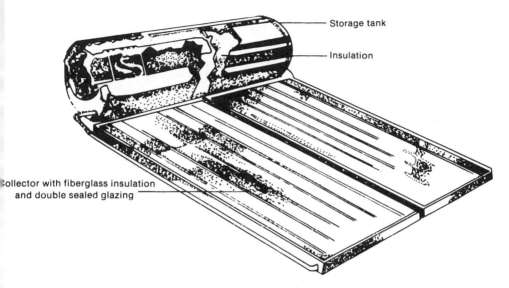

Storage tank

Insulation

Collector with fiberglass insulation
and double sealed glazing

FIG. 4. Schematic diagram of a flat-plate thermosiphon collector.

Collectors for passive solar thermal applications range from small flat-plate thermosiphon units (Fig. 4) to various types of solar building components. The most common types of solar building components are the Trombe wall, the atrium or sunspace, and direct gain windows. Solar buildings can also incorporate vents or windows for ventilation cooling. In a solar building, natural convection is the major heat distribution mechanism between building zones.

From an examination of the diagrams of the main types of solar buildings in Fig. 5, it is apparent that their geometries and thermal boundary conditions can vary widely. In addition, buildings often are complicated by the presence of furniture, draperies, and nonrectangular wall intersections.

The enclosure aspect ratio (room height/room length) in building applications can be large or small. The Trombe wall is a large aspect ratio configuration with parallel flat plates in a vertical position. Other types of solar building configurations (sunspace, direct gain) have aspect ratios close to one.

The Rayleigh numbers in building applications are on the order of 1×10^{10}, indicating that the natural convection flow next to heated and cooled surfaces will have a boundary-layer structure. Depending upon the specific application, this natural convection flow may be laminar, partially turbulent, or fully turbulent. Because of the complexity of the problem, it is important to carefully integrate results from laboratory experiments with results from measurements in full-scale buildings. This integration process

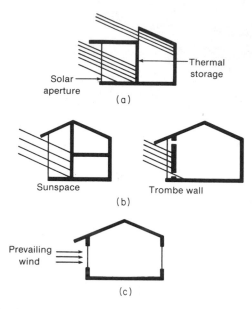

FIG. 5. Schematic diagrams of various types of passive heated and cooled solar buildings: (a) direct gain, (b) indirect gain, (c) ventilation cooling.

is summarized in Table I. Results gained from laboratory research must be reduced to reliable design information before they become useful tools to builders and architects.

B. EFFICIENCY AND OVERALL HEAT-LOSS COEFFICIENTS OF ACTIVE COLLECTORS

The useful heat output q_u of a thermal collector using a liquid or gaseous working fluid is proportional to the rate of increase of the temperature of the working fluid [3]:

$$q_u = \dot{m}c_p(T_{out} - T_{in}) \tag{2}$$

where $\dot{m}$ is the working fluid mass flow rate through the collector, c_p is the specific heat at constant pressure, T_{in} is the inlet temperature of the working fluid, and T_{out} is the outlet temperature.

It should be noted that the heat transferred to the working fluid in the collector does not equal the useful energy delivered by the solar system. Heat losses from connecting pipes and valves and transient effects at start-up in the morning and after insolation has been interrupted by clouds decrease system heat delivery [4].

The instantaneous thermal efficiency η for a thermal collector is given by the relation [3]

$$\eta = q_u/A_a I_c = \eta_o - U(T_H - T_{amb})/I_c C \tag{3}$$

where η_o is the optical efficiency, T_H the average receiver surface temperature in degrees Kelvin, T_{amb} the ambient temperature in degrees Kelvin, U the total heat-transfer (or thermal loss) coefficient between the collector surface and the surrounding in watts per square meter per degree Kelvin, based on the receiver area A_r and average surface temperature T_r, and I_c the solar irradiance (or insolation) on the collector aperture area in watts per square meter.

The insolation I_c comprises different elements of the solar radiation, depending on the type of solar collector [5].

1. For flat-plate collectors, I_c is the total hemispherical irradiation, $I_b + I_d$.

2. For tracking collectors with low concentration ($C < 10$), I_c is the radiation within the acceptance angle $[I_b + (I_d/C)]$.

3. For tracking collectors with high concentration ($C > 10$), I_c is the solar beam radiation I_b.

Here I_d is the diffuse insolation and I_b is the solar beam radiation.

TABLE I

INTEGRATION OF LABORATORY-SCALE AND FULL-SCALE RESEARCH RESULTS

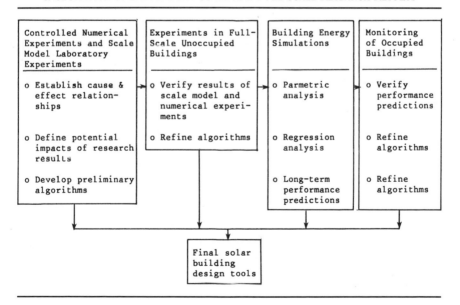

Controlled Numerical Experiments and Scale Model Laboratory Experiments	Experiments in Full-Scale Unoccupied Buildings	Building Energy Simulations	Monitoring of Occupied Buildings
o Establish cause & effect relationships	o Verify results of scale model and numerical experiments	o Parmetric analysis	o Verify performance predictions
o Define potential impacts of research results	o Refine algorithms	o Regression analysis	o Refine algorithms
o Develop preliminary algorithms		o Long-term performance predictions	o Refine algorithms

Final solar building design tools

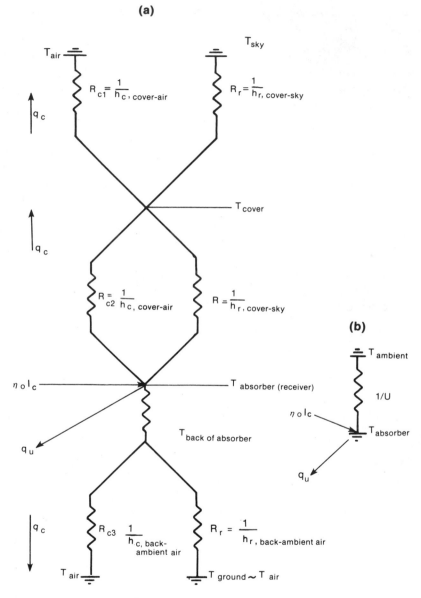

FIG. 6. Thermal network for the overall heat-loss coefficient U, from a receiver with a single transparent cover. (a) Detailed network for convection and radiation. (b) Simplified network.

The efficiency analysis, Eq. (3), shows that the three most important factors affecting the efficiency of solar thermal conversion for a given temperature output and insolation are the concentration ratio, the optical efficiency, and the overall heat-loss coefficient of the absorber. Of these three, the overall heat-loss coefficient is critically dependent upon natural convection phenomena.

The overall heat-loss coefficient is a simplified concept since the heat loss from a solar absorber occurs by complex interactions between radiation and convection. Figure 6a shows a thermal network for the heat loss from a collector with a single cover that is transparent to solar radiation, but opaque in the infrared. The overall heat-transfer loss coefficient U can be obtained from a detailed thermal analysis if the individual convection heat transfer coefficients and the radiation exchange coefficients are known. Such an analysis leads to the simplified circuit with U as the loss coefficient used in the efficiency equations (Fig. 6b). The radiation exchange coefficient can be calculated from available information with relatively good accuracy [3, 5, 6], but the free convection heat-transfer coefficient requires knowledge of the operating conditions and geometry. Figure 7 shows in simplified fashion the various heat-flow resistances for calculating heat-transfer coefficients for flat-plate collectors, compound parabolic collectors, and single-axis-tracking parabolic troughs. The schematic diagrams in Fig. 3 show the geometries applicable for a central receiver with an external

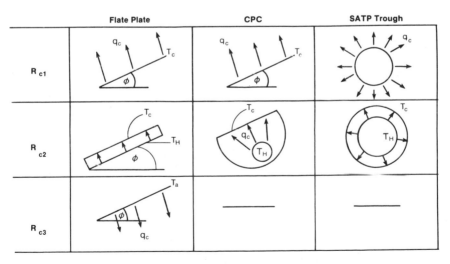

FIG. 7. Geometries and orientation for natural convection losses from flat plate collectors, compound parabolic collectors, and single-axis-tracking parabolic troughs; R_{c1}, R_{c2}, and R_{c3} as in Fig. 6.

or cavity design. In all of these cases, free convection plays an important role. In Section II we will present the available information about free convection heat transfer for the geometries relevant to all of these active solar systems.

C. THERMAL DESIGN CHARACTERISTICS OF SOLAR BUILDINGS

Buildings rank with transportation and industry as one of the three main users of energy in the U.S. economy. Building energy consumption can be reduced by designing buildings to use solar energy for a portion of their heating and cooling needs.

In heating applications, solar buildings can be categorized depending on whether the solar energy that enters through the windows is used directly in the building zone in which it is absorbed (direct gain) or is transported from the direct gain zone to another building zone (indirect gain) (Fig. 5).

In cooling applications, prevailing winds are used in combination with stack driven ventilation to move cool outside air through the building. These types of solar applications have come to be known as passive solar to distinguish them from the active solar applications that have been discussed in the previous section.

A passive system has been defined by Balcolmb [7] as "one in which the thermal energy flow is by natural means." Most passive designs use south-facing glass in the building as the solar collection element and structural mass in the building as the thermal storage element. A successful passive design requires integration of the solar collection, thermal storage, and heat distribution functions into the architecture of the building. Obviously, a thorough understanding of natural convection in enclosed spaces is absolutely necessary to successfully achieve this integration without sacrificing the comfort standard of conventional heating systems.

In practice, solar buildings can consist of any combination of direct gain, indirect gain and ventilation cooling components. A fundamental understanding of the heat-transfer aspects of each of these generic configurations is necessary if they are to be successfully integrated into an overall building design. In Section III we present available information about free convection relevant to the design of passive solar systems.

II. Natural Convection in Active System Configurations

As mentioned previously, natural convection is important for the performance of solar collectors because it directly affects the thermal losses between the absorber surface and its surroundings.

A. FLAT-PLATE COLLECTORS

The natural convection losses from flat-plate collectors have commonly been analyzed by modeling the collector as a large-aspect-ratio, two-dimensional air gap with isothermal hot and cold walls (Fig. 8). This air gap may exist between the absorber and the transparent cover as well as between two transparent covers above the absorber if a double-glazed collector design is used. In both cases free convection occurs jointly with radiation. Flat-plate collectors usually operate without concentration and are inclined to the horizontal by an angle ϕ. There have been a number of reviews in recent years, including those by Buchberg [8], Catton [9], and Ostrach [10, 11] that cover various aspects of convection in enclosures. The reviews that most directly deal with the relation of enclosure research to heat losses from solar collectors are those of Buchberg *et al.* [8] and Catton [9].

A large fraction of the energy losses from flat-plate solar collectors is due to a recirculating convective cell that forms in the cavity between the hot absorber surface and transparent cover plate. The magnitude of the heat leak produced by the buoyant circulation cell between the absorber and cover plates of the collector is proportional to the overall temperature difference between the absorber and cover. It also depends upon cavity geometry, thermal boundary conditions, and physical properties of the cavity fluid.

The structure of the natural convection flow in a flat-plate collector is a strong function of the tilt angle of the collector. As the collector tilt angle

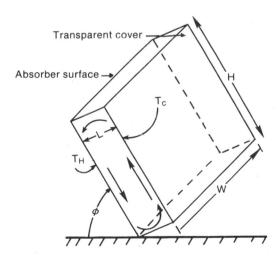

FIG. 8. Schematic diagram of gap in a flat-plate collector with a single transparent cover.

decreases from vertical ($\phi = 90°$) to horizontal ($\phi = 0°$) the flow changes from convection in a vertical slot with differentially heated side walls to Benard convection in a horizontal fluid layer heated from below and cooled from above. Hart [12] observed that for $0 \leq \phi \leq 80°$ thermal instabilities lead to the formation of longitudinal rolls with the axis of each roll oriented up the slope, and superimposed upon the primary flow. For near vertical orientations ($80 \leq \phi \leq 90$) secondary instabilities take the form of transverse rolls with axes oriented across the slope. The secondary flows cause an increase in convective heat transfer across the collector air gap.

Additional experimental studies of this phenomenon have been conducted by Ozoe et al. [13], Ruth et al. [14], Linthorst et al. [15], Inaba [16], and Goldstein and Wang [17]. Three-dimensional numerical calculations have been carried out by Ozoe et al. [18]. Schematic drawings visualizing natural convection in flat-plate collector geometries are shown in Fig. 9. Linthorst et al. [15] report that the characteristics of natural convective flow in an inclined rectangular cavity are a strong function of the aspect ratio of the cavity. Their experimentally derived curves for transition from stationary to nonstationary and two-dimensional to three-dimensional flow are shown in Fig. 10 as a function of aspect ratio and Rayleigh number Ra. For large values of the aspect ratio (AR), the flow quickly

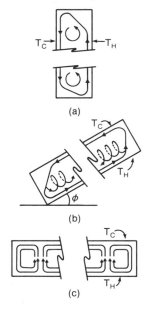

(a)

(b)

(c)

FIG. 9. Flow structure inside a flat-plate enclosure for (a) $\phi = 90°$ (transverse rolls), (b) $\phi = 30°$ (longitudinal rolls), (c) $\phi = 0$ (Benard convection). (Adapted from Linthorst et al. [15].)

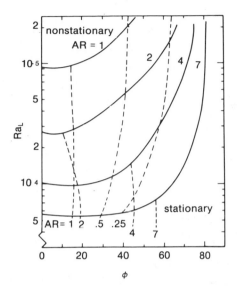

FIG. 10. Stability diagram showing transition from stationary to nonstationary (————) and from two-dimensional to three-dimensional flow (---------). (From Linthorst *et al.* [15])

becomes nonstationary and/or three dimensional for inclined orientations. For AR = 7 and $0° \leq \phi \leq 60°$, the critical Rayleigh number for transition to nonstationary three-dimensional flow is about 6,000. This means that for conditions representative of a flat-plate solar collector (AR $\geqslant$ 1 and $Ra_L \sim 2 \times 10^5$) the flow is always strongly three dimensional and nonstationary.

ElSherbiny *et al.* [19] conducted a comprehensive experimental investigation of the heat transfer in air-filled, high-aspect-ratio enclosures with isothermal walls, covering the ranges

$$10^2 \leq Ra_L \leq 2 \times 10^7, \quad 5 \leq H/L = AR \leq 110, \quad \text{and} \quad 0° \leq \phi \leq 90°$$

where $Ra_L = g\beta L^3 \, \Delta T/\nu\alpha$ and ΔT is the temperature difference between the lower and upper surface of the air gap. A comparison of the ranges of these parameters with previous experimental studies is shown in Fig. 11 [19a–19d].

ElSherbiny *et al.* found that the transition from the conduction to convection regime in vertical enclosures is a strong function of the aspect ratio when AR < 40. The following heat-transfer correlations are recommended:

1. For vertical layers ($\phi = 90°$):

$$\overline{Nu_L} = [\overline{Nu_1}, \overline{Nu_2}, \overline{Nu_3}]_{max} \tag{4}$$

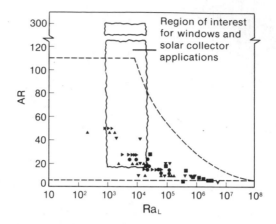

FIG. 11. Summary of experimental investigations of natural convection in high-aspect-ratio enclosures. ▶, Degraaf and Van der Held [19a]; ▲, Eckert and Carlson [19b]; ■, Schinkel and Hoogendoorn [19c]; ●, Randall, Mitchell and El-Wakil [19d], ----------, ElSherbiny, Raithby, and Hollands [19] (adapted from ElSherbiny, Raithby, and Hollands [19].)

where

$$\overline{Nu_1} = 0.0605 \ Ra_L^{1/3} \tag{5}$$

$$\overline{Nu_2} = [1 + \{0.104 \ Ra_L^{0.293}/(1 + (6310/Ra_L)^{1.36})\}^3]^{1/3} \tag{6}$$

$$\overline{Nu_3} = 0.242(Ra_L/AR)^{0.272} \tag{7}$$

2. For inclined layers ($\phi = 60°$):

$$\overline{Nu_L} = [\overline{Nu_1}, \overline{Nu_2}]_{max} \tag{8}$$

where

$$\overline{Nu_1} = [1 + \{0.0936 \ Ra_L^{0.314}/(1 + G)\}^7]^{1/7} \tag{9}$$

$$G = 0.5/[1 + (Ra_L/3160)^{20.6}]^{0.1} \tag{10}$$

$$\overline{Nu_2} = (0.104 + 0.175/AR)Ra_L^{0.283} \tag{11}$$

The notation in Eqs. (4) and (8) indicates that the maximum value of the average Nusselt number calculated from the correlations for Nu_i (where $i = 1$ to 3) should be used.

For tilt angles between 60° and 90° ElSherbiny *et al.* [19] suggest a linear interpolation between the limiting correlations given above:

$$\overline{Nu_\phi} = [(90° - \phi) \ Nu_{60} + (\phi - 60°) \ Nu_{90}]/30° \tag{12}$$

Extrapolating these equations beyond the experimental range of variables is not recommended.

For tilt angles between 0° and 75°, Hollands *et al.* [20] recommend the following correlation for the average Nusselt number:

$$\overline{\mathrm{Nu}} = 1 + 1.44\left(1 - \frac{1708}{\mathrm{Ra}\cos\phi}\right)^{\bullet}\left[1 - \frac{1708\,(\sin 1.8\phi)^{1.6}}{\mathrm{Ra}\cos\phi}\right]$$
$$+ \left[\left(\frac{\mathrm{Ra}\cos\phi}{5830}\right)^{1/3} - 1\right]^{\bullet} \tag{13}$$

where L is the distance between the plates at temperatures T_H and T_C, respectively, and the Rayleigh number Ra is given by

$$\mathrm{Ra} = \frac{2g(T_H - T_C)L^3}{v^2(T_H + T_C)}$$

A dot to the right of a bracket denotes that $[\chi]^{\bullet} = (|\chi| + \chi)/2$. Thus when Ra $< (1708)/(\cos\phi)$ the Nusselt number in Eq. (13) is exactly equal to unity. The condition Nu $= 1$ implies that the heat transfer across the air cavity is by pure conduction.

The natural convection circulation in the cavity between the cover plate and the absorber can be suppressed by making the aspect ratio of the collector very large or very small. The former approach is used in the design of double-pane windows. The latter approach is used in the design of various types of internal partitions (honeycombs, horizontal slats, vertical slats) that are placed in the collector cavity (Fig. 12). A summary of the aspect ratio dependence of natural convection in vertical rectangular enclosures as reported by Bejan [94] is shown in Fig. 13. The Nusselt number reaches a peak for values of the aspect ratio between 0.1 and 1.0, and drops rapidly in value for very large and very small values of the aspect ratio.

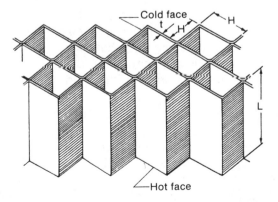

Fig. 12. Schematic diagram of honeycomb structure used to suppress convection.

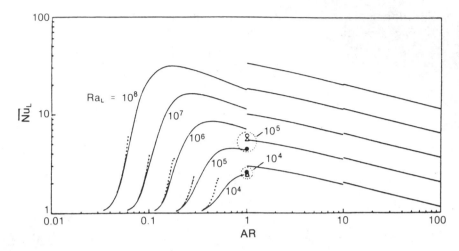

FIG. 13. Aspect ratio dependence of heat-transfer coefficient (from Bejan [92]).

A review of the theory and application of honeycombs for suppressing natural convection in flat-plate solar collectors is given by Buchberg *et al.* [21, 22] and by Hollands [23]. For an inclined, square honeycomb, the Nusselt number depends on the Rayleigh number, the inclination, and the AR of the honeycomb. For the range $0 < \text{Ra} < [6000 \, \text{AR}^4]$, $30° < \phi < 90°$, and $1/\text{AR} = 3$, 4, or 5, the Nusselt number for air is given by Cane *et al.* [24] in the form

$$\overline{\text{Nu}_L} = \frac{\overline{h_c} L}{k} = 1 + 0.89 \cos(\phi - 60°)$$
$$\times \left(\frac{\text{Ra}_L \, \text{AR}^4}{2420} \right)^{(2.88 - 1.64 \sin \phi)} \tag{14}$$

For minimum heat loss the honeycomb should be designed to give a Nusselt number of 1.2, according to Hollands *et al.* [25]. For air at atmospheric pressure and moderate temperatures, $370 \, \text{K} > T_m > 280 \, \text{K}$, the aspect ratio for minimum heat loss can be found from

$$\frac{1}{\text{AR}} = C(\phi) \left(1 + \frac{200}{T_m} \right)^{1/2} \left(\frac{100}{T_m} \right) (T_H - T_C)^{1/4} L^{3/4} \tag{15}$$

where L is the thickness of the honeycomb in centimeters, T_m is the average of the coverplate and absorber temperature in degrees Kelvin, and the function $C(\phi)$ is plotted in Fig. 14.

Hollands *et al.* [26] and Hoogendoorn [27] independently calculated total heat transfer in honeycomb structures including radiation effects.

They found that conduction heating of the honeycomb increases the radiative losses from the collector and reduces the benefits associated with the convection suppression that is provided by the honeycomb if the honeycomb is allowed to come into direct contact with the hot absorber plate. Hollands and Iynkaran [28] recommend that a 10-mm air gap be left between the honeycomb and the absorber plate to reduce conductive heating of the honeycomb structure.

Two-dimensional slats are an alternative method of convection suppression. These slats can be oriented horizontally (forming transverse slots) or vertically (forming longitudinal slots). Transverse slats have been investigated by Arnold *et al.* [29, 30], Meyer *et al.* [31], and Smart *et al.* [32]. These studies demonstrate that for transverse rectangular cell aspect ratios less than 0.1, convection is suppressed for Ra_L up to 4×10^5 for a tilt angle of 60°. Meyer *et al.* [31] found that heat transfer increased, compared to an enclosure without transverse slats, for transverse rectangular cell aspect ratios in the range $0.5 \leq Ar \leq 4$. Symons and Peck [33] and Symons [34] have investigated the reduction in convective heat transfer produced by longitudinal slats. Symons and Peck [33] compared the heat transfer across a transverse slot with AR = 0.17 to the heat transfer across a longitudinal slot with the same cross-sectional dimensions. They found that the heat transfer across the longitudinal slot was less than the heat transfer across the transverse slot for tilt angles in the range $24° \leq \phi \leq 75°$. For tilt angles in the range $0° \leq \phi \leq 24°$, the heat transfer was the same, regardless of slot orientation. For tilt angles in the range $75° \leq \phi \leq 90°$, the heat transfer across the transverse slot was less than the heat transfer across the longitudinal slot. The effectiveness of longitudinal slots appears to be related to their ability to damp out the longitudinal convection cells that form when the collector is tilted from the vertical.

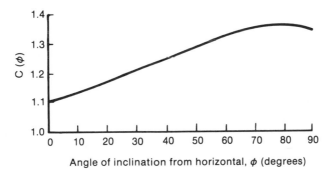

FIG. 14. Plot of $C(\phi)$ versus ϕ for use in Eq. 15 (from Hollands *et al.* [25]).

Some aspects of heat transfer in flat-plate solar collectors require more sophisticated models than those based on enclosures with isothermal walls. Balvanz and Kuehn [35] have shown that the effect of finite wall conductivity can be significant, particularly for a constant-flux boundary condition where wall conduction can reduce the wall-temperature gradient and produce a corresponding decrease in heat transfer from the wall. Figure 15a shows the wall temperature profile measured by Anderson and Bohn [36] in an enclosure with a finite conductivity wall. Figure 15b shows the corresponding decrease in heat transfer predicted by Balvanz and Kuehn [35]. MacGregor and Emery [37] calculated natural convection flow in a vertical enclosure while varying thermal boundary conditions at the heated wall. They found 30% higher convective losses for a constant flux condition than for an isothermal condition. In their work the average temperature difference between the hot and cold walls was used in the definition of the Nusselt number for the constant flux surface. Schinkel and Hoogendoorn [38] repeated the calculations of MacGregor and Emery for $Ra_L = 5.8 \times 10^4$ and found the Nusselt number for a constant flux surface to be more than 20% more than for an isothermal surface. This prediction was found to agree with experiments in air. Schinkel and Hoogendoorn also

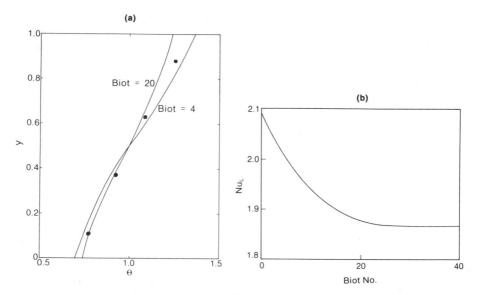

FIG. 15. (a) Experimental measurements of wall surface temperature by Anderson and Bohn, ●, Bi = 10 [36] for finite wall conductance compared to theoretical predictions by Balvanz and Kuehn [35]; Bi = $K_w t/K_f H$. $\theta = (T - T_C)/(T'_H - T_C)$. (b) Effect of finite Bi on overall Nusselt number (after Balvanz and Kuehn [35]).

did experimental comparisons at $\phi = 60°$, $40°$, and $20°$ and found increases of 14%, 11%, and 9%, respectively.

In an actual flat-plate collector, the absorber will be at some condition in between an isothermal and a constant-flux surface. This overlapping state is caused by the coupling between the free-convective flow in the cavity separating the absorber and cover and the forced-convective flow of the heat transfer fluid through tubes in the absorber plate. Chao et al. [39] numerically analyzed the effect of spatial sawtooth variations in the temperature of an inclined enclosure with AR = 2. They found that surface-temperature variations produced a stronger circulation and a higher overall Nusselt number than did a uniform temperature. The sawtooth variation in surface temperature studied by Chao et al. [39] is particularly relevant to liquid collectors, where heat-exchanger tubes are connected at regular intervals along the surface of the absorber plate.

The thermal boundary condition at the cover plate results from the interaction of internal and external convection and thus is not known a priori as in the case of the isothermal wall model. A number of researchers have looked at cases where a free-convection boundary layer exists on the outside surface as well as on the inside surface of the cover plate. Such a condition would exist when the external wind velocity is zero. Lock and Ko [40] demonstrated that the resulting interaction produces a nearly constant-flux surface with a linear temperature profile except for regions near the top and bottom of the plate. Anderson and Bejan [41, 42] extended the analysis of Lock and Ko to wider range of plate thermal resistance and included the effects of thermal stratification in the fluid on either side of the plate. Viskanta and Lankford [43] conducted experiments in an air-filled enclosure that was separated into two zones by a conducting partition. Sparrow and Prakash [44] conducted a numerical investigation of an enclosure with $H/L = 1$ which was coupled via a conductive wall to an external natural convection flow.

B. Line Focusing Collectors

Line-focus collectors have recently received a great deal of attention for industrial heat applications at intermediate temperatures [44, 45]. There are two commercial types of line focusing collectors: the compound parabolic trough and the tracking parabolic trough. Both are usually deployed in a horizontal position. The parabolic trough collector must track the sun continuously. The optical design of the compound parabolic type requires no tracking at concentration ratios below 2 and only biyearly tilt adjustment at concentration ratios below 5.

1. *Compound Parabolic Collector*

Compound parabolic collectors have been studied by Winston [46], Rabl [47], and O'Gallager *et al.* [48]. In a CPC, heat loss is by natural convection from the cylindrical receiver surface in the space formed by the aperture cover and the reflector. Typical CPC configurations are shown in Fig. 16a. Generalizing the shape of CPC cavities and receivers is difficult because the detailed geometry of these designs depends upon the manner that the CPC shape is designed.

There are two approaches to the calculation of natural convection losses from CPC collectors. The losses can be approximated by replacing the collector with an equivalent eccentric cylindrical annulus, or the losses can be calculated directly. Examination of the shapes presented by Rabl [47], O'Gallager *et al.* [48], and Ortabasi and Fehlner [49] shows that with concentration ratios between 1.6 and 3.0 the heat loss and flow can be approximated by natural convection in an eccentric cylindrical annulus.

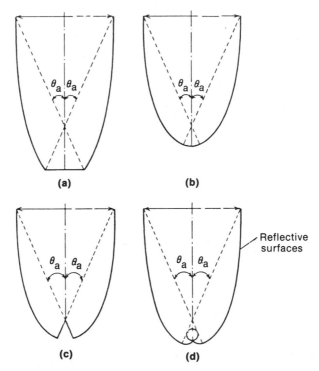

FIG. 16. Schematic diagrams of four typical compound parabolic collectors. Four possible absorber configurations are shown. θ is the acceptance angle of the collector.

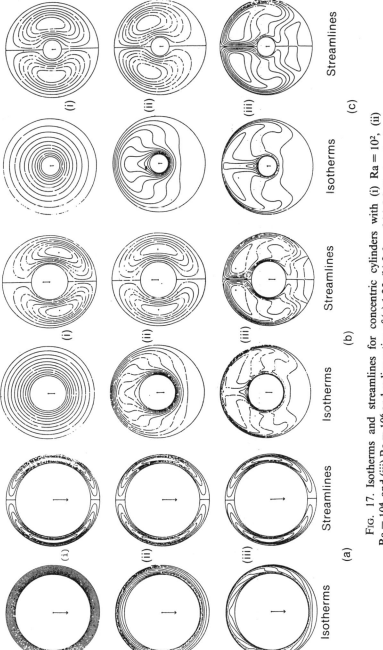

FIG. 17. Isotherms and streamlines for concentric cylinders with (i) Ra = 10^2, (ii) Ra = 10^4, and (iii) Ra = 10^6 and radius ratios of (a) 1.25, (b) 2.6, and (c) 5.0. Ra is calculated based upon the average gap between cylinders. (From Lee *et al.* [50]).

Correlations for the evaluation of the heat loss in an eccentric cylindrical annulus geometry are given by Lee *et al.* [50]. These authors determined, both experimentally and numerically, the characteristics of natural convection heat transfer in concentric as well as eccentric cylindrical annuli. The results cover a range of inner-to-outer diameter ratios from 1.25 to 5.0 and eccentricity ratios up to ±0.9 for air (Pr = 0.7). Figure 17 shows the isotherms and streamlines for concentric cylinders for D_i/D_o ratios of 1.25, 2.6, and 5.0 at Ra of 10^2, 10^4, and 10^6. Figure 18 shows the isotherms and streamlines at eccentricity ratios e of -0.9, 0.67, 0.9 for Ra = 5 × 10^5 and D_i/D_o of 2.6. Figure 19 shows the results for the average convection heat-transfer coefficient between the inner and outer surface versus Ra for eccentricity ratios of 0, -0.67, 0.67, and 0.33. Lee *et al.* based their definition of Rayleigh number on the average size of the gap between cylinders.

A CPC collector configuration, representative of commercial designs, has been studied by Hsieh [51] and Hsieh and Mei [52]. His configuration (see Fig. 20) can be approximated by an annular space between the receiver envelope and the surface formed by the reflector and the transparent aperture cover at the top. When the outer surface of the structure is treated

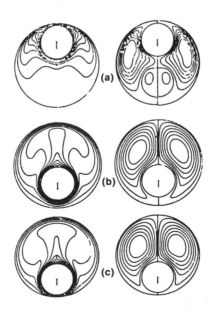

Fig. 18. Isotherms and streamlines at eccentricity ratios of (a) -0.9, (b) $\frac{2}{3}$, (c) 0.9, for Ra = 5 × 10^5 and radius ratio of 2.6. The Rayleigh number is calculated from the average gap between cylinders. (From Lee *et al.* [50]).

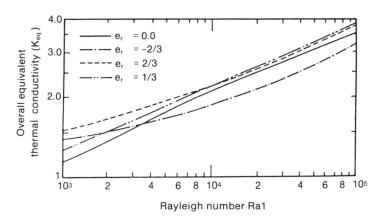

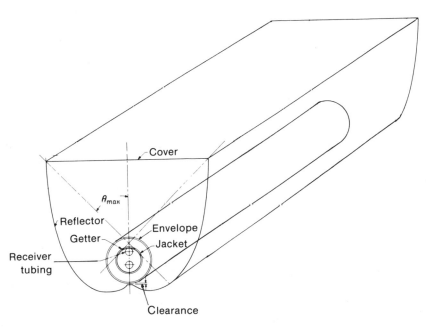

FIG. 19. Overall heat-transfer coefficient versus Rayleigh number for radius ratio of 5. Rayleigh numbers are calculated from the average gap between cylinders. Here, $K_{eq} = q_c \ln(S)/2\pi k \, \Delta T$, where

$$S = \frac{\sqrt{(r_o + r_i)^2 - e^2} + \sqrt{(r_o - r_i)^2 - e^2}}{\sqrt{(r_o + r_i)^2 - e^2} - \sqrt{(r_o - r_i)^2 - e^2}}$$

(From Lee *et al.* [50].)

FIG. 20. Commercial design of a CPC according to Hsieh [51].

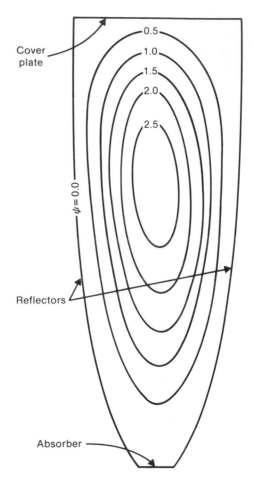

FIG. 21. Convective pattern in a compound parabolic collector with a coverplate and planar absorber (from Abdel-Khalik *et al.* [53]).

as an equivalent cylinder, the heat-loss mechanism from the receiver can be approximated as natural convection through an annulus with a diameter ratio of about 3 and an eccentricity e of about 0.75.

Abdel-Khalik *et al.* [53] carried out a finite-element analysis of a CPC collector with its axis oriented horizontally and developed heat-loss equations for concentration ratios of 2 to 10 and Rayleigh numbers, based upon cavity thickness, of 2×10^3 to 1.3×10^6. The convection pattern in this Rayleigh number range was found to be unicellular (Fig. 21). The reflector walls were assumed to be adiabatic while the absorber and cover plate were isothermal.

Meyer *et al.* [54] determined the losses from a V-trough collector as a function of Rayleigh number and tilt angle. Their collector had straight rather than parabolic sides, but it approximated closely the behavior of a CPC collector. They found that the collector tilt affects heat transfer similarly to that for rectangular enclosures.

Convective losses from a CPC collector can be reduced when the absorber is cylindrical by surrounding the absorber with a concentric glass tube. Collares Pereira *et al.* [55] and Woo [56] have conducted experiments with glass tubes surrounding the CPC collectors. The available experimental and analytic results, for the overall loss coefficient in CPC collectors with and without a transparent cylindrical cover over the receiver, are summarized in Table II [56a–56c]. The overall loss coefficients found in experimental tests of various CPC designs are in reasonably good agreement with each other and they indicate that surrounding the absorber with a transparent cylinder reduces losses by about a factor of two.

TABLE II

SUMMARY OF HEAT LOSSES IN NONEVACUATED CPC COLLECTORS[a]

Reference	Heat-loss coefficient Normalized to collector aperture (W/m² °C)	Geometric concentration	Heat-loss coefficient normalized to absorber (W/m² °C)	Remarks
I. No envelope				
Rabl *et al.* [56a]	1.85 ± 0.1	5.2	9.6 ± 0.5	Tube absorber
	2.7 ± 0.2	3.0	8.1 ± 0.6	Fin absorber
Meyer *et al.* [54]	1.8–2.5	2–3	3.6–7.5	Flat absorber, measured experimentally
II. Tube with Glass Envelope				
Rabl *et al.* [56a]	2.2	1.5	3.3	Calculated
Patton [56b]	1.82	1.6	2.9	Measured experimentally
Woo [56]	2.8 ± 0.1	1.65	4.6 ± 0.2	Measured experimentally
Collares Pereira *et al.* [55]	2.64	1.5	4.0	Measured experimentally
Prapas *et al.* [56c]	—	—	4.4	Calculated for absorber only

[a] Well-designed CPCs with selective absorber coating, and absorber thermally isolated from mirrors so that losses are dominated by convection.

2. *Single-Axis-Tracking Parabolic Trough Collectors*

In a single-axis-tracking parabolic trough concentrator, which operates at concentration ratios from 25 to 50, the incident solar radiation is reflected onto a planar or a cylindrical absorber, as shown in Fig. 22. This absorber generally has a transparent cover to reduce heat losses from its surfaces. Unless the space surrounding the absorber is evacuated, the convection heat loss is by natural convection in the space between flat plates or between concentric tubes in conjunction with radiation. When the absorber has a selective surface, natural convection is the dominant mechanism. Rabl, Bendt, and Gaul [57] have shown that for conventional, unevacuated line-focusing parabolic trough collectors with tubular receivers, the ratio of outer to inner diameter (D_o/D_i) varies typically from 1.5 to 2.6.

Natural convection flow and heat transfer under geometric conditions similar to those encountered in the tubular receivers of single-axis-tracking line-focusing collectors have been studied by several investigators, most recently by Kuehn and Goldstein [58–61] and Lee *et al.* [50]. Equations for heat transfer by natural convection between isothermal long concentric horizontal circular cylinders have been developed by Kuehn and Goldstein and compared with experimental data obtained by various investigators. They recommend for the average Nusselt number between isothermal concentric cylinders with inner and outer surface temperatures T_i and T_o,

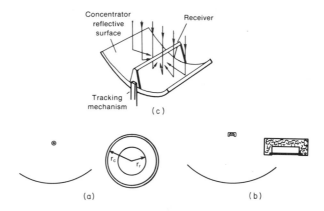

FIG. 22. (a) Schematic front view of parabolic trough concentrating collector with (a) cylindrical and (b) planar receiver. (c) Three dimensional schematic view of a paraboloid trough collector.

respectively, the relation

$$\overline{\mathrm{Nu}}_{D_i} = \frac{2}{\ln\{(1 + 2/E^{1/15})/(1 - 2/F^{1/15})\}} \tag{16}$$

where

$$E = [0.518 \, \mathrm{Ra}_{D_i}^{1/4} \, (1 + (0.559/\mathrm{Pr})^{3/5})^{-5/12}]^{15} + (0.1 \, \mathrm{Ra}_{D_i}^{1/3})^{15} \tag{17}$$

$$F = \left(\left[\left(\frac{2}{1 - e^{-0.25}}\right)^{5/3} + G^{5/3}(0.587 \, \mathrm{Ra}_{D_o}^{1/4})^{5/3}\right]^{3/5}\right)^{15}$$

$$+ (0.1 \, \mathrm{Ra}_{D_o}^{1/3})^{15} \tag{18}$$

$$G = \left[\left(1 + \frac{0.6}{\mathrm{Pr}^{0.7}}\right)^{-5} + (0.4 + 2.6 \, \mathrm{Pr}^{0.7})^{-5}\right]^{-1/5} \tag{19}$$

and

$$\overline{\mathrm{Nu}}_{D_i} = \bar{h}_c D_i / k \tag{20}$$

$$\bar{h}_c = \frac{q_c}{\pi D_i L (T_i - T_o)} \tag{21}$$

$$\mathrm{Ra}_{D_i} = g\beta D_i^3 (T_i - T_o)/\nu\alpha, \qquad \mathrm{Ra}_{D_o} = g\beta D_o^3 (T_i - T_o)/\nu\alpha \tag{22}$$

Equation (16) correlates available experimental data over the ranges of Rayleigh numbers from 10^2 to 10^{10} and Prandtl number between 0.01 and 1,000 as shown in Fig. 23.

Evaluation of T_i in the above relations requires iteration. To avoid this, Kuehn and Goldstein [61] suggest the simple relation, valid for $\mathrm{Pr} = 0.71$ and laminar flow,

$$\mathrm{Nu}_{D_i,\mathrm{conv}} = \frac{2}{(1 + 2/0.4 \, \mathrm{Ra}_{D_i}^{1/4})/(1 - 2/0.587 \, \mathrm{Ra}_{D_o}^{1/4})} \tag{23}$$

The above correlation is considered satisfactory for horizontal trough collectors with cylindrical receivers, although it does not take into account the nonuniform temperature of the receiver surface and is therefore subject to an unknown error under practical operation conditions in the sun.

C. POINT-FOCUSING SYSTEMS

The earliest reported application of a point-focusing system was Archimedes' use of reflecting mirrors in 212 B.C. to set the ships of the invading Romans on fire at Syracuse. Many inventors, alchemists, and scientists have proposed point focusing of solar radiation as a means of achieving

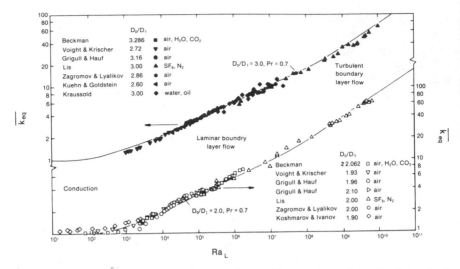

FIG. 23. Comparison of correlating equations with experimental results for natural convection between horizontal concentric cylinders; $\overline{K}_{eq} = Nu_{Di} \cosh^{-1}[D_i^2 + D_o^2 - 4\varepsilon^2/2D_i D_o]/2$, where ε is the normalized distance that the inner cylinder is displaced from concentricity. The Rayleigh number is based on the average gap width. (From Kuhn and Goldstein [60]).

extremely high temperatures for many applications [62]. But use of point focusing for large-scale installations has only taken place within the past decade. Since 1980 several central receiver systems in sizes up to 10 mW electric capacity have been built for electric power production and industrial heat and dual-axis-tracking paraboloid concentrators have been employed for heat and electric power generation. The distributed total energy system at Shenandoah, Georgia is shown in Fig. 24. The collector field for this paraboloid dish system (PDS) consists of 114 parabolic dishes; each of them is 7 m in rim diameter with a cavity receiver located at the focus as shown schematically in Fig. 1b. Paraboloid dishes are generally less than 10 or 20 m in diameter because larger dishes are difficult to build and are subject to excessive wind loading. Hence, receivers for dish systems are relatively small. Natural convection for this type of receiver will be discussed in Section II,C,3.

Figure 1e is a simple diagram of a central receiver solar power system. Dual-axis-tracking heliostats concentrate direct solar radiation onto a tower-mounted central receiver, which can be a cavity or external design. There the radiant energy is used to heat a working fluid to high temperatures for piping to the bottom of the tower and subsequent use as a high-temperature heat source for industrial processes, operating a turbine, or storing for future use (63).

Central receivers generally are large and have high surface temperatures and complex geometries. Heat transfer and fluid flow occur in regimes for which experimental data are scant and predictive methods are uncertain. The lack of steady flow in atmospheric boundary layers further complicates analysis in the case of the external receiver. Abrams [64] developed a map of available natural convection data in the Grashof number versus Reynolds number regimes as well as the operating regime of typical solar central receivers (Fig. 25). Progress has been made in providing information upon which the convective losses of central receivers can be based, but uncertainties remain in understanding and predicting convection losses from central receivers.

To predict the efficiency of any receiver in a point-focusing system, it is necessary to calculate the amount of incident solar radiation that is intercepted, reflected, and emitted by the receiver and the losses by convection and conduction. In many cases, natural convection is a dominant factor, and in this section we will summarize the available information for the three industrially most important receiver types: external and cavity receivers for CRS applications and cavity receivers for PDS applications.

FIG. 24. Two-axis-tracking parabolic dish collectors in Shenandoah Field (courtesy of Sandia National Labortories).

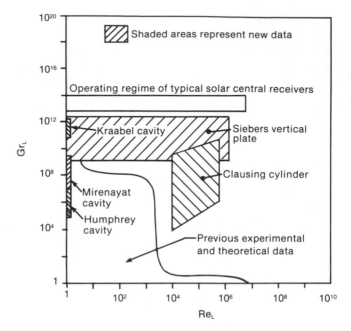

FIG. 25. Range of scaling variables for which convection data are available for solar central receivers (from Abrams [64]).

1. *External Receivers—Central Receiver Systems*

An external receiver essentially is a cylinder formed of vertical flow tubes. The solar radiation reflected from the heliostat field falls on tubes arranged outside the cylinder. The working fluid passes through the radiatively heated tubes. The receiver for the 10-MW Solar 1 prototype solar power plant in Barstow, California is a once-through boiler located atop a 76-m-tall tower that is 7 m in diameter and 12.5 m high (Fig. 26). The average outside temperature of this receiver is approximately 600°C, and wind velocities perpendicular to this receiver range between 0 and 25 m/sec. For these conditions and cylinder diameter, Reynolds numbers are between 0 and 10^8 and Grashof numbers between 10^{12} and 10^{14}.

Heat-transfer experiments simulating these conditions have been performed by Siebers *et al.* [65] using a 3 m by 3 m electrically heated plate located in a wind tunnel. Air velocities ranged from 0 to 6 m/sec and plate temperatures went to 600°C under conditions of forced, natural, and mixed convection. In these experiments the wind was parallel to the plate, whereas in a cylindrical central receiver, winds were perpendicular as well as parallel to the surface. Experimental data were obtained in the range of

Grashof numbers between 1×10^9 and 2×10^{12} and Reynolds numbers between 0 and 2×10^6 for plate-to-ambient absolute temperature ratios in the range of 1 to 2.7. The local Nusselt number, based on the vertical distance from the bottom y and obtained from natural convection experiments, are given by Eqs. (24) and (25).

1. For turbulent natural convection

$$Nu_y = 0.098 \, Gr_y^{1/3}(T_s/T_{amb})^{-0.14} \quad \text{for} \quad 10^9 < Gr_y < 2 \times 10^{12} \quad (24)$$

2. For laminar natural convection

$$Nu_y = 0.404 \, Gr_y^{1/4} \quad \text{for} \quad Gr_y < 10^9 \quad (25)$$

Equation (24) applies when either the wall temperature or the heat flux is uniform, whereas Eq. (25) applies only when the heat flux is uniform. Transition from laminar to turbulent flow occurs over a range of Grashof numbers, which depends on the temperature ratio (T_s/T_∞) [65]. The experimental results are compared with Eqs. (24) and (25) in Fig. 27.

For conditions with forced convection dominant and the free-stream velocity parallel to the plate the following equations apply.

FIG. 26. Photograph of Solar 1, prototype solar power plant at Barstow, California.

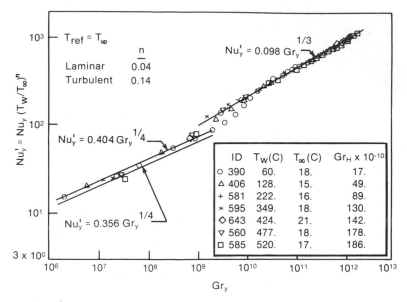

FIG. 27. Correlation of effects of variable properties on natural convection from a vertical surface in air (from Siebers *et al.* [65]).

1. For laminar forced convection

$$\mathrm{Nu}_x = 0.453\,\mathrm{Re}_x^{1/2}\,\mathrm{Pr}^{1/3} \qquad \text{for} \quad \mathrm{Re}_x < 2 \times 10^5 \qquad (26)$$

2. For turbulent forced convection

$$\mathrm{Nu}_x = 0.307\,\mathrm{Re}_x^{-0.8}\,\mathrm{Pr}^{0.6}(T_s/T_{\mathrm{amb}})^{-0.4} \qquad \text{for} \quad \mathrm{Re}_x > 2 \times 10^5 \quad (27)$$

In Eqs. (26) and (27), x is the distance from the leading edge of the plate. All fluid properties in Eqs. (24) to (27) should be evaluated at the ambient temperature T_{amb}.

The length-averaged heat-transfer coefficient for natural convection $\bar{h}_{\mathrm{nat}}$ is determined from Eqs. (24) and (25). The length-averaged heat-transfer coefficient for forced convection $\bar{h}_{\mathrm{for}}$ is determined from Eqs. (26) and (27). For $\mathrm{Gr}/\mathrm{Re}^2 > 10.0$, the heat transfer should be determined from the equations for natural convection while for $\mathrm{Gr}/\mathrm{Re}^2 < 0.7$, the forced convection equations are applicable. It is important to note that these equations have been validated only with ambient wind parallel to the plate and orthogonal to the buoyant natural convective flow. Errors resulting from the application of these equations to external receivers not meeting these conditions are not known.

For mixed-convection in the flow regime defined by $0.7 \leq \mathrm{Gr}/$

$Re^2 \leq 10.0$, heat-transfer coefficients are calculable to within $\pm 10\%$ by

$$\bar{h}_{mix} = (\bar{h}_{nat}^3 + \bar{h}_{for}^3)^{1/3} \tag{28}$$

Analysis of the experimental data [65] showed some scatter in the transition region (Fig. 28). A correlation for the average location of transition is

$$Re_{crit} = \frac{4 \times 10^5}{1 + 6.4(Gr_x/Re_x^2)^{1.5}} \tag{29}$$

Because it is difficult to achieve high Grashof numbers in air, Clausing [66] built a cryogenic wind tunnel and used nitrogen at 80 K as the working fluid. The setup achieved simultaneously values of $Gr \approx 3 \times 10^{10}$ and $Re \approx 3 \times 10^6$. Abrams [64] expressed concern about the forced-convection data obtained in the tunnel because turbulence is intense and the velocity distribution is asymmetric, with variations being approximately 23%. But the natural convection correlations proposed by Clausing [67] and Siebers et al. [65] agree within 20%.

Natural convection data obtained in the cryogenic tunnel wind a 14-cm-diameter, 28-cm-tall cylinder [67] yielded the following empirical correlations for the average-Nusselt-number base on the height H:

$$Nu_H = 0.082 \, Ra_H^{1/3} \times f(T_s/T_{amb}) \quad \text{for} \quad 1.6 \times 10^9 < Ra_L < 10^{12} \tag{30}$$

where $f(T_s/T_{amb}) = -0.9 + 2.4 \, (T_s/T_{amb}) - 0.5(T_s/T_{amb})^2$ for $1 < (T_s/$

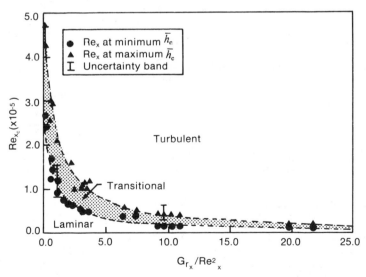

FIG. 28. Effect of Re and Gr on transition in mixed convection from a vertical heated surface (from Siebers et al. [65]).

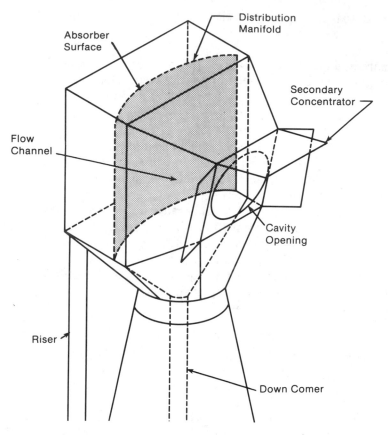

FIG. 29. High-temperature direct absorption receiver cross section for a falling-film molten salt system. The shaded region indicates the location of falling film. From Wang and Copeland [69a]).

$T_{amb}) < 2.6$. In Eq. (30) properties are based on the average temperature $(T_s + T_{amb})/2$.

A computer code to predict the heat loss from external receivers in a steady ambient wind has been developed at Sandia National Laboratory for the cylindrical receiver of Solar 1 [68]. The code can determine the laminar and turbulent mixed-convection heat-transfer coefficient on an external receiver up to the line of separation. Availability of heat-transfer data in the wake region is scant, but measurements in pure forced convection from cylinders [69] have shown that between 30 and 50% of the total convective losses can occur in the wake region. Atmospheric turbulence differs from that in a wind tunnel. The effect of the wake region as well as of atmospheric turbulence on the convection loss has not yet been investigated.

2. Cavity Receivers—Central Receiver Systems

In a cavity-type receiver, solar radiation passes through an aperture into the interior. In current designs the interior is lined with flow passages in which the working fluid is heated, but direct contact heat exchangers have also been proposed for use in cavity receivers. Figure 29 shows an advanced design currently under development at the Solar Energy Research Institute (SERI), using direct radiation absorption [69a]. In this design radiation is absorbed by a thin layer of molten salt that flows down an inclined absorber surface. Another direct absorption concept under development at Sandia makes us of solid particles falling through the cavity receiver to directly absorb incoming solar radiation.

The analysis and interpretation of experimental data on the convection heat loss from cavity receivers are more complex than those for external receivers. Clausing [67] postulated a convective flow pattern for cavity receivers and devised a network representation of the loss mechanism. Although the analytic model is oversimplified, it does cover some of the key elements that govern convective losses from cavity receivers. The density of the air entering a cavity solar receiver is typically a factor of three or four larger than the density of the air at the temperature of the refractory surfaces inside the receiver. If the aperture is in the lower portion of the cavity, the air inside the cavity will be stratified and relatively stagnant in the upper region. Thus it is reasonable to divide the volume into a convective and a stagnant zone (Fig. 30). The convective heat loss from a cavity receiver depends on two factors: (1) the ability of buoyancy to transfer mass and energy across the aperture and (2) flow across the aperture because of wind. Preliminary experimental results indicate that the thermal resistance between the interior cavity walls and the air inside the cavity

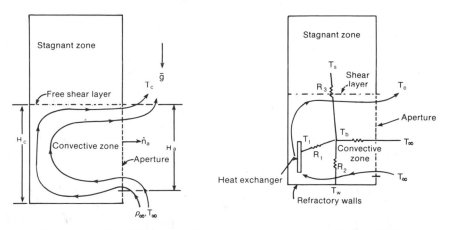

FIG. 30. Postulated flow pattern for cavity-type receiver (from Clausing [67]).

controls the convective heat loss and that the external wind has relatively little effect. By performing a simple energy balance on the flow through the aperture of the receiver, the convective loss q_c can be expressed as

$$q_c = \dot{m}c_p(T_c - T_{amb}) = (\rho_\infty V_a A_a)c_p \,\Delta T \qquad (31)$$

In Eq. (31) T_c is the average gas temperature inside the cavity, A_a the aperture area, and V_a the average velocity of the inflow, which can be expressed according to Clausing [67] by

$$V_a = \frac{1}{2}\left[V_b^2 + \left(\frac{V_{wind}}{2}\right)^2 \right]^{1/2} \qquad (32)$$

where

$$V_b = [g\beta(T_c - T_{amb})H]^{1/2} \qquad (33)$$

and L equals the projected vertical aperture height.

Another expression for q_c, derived from the thermal network model in Fig. 30, is

$$q_c = \bar{h}_t A_t(T_t - T_b) + \bar{h}_w A_w(T_w - T_b) + \bar{h}_s A_s(T_s - T_b) \qquad (34)$$

The subscripts t, w, and s refer to the heat exchanger surface, wall, and stagnation region, respectively (Fig. 30).

The average heat-transfer coefficients between the inner surfaces and the air can be estimated from the semiempirical relation given by Clausing [67],

$$\overline{Nu_H} = 0.082 \, Ra_H^{1/3}\,[-0.9 + 2.4\,(T_s/T_{amb}) - 0.5(T_s/T_{amb})^2]\,f(\phi) \qquad (35)$$

where

$$f(\phi) = 1 \qquad\qquad\qquad \text{for} \quad 0° < \phi < 135°$$

and $\qquad\qquad\qquad\qquad\qquad\qquad\qquad\qquad\qquad\qquad\qquad (36)$

$$f(\phi) = 0.66\,[1 + (\sin \phi)/\sqrt{2}] \qquad \text{for} \quad \phi > 135°$$

The angle ϕ in Eq. (35) is the zenith angle between the normal to the heat-transfer surface and the zenith. For a heated downward facing surface $\phi = 180°$ and $f(\phi) = 0.66$. Equation (35) holds for $Ra_H > 1.6 \times 10^9$ and $1 < (T_s/T_{amb}) < 2.6$. The above predictions by Clausing [67] were found to be in close agreement with experimental data obtained by McMordie [70] on a full-scale cavity receiver and Mirenayat [71] in a laboratory scale electrically heated cavity.

Experimental studies of natural convection in two-dimensional open cavities have been conducted by Sernas and Kyriakidas [72], Chen and Tien [73], Hess and Henze [74], and Humphrey et al. [75] and in a three-dimensional cavity by Kraabel [76]. Sernas and Kyriakidas, Chen and Tien, and Hess and Henze studied convection from cavities in which only the back wall was heated. Sernas and Kyriakidas and Chen and Tien found that the heat transfer from a heated back wall of height H approached that for the heat transfer from a heated vertical plate of similar height in a semi-infinite medium for $Ra_H > 10^7$. Hess and Henze examined the effect of baffles on the heat transfer from an open cavity in the Rayleigh range $3 \times 10^9 \leq Ra_H \leq 3 \times 10^{11}$. The baffles were placed at the top and bottom of the cavity aperture and extended one quarter of the height of the cavity. Even though the baffles reduced the effective aperture area of the cavity by 50%, the heat transfer from the cavity was only reduced by 10%. Humphrey, Sherman, and Chen conducted experiments in a cavity in which the floor and back wall of the cavity were both heated. Their experiments were conducted for $Ra_H = 2.9 \times 10^7$ and included studies of the effects of cavity tilt angle, external forced convection flow, and cavity aspect ratio. They found the natural convection flow to be unsteady with periodic oscillations at frequencies of 2–5.5 Hz.

Kraabel [76] conducted his experiments using a 2.2 m cubical cavity with electrical heating elements on all of the interior surfaces of the cavity. The aperture was vertical and was one face of the cube. Automatically positioned probes in the aperture plane measured velocity and temperature distributions. The convective loss was determined by (1) integrating the product of the temperature and velocity distributions and by (2) calculating the difference between the electric power input and the power radiated from the cavity. The two results were in good agreement. The second method, being more rapid, was used for most of the convective loss determinations. The cavity wall temperature was varied from 90° to 750°C, corresponding to Grashof number variations in the range from 9.4×10^{10} to 1.2×10^{12}. Flow visualization revealed secondary flow patterns characterized by a pair of counter-rotating vortices that fully occupied the upper half of the cavity. The total convective losses were correlated by

$$\overline{Nu_L} = 0.088 \, Gr_L^{1/3}(T_s/T_{amb})^{0.18} \tag{37}$$

where the physical properties in the Nusselt and Grashof numbers are evaluated at the ambient temperature. Kraabel found that Eq. (37) also fitted Mirenayat's [71] measurements of the convective losses from 0.2-m and 0.6-m cavities, thus being valid to Gr_L as low as 5×10^7. In Eq. (37), the appropriate heat-transfer area to calculate convective loss is the entire interior surface area of the cavity. Kraabel's experiments were conducted with the cavity protected from environmental winds by large curtains. However, on occasion winds arose during tests without discernible effect upon the convective loss. This finding is consistent with McMordie's

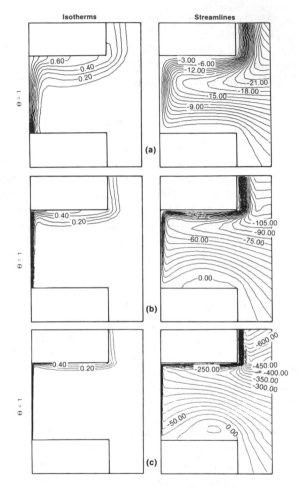

FIG. 31. Calculated flow pattern in an open cavity: (a) $Ra_H = 10^5$, (b) $Ra_H = 10^7$, (c) $Ra_H = 10^9$ (from Chen and Tien [78]).

experience at the Central Receiver Test Facility [70]. But so far, there are not sufficient measurements to conclude that cavity convective losses are unaffected by ambient wind.

A computer model that predicts laminar natural convection heat loss from two dimensional cavities has been developed by LeQuere *et al.* [77]. They found that the flow in a bottom heated cavity was unsteady for $Ra_H > 10^6$. Calculations have also been performed by Chen and Tien [78] for a two-dimensional cavity with a heated back wall in laminar flow. They calculated the steady flow characteristics over the range $1 \times 10^3 \leq Ra_H \leq 1 \times 10^9$.

Humphrey, Sherman, and To [79] extended the work of LeQuere, Humphrey, and Sherman to turbulent flow and compared their calculations with the experimental measurements from a small air-filled cavity. Good agreement was found between measurements and predictions of the velocity and temperature fields. Calculated results that show the flow patterns in the cavity studied by Chen and Tien [78] are shown in Fig. 31.

Boehm [80] summarized and evaluated available data of thermal losses from central receivers and also recommended Eq. (37) to predict the natural convection heat loss from cavity receivers, for the range $10^5 <$ Gr $< 10^{12}$ with the height of the cavity as the significant length and with fluid properties based on the ambient temperature T_∞. Boehm [80] also reported that Kraabel compiled natural convection results in air for flat plates, vertical cylinders, and cavities whose apertures are equal in height to the inner back panels and recommended the equation

$$\overline{\mathrm{Nu}}_H = 0.052 \, \mathrm{Gr}_H^{0.36} \tag{38}$$

with properties evaluated at T_∞. Available experimental data are compared with Eq. (38) in Fig. 32. A comparison of the correlation given by Eq. (38) with experimental field data from cavity receivers is shown in Fig. 33. Data for the molten salt alternative central receiver (ACR), the molten salt electric equipment (MSEE), and the sodium-cooled cavity receiver (SCCR) are shown in Fig. 33. The maximum estimated wind effects are shown by an error bar. A similar comparison of Eq. (38) with available data from external receivers is shown in Fig. 34. Here are shown data for the Advanced Sodium Receiver (ASR) of the International Energy Agency (IEA).

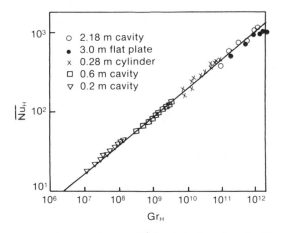

FIG. 32. Results of natural convection studies in air for length scale, H, measured significant length in vertical direction and fluid properties evaluated at ambient temperature (from Boehm [80]).

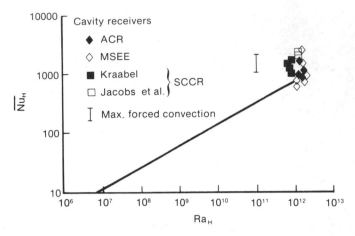

FIG. 33. Experimental convective loss results for cavity receivers (from Boehm [80]).

The relative importance of natural convective losses from cavity receivers can be reduced by operating the receivers at high flux levels. Anderson [81] has examined convective heat and mass transfer in a high-flux receiver with a falling-film on the absorber panel (Fig. 29). The performance characteristics of a carbonate molten-salt film with $(T_i + T_o)/2 = 700°C$ are shown in Fig. 35. The temperature difference between the falling film and the heated wall is plotted as a function of film Reynolds number and solar flux in Fig. 35a, and the relative importance of natural convective losses from the surface of the film is plotted in Fig. 35b. As the level of the solar flux falling on the film is increased i.e., as the film

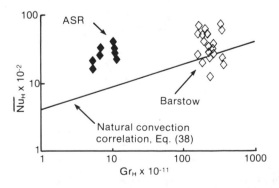

FIG. 34. Existing convective loss results for external receivers (from Boehm [80]).

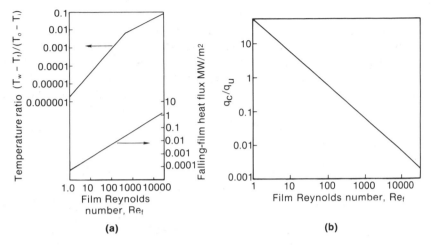

FIG. 35. Temperature distribution and heat flux in a falling-film receiver with Pr = 1 and $(T_o - T_i) = 400°C$. (a) Film temperature and heat flux versus Reynolds number Re_f. (b) Convective-to-nonconvective loss ratio versus Reynolds number Re_f (from Anderson [81]).

Reynolds number is increased) the relative importance of natural convection losses decreases substantially.

An interesting concept for suppressing natural convection into and out of a central receiver has been proposed by Taussig [82]. Called the aerowindow, it injects a transparent gas stream across the receiver aperture, thereby insulating the cavity from the surroundings. Aerowindows generate a vortex inside the receiver when Gr/Re^2 is between 1 and 10. In this regime both natural and forced convection are important, but no experimental data are available. A numerical analysis of the laminar regime by Humphrey and Jacobs [83] suggests that small downward flows are more effective in reducing thermal loss than comparable upward flows.

3. Cavity Receivers — Parabolic Dish Systems

In contrast to central receiver systems, which focus the energy collected by a large heliostat field upon a single receiver, distributed solar energy collection systems utilize many small receivers connected together to collect the energy delivered from each focusing reflector. The convective transport process in distributed collection systems is more complicated than in central receiver systems because the orientation of each receiver changes as the collector tracks the sun throughout the day. The Shenandoah project, located at Shenandoah, Georgia consists of a field with 120 parabolic dishes, each with a cavity-type receiver [84]. The basic configura-

tion of the parabolic dish/receiver system is shown in Fig. 36. A detail of
the cavity receiver design for the Shenandoah project is shown in Fig. 37.
The interior wall of the cavity is covered by an elliptical coil assembly. The
outer portion of the cavity is shrouded by a wind shield to reduce convective losses. Kugath *et al.* [85] measured thermal losses from this receiver by
pumping hot fluid through the receiver when the collector was not tracking
the sun. Conduction losses were determined by blocking the aperture
opening, and radiative losses were calculated based upon the cavity geometry and temperature distribution. Natural convection losses were determined by subtracting conduction and radiative losses from the total losses
measured during the test. Natural convection losses from the receiver are
shown as a function of the receiver declination angle in Fig. 38. The
minimum losses occur when the cavity is pointing directly downward with
a declination angle of 90°. Kugath *et al.* [85] also measured the effects of a
10 mph wind upon receiver losses and found that the total heat loss was
strongly dependent upon cavity orientation. The maximum heat losses
occurred when the forced flow was directed at the aperture of the cavity
and were four times the magnitude of the pure natural convection losses
that were measured in the absence of any wind. The following empirical
correlation for free-convection loss from a cavity as a function of cavity

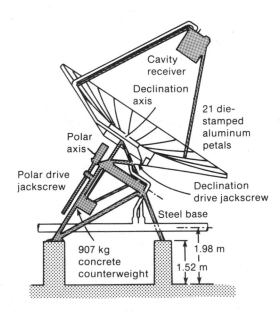

FIG. 36. Schematic diagram showing receiver and parabolic dish in use in the solar thermal
cogeneration project at Shenandoah, Georgia.

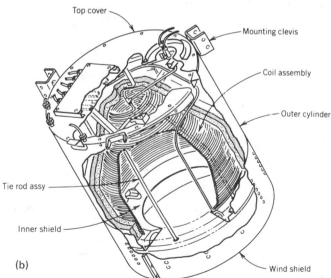

FIG. 37. Cutaway view of a cavity (focal plane) receiver (courtesy of Sandia National Laboratories).

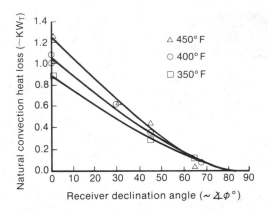

FIG. 38. Curves of natural convection losses versus receiver declination angle (from Kugath *et al.* [85]).

orientation has been proposed by Koenig and Marvin [86]:

$$\overline{\mathrm{Nu}}_L = 0.52 P(\phi) K^{1.75}\, \mathrm{Ra}_L^{0.25} \tag{39}$$

$$P(\phi) = \cos^{3.2}\phi \qquad \text{when}\quad 0° \le \phi \le 45° \tag{40}$$

$$P(\phi) = 0.707\,\cos^{2.2}\phi \qquad \text{when}\quad 45° \le \phi \le 90°$$

$$K = R_{\mathrm{aper}}/R_{\mathrm{cav}} \qquad \text{when}\quad R_{\mathrm{aper}} \le R_{\mathrm{cav}} \tag{41}$$

$$K = 1, \qquad \text{when}\quad R_{\mathrm{aper}} = R_{\mathrm{cav}}$$

and

$$L = \sqrt{2}\,R_{\mathrm{cav}} \tag{42}$$

The characteristic length used in the evaluation of the Nusselt number and Rayleigh number in (39) is $\sqrt{2}\,R_{\mathrm{cav}}$, where R_{cav} is the radius of the cavity.

Harris and Lenz [87] calculated the performance of distributed dish cavity receivers as a function of cavity geometry. Cavity geometries considered in their study are shown in Fig. 39. Natural convection losses were calculated by using Eq. (39). They found cavity losses to be 12% of the energy entering the cavity. For a cavity temperature of 550°C these cavity losses were due to equal contributions from radiation and natural convection. It was also found that, for the same cavity aperture and insulation thickness, cavity geometry had almost no effect on system efficiency.

Because of the complexity of the convective heat transfer process in open cavities, Somerscales and Kassemi [88] have suggested the use of an electrochemical technique that measures mass transfer rather than heat

transfer to determine the convective-loss characteristics of the cavity. Somerscales *et al.* suggest the use of the electrochemical technique because it
has the potential to be simpler, cheaper and faster than heat transfer
measurements. Heat transfer can then be inferred based upon the analogy
between mass and heat transfer. Somerscales and Kassemi [88] examined
mass transfer in nine cylindrical cavities with diameter to height ratios in
the range $\frac{1}{2} \le D/H \le 2$. They found that the comparison of their mass
transfer results with heat transfer results was not entirely satisfactory because:

1. There was a considerable difference in the range of values of D/H
used in the heat-transfer and mass-transfer experiments that were compared in their study. The heat-transfer measurements were made with deep
cavities (D/H small), whereas in the mass-transfer tests the cavities were
shallow (D/H large).

2. The Schmidt number of the fluids was many times greater than the
Prandtl number of the fluids used in the heat-transfer tests.

3. It was not certain that comparable flow regimes were being considered.

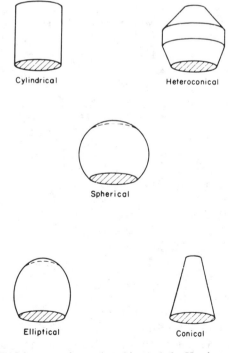

FIG. 39. Cavity geometries analyzed in study by Harris and Lenz [87].

Additional research is necessary to clarify their results and provide correlations for natural convection losses from cavity-type receivers with changing orientation relative to the force field.

III. Natural Convection in Solar Buildings

Solar energy can be used to provide a large fraction of a buildings energy requirements by designing buildings to act as efficient solar collectors, (Balcomb et al. [89], Jones, and McFarland [90]). In heating applications, sunshine enters south facing windows and strikes absorbing surfaces, which heat up and warm the air in the adjacent room. This warm air is then moved largely by natural convection to the remainder of the house where it is either used immediately for heating or stored for later use. In ventilation cooling applications, the direction of heat transfer is essentially reversed. Cold night air is introduced through open windows or vents and used to remove heat from the interior of the building. Natural convection plays an important role in the transport of energy within the building, in both heating and cooling applications. Thermal network models for solar buildings are similar to those for flat-plate collectors with the exception that horizontal surfaces play an important role in determining the heat-transfer characteristics of the building. In addition, a successful design for a solar building includes a unique performance criterion that is not required of active solar energy systems. A solar building should not only provide a high level of thermal efficiency, it must also provide for the thermal comfort of the occupants of the building. A fundamental understanding of natural convection in enclosures with complicated geometries and complicated thermal boundary conditions is important to designing efficient and comfortable solar buildings.

Common types of solar building components have been described previously (Fig. 5) and include direct gain, indirect gain and ventilation cooling applications. Simple thermal models for these configurations are shown in Fig. 40. In direct gain applications, one must predict thermal stratification, temperature distributions, and heat transfer in a single zone enclosure as a function of the location (horizontal or vertical wall) at which heating or cooling occurs. In indirect gain applications, one would like to know how the geometry of openings affects the transport of energy between building zones. Finally, in ventilation cooling applications one would like to know how opening geometry affects the transport of energy between the interior and exterior of the building. An understanding of these transport processes is necessary for the sizing and orientation of

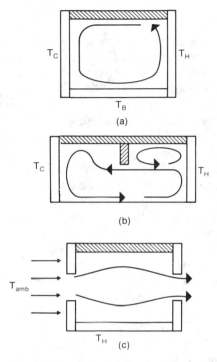

FIG. 40. Simple thermal models for solar building applications; see also Fig. 3. (a) Direct gain, (b) indirect gain, (c) ventilation cooling.

thermal storage and apertures that maximize the performance of the building. In this section we will examine the contribution of natural convection to heat transfer through the building envelope, heat transfer within a single zone, and heat transfer between zones in solar buildings. Infiltration, and forced convection heat transfer also are important in many building applications but are not considered here.

A. BUILDING ENVELOPE

The envelope of a building consists mostly of planar surfaces such as walls and windows. The total heat loss under given climatic conditions depends on infiltration rates, heat losses through walls and their insulation, and heat losses and gains through windows. Natural convection plays an important role in all these loss mechanisms.

1. *Single-Pane Windows*

Single-pane windows constitute one of the largest sources of heat loss from buildings, especially when they face north. A single-pane window can be idealized as a vertical surface between two air reservoirs.

Lock and Ko [40], Anderson and Bejan [41], and Viskanta and Lankford [43] have examined combined natural convection/conduction heat transfer through a vertical plate separating two semi-infinite fluid reservoirs. Sparrow and Prakash [44] conducted a numerical analysis of an enclosure with AR = 1, which was coupled through one conductive wall to an external natural convection flow. The results of these studies can be used to develop a general description of heat transfer through a single-pane window. Figure 41 compares the relative importance of radiation losses to internal convection heat losses through a single pane window as a function of external wind velocity. It can be seen that the radiation and convection heat losses are of the same order of magnitude in many building applications.

The natural convection heat-transfer coefficient used in Fig. 41 for zero external wind is based upon the conjugate conductive/convective analysis of Anderson and Bejan [41]. The natural convection heat-transfer coefficient for high external wind is based upon the boundary layer analysis of Gill [91] and Bejan [92], assuming that the window's surface temperature approaches the external air temperature. The gradual decrease of $\bar{h}_{rad}/\bar{h}_{conv}$ with increasing temperature difference results from the dependence of h_{conv} upon $(T_H - T_C)^{1/4}$. The shift in the two curves for different wind velocities also results from the dependence of $\bar{h}_{conv}$ upon temperature difference.

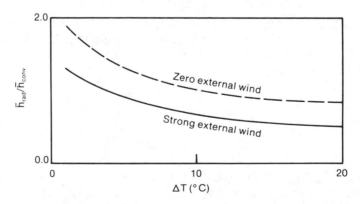

FIG. 41. Relative contributions of convective and radiative heat transfer through a single-pane window: $\bar{h}_{rad} = \sigma(T_H^4 - \bar{T}_{window}^4)/(T_H - T_C)$, $T_C = 0°C$.

2. Double-Pane Windows

Most new buildings in the U.S. have double pane windows because of their superior insulating qualities. Korpela *et al.* [93] carried out a series of numerical experiments on high-aspect-ratio enclosures with the specific application of double-pane windows in mind. Their primary goal was to determine the optimum spacing between panes for minimum heat loss. They found that as the AR was increased at constant window height, the flow in the cavity exhibited, in succession, a conduction-dominated unicellular flow, a multicellular flow, and finally reversion to a unicellular flow. The final unicellular flow was not a boundary-layer flow but was in the transition regime between conduction- and convection-dominated heat transfer. The minimum heat transfer was found at the onset of multicellular flow. Korpela *et al.* [93] suggest the following formula for calculating the optimum AR for minimum heat transfer as a function of Gr_H:

$$AR^3 + 5 \, AR^2 = 1.25 \times 10^{-4} \, Gr_H \qquad (43)$$

where Gr_H is the Grashof number and equals $g\beta H^3 \, \Delta T / v^2$ and H is the height of the cavity. ElSherbiny *et al.* [19] conducted an extensive series of experiments in large aspect ratio enclosures. They found that for AR > 40, the transition between conduction and convection dominated heat transfer is independent of AR. Therefore, the optimum spacing for large AR, double-pane windows depends only on the temperature difference across the window and is independent of the height of the window. For large AR Eq. (43) reduces to

$$L_{opt} = 20(v\alpha/g\beta \, \Delta T)^{1/3} \qquad (44)$$

It is important to note that the (constant H/variable L) case considered by Korpela, Lee, and Drummond [93] is different from the (constant L/variable H) case previously considered by Bejan [94]. Bejan's analysis assumes boundary-layer flow and considers the dependence of Nu_L on AR with fixed plate spacing L and variable height H. Korpela *et al.* limited their study to Gr_H numbers in the transitional region between conduction- and boundary-layer dominated heat transfer and considered the dependence of Nu_H upon AR with fixed H and variable L. The analysis by Korpela *et al.* predicts the combination of Gr_H and AR that produces minimum heat transfer at fixed H, whereas the analysis by Bejan gives the combination of Gr_L and AR that produces maximum heat transfer at fixed L.

B. SINGLE-BUILDING ZONES

An understanding of the impact of complicated boundary conditions upon the natural convection pattern in enclosures resembling a room (i.e., enclosures with aspect ratios of the order of one) is important to predict natural convection heat transfer within a building. In building applications one commonly encounters nonrectangular geometries with nonuniform thermal boundary conditions with surfaces that are often not perfectly smooth or straight. Flow patterns can be complicated by the presence of internal obstructions such as wall hangings, window coverings, and furniture. The thermal boundary conditions in building applications are generally three dimensional and can involve heating and cooling of several surfaces. In comparison with a flat-plate collector that has only one heated surface at the bottom and one cooled surface at the top, any building surface can be thermally active, regardless of its orientation.

Heating and cooling of vertical surfaces produce an overall circulation pattern similar to that seen in vertical flat-plate collectors (Fig. 9a). Floor heating generates unstable vertical temperature gradients that can lead to the formation of thermal plumes or Benard circulation similar to that found in a horizontal flat-plate collector (Fig. 9c). In buildings, combinations of horizontal and vertical temperature gradients can occur within a single building zone. The flows generated by these two types of temperature gradients compete with each other and their combined action determines the heat transfer, temperature distributions, and thermal stratification levels in the buildings.

1. *Triangular Spaces*

Most studies of single-zone building heat transfer have been for rectangular enclosures, whereas many building applications include nonrectangular spaces. Notable exceptions to the rectangular studies are experimental studies of triangular enclosures by Flack et al. [95], Flack [96], and Poulikakos and Bejan [97]. In addition, Akinsete and Coleman [98] and Poulikakos and Bejan [99] conducted numerical studies of natural convection in triangular enclosures. Triangular spaces occur in A-frames, attics, and rooms with cathedral ceilings or clerestories (Fig. 5). Flack et al. [95] measured the heat transfer in an air-filled horizontal isosceles triangular enclosure by using a Wollaston prism/Schlieren interferometer. The base of the enclosure was insulated and the upper enclosure had one isothermal heated side and one isothermal cooled side. Their experiments showed that the average Nusselt number was within 20% of that for a rectangular enclosure, if the heated side of the triangle is used as the characteristic length dimension. However, a strong conduction-dominated region formed near the apex of the enclosure due to the physical proximity of the hot and cold surfaces in that region. This conduction-dominated region resulted in a sharp increase in heat transfer on the hot wall near the apex.

Flack [96] used the same apparatus previously used by Flack *et al.* [95] to examine the case when both upper sides of the enclosure were at the same temperature, while the floor of the enclosure was maintained at a different temperature. For stable heating (hot ceiling, cold floor) the heat transfer varied at most by 10% from a pure conduction solution. Four horizontally aligned Bernard cells formed along the axis of the enclosure during unstable heating (hot base, cold sides). The axis of rotation of the Benard cells was perpendicular to the long axis of the enclosure. Poulikakos and Bejan [97] considered natural convection in an air- or water-filled unstably heated right-triangular enclosure. The flow was found to be turbulent during the water experiments because of the high Rayleigh numbers ($Ra_H \sim 10^8$) that were reached during the water experiments. Poulikakos and Bejan [97] report the following correlation for their air experiments

$$\overline{Nu}_H = 0.345\, Ra_H^{0.3} \quad \text{for} \quad H/L = 0.207 \quad \text{and} \quad 10^6 \geq Ra_H \geq 10^7 \quad (45)$$

Akinsete and Coleman [98] conducted a numerical study on a stably heated right-triangular enclosure with $800 < Gr_L < 6400$, $0.0625 < AR \leq 1$ and $Pr = 0.733$. Like Flack [96], they determined that heat transfer was conduction dominated for high-aspect-ratio enclosures, indicated by a drastic drop in heat transfer as AR was increased. Poulikakos and Bejan [99] conducted a two-dimensional transient numerical study of an isosceles triangular enclosure with cold upper sides and a warm base. They assumed that the fluid in the enclosure was initially isothermal at the base temperature T_H, and at time $t = 0$ the upper sides of the enclosure were suddenly cooled to T_G. For large AR enclosures, the transient Nusselt numbers initially overshot their steady state values. This numerical solution assumed the presence of two symmetrical axially oriented rolls in contrast to the transversely oriented rolls observed by Flack [96].

2. Heating and Cooling of Vertical Surfaces in Enclosures

The thermal boundary conditions in buildings heated by direct solar gain are generally three dimensional, involve both horizontally and vertically imposed temperature gradients, and include local and uniformly distributed heat sources. Adjacent surfaces at different temperatures can produce interactive flows. One recent study by Sparrow and Azevedo [100] investigated whether heat transfer from a vertical plate would be different if the edges were unshrouded (permitting possible lateral inflow of fluid toward the plate) or if they were shrouded (thereby blocking the possible inflow). Figure 42 illustrates the four lateral-edge configurations investigated as well as the basic vertical flat-plate assembly. Configuration I corresponds to the case of unshrouded and insulated lateral edges, configu-

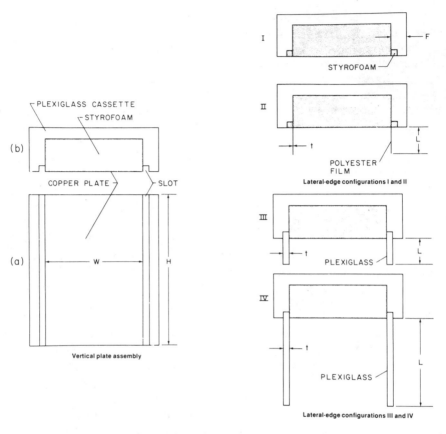

Fig. 42. Lateral-edge configurations used by Sparrow and Azevedo [100].

ration II corresponds to a hydrodynamic blockage with low thermal mass, while configurations III and IV are similar to II but with a different material. The dimensions defining the lateral-edge configurations are displayed in Table III, corresponding to the symbols in Fig. 42.

Experiments were performed under conditions corresponding to Rayleigh numbers from about 8.5×10^7 to 10^9 with Prandtl numbers essentially constant at about 5. The results of this study are plotted in Fig. 43; the data are correlated by the equation

$$\overline{Nu}_H = 0.623 \, Ra_H^{1/4} \tag{46}$$

where the pertinent length dimension in the Rayleigh number is the height of the plate H. The key conclusion drawn by the authors is that lateral-edge effects are negligible for plates with ratios of height to width of less than 1.5

TABLE III

DIMENSIONS DEFINING THE LATERAL-EDGE
CONFIGURATIONS[a]

Case	L	t	F
I	—	—	1.4
II	2.54	0.011	1.4
III	2.54	0.635	1.4
IV	8.50	0.635	1.4

[a] All dimensions are in centimeters.

and insensitivity to lateral-edge effects may be extrapolated to a height to width ratio of as much as 4.

Bohn *et al.* [101] examined heat transfer in a cubical enclosure with three-dimensionally heated and cooled vertical walls and found that the temperature in the middle of the enclosure was a strong function of the temperature distribution on the vertical walls of the enclosure. By using the area-weighted bulk-temperature difference defined by

$$\Delta T_b = T_H - T_b, \qquad T_b = \sum_i T_i A_i / \sum_i A_i \qquad (47)$$

in the definition of the heat-transfer coefficient it was possible to express the heat-transfer results for any combination of hot and cold vertical walls by a single correlation. Bohn and Anderson [102] subsequently found that the bulk temperature defined by Eq. (47) closely predicted the average core temperature in a cubical enclosure with three-dimensional thermal bound-

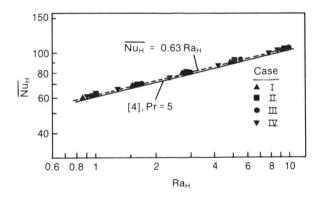

FIG. 43. Nusselt number results for the various lateral-edge configurations (from Sparrow and Azevedo [100]).

ary conditions on its vertical walls. This bulk-temperature-difference scaling demonstrates that the heat transfer in the boundary-layer regime of an enclosure with complicated thermal boundary conditions on vertical surfaces is driven by the temperature difference between a given surface and the bulk fluid temperature in the core of the enclosure. If this scaling is not observed, the heat transfer from a vertical surface in an enclosure with complicated thermal boundary conditions on the vertical walls can appear to be drastically different than that in an enclosure where one surface is uniformly heated and one surface is uniformly cooled.

This point can be demonstrated by considering the example of an enclosure with three heated vertical walls and one cooled vertical wall. For this case, the bulk temperature difference is

$$\Delta T_b = (T_H - T_C)/4 \qquad (48)$$

and since $T_H - T_C = \Delta T$, $\Delta T/\Delta T_b = 4$.

Because the overall heat transfer to the cold surface for this example has to be the same regardless of the temperature difference used to define the heat-transfer coefficient, this implies

$$\bar{h}_b \, \Delta T_b = \bar{h} \, \Delta T \qquad \text{or} \qquad \bar{h}/\bar{h}_b = \tfrac{1}{4} \qquad (49)$$

Thus, the heat-transfer coefficient based upon ΔT will be one-quarter that of the heat-transfer coefficient based upon ΔT_b for the three-heated- and one-cooled-wall geometry described above. This apparent discrepancy results from the choice of the temperature difference used to define the heat-transfer coefficient; it does not indicate a change in the convective heat-transfer mechanism. If ΔT_b is used rather than ΔT, the heat-transfer coefficient remains constant regardless of the thermal boundary conditions at the vertical walls of the enclosure. Bohn *et al.* [101] recommend the following correlation for natural convection heat transfer in enclosures with multiple heated and cooled vertical walls

$$\bar{h}_b L/k = 0.62 \, \mathrm{Ra}_H^{0.250} \qquad (50)$$

Depending upon window location, direct solar gain may produce many localized, heated areas rather than a uniformly heated wall. Jaluria [103] conducted a numerical study of the interaction of multiple horizontal heated strips on a vertical surface in the boundary layer regime. He found that the velocity increased and the temperature decreased as the fluid moved downstream from the region being heated. Downstream heaters experienced heat-transfer enhancement provided they were far enough downstream to benefit from the added velocity induced by upstream heaters, without being exposed to hot fluid.

A majority of experiments with Rayleigh numbers in the range corresponding to full-scale buildings' interior spaces ($\sim 10^{10}$) have been performed with water and Freon using small-scale laboratory test cells. Workers at Los Alamos National Laboratory (see Yamaguchi [104]) have used freon in small-scale tests so that high Rayleigh numbers can be achieved in small enclosures. Small-scale testing can be simpler, quicker, and cheaper to accomplish than equivalent full-scale testing, but there is always some uncertainty regarding the degree to which the small-scale test actually models the full-scale situation. Olsen et al. [105] recently attempted to answer this question by conducting simultaneous small-scale and full-scale experiments over the range $1 \times 10^{10} \leq \mathrm{Ra}_H \leq 5 \times 10^{10}$. Olsen et al. [105] used Freon® 114 gas in the scale model. The physical capabilities of the full-scale and small-scale experiments are shown in Table IV. A comparison between temperature measurements in the full-scale and small-scale experiment is shown in Figs. 44 and 45. Figure 44 shows a comparison of vertical temperature profiles taken midway between the hot and cold vertical walls, and Fig. 45 shows a comparison of horizontal temperature profiles taken near the heated vertical wall.

Olsen et al. performed flow visualization experiments by injecting a neutrally buoyant smoke tracer. The vertical boundary layers on the heated and cooled surfaces were turbulent, characterized visibly by random eddy motion. Once the hot boundary layer reached the ceiling, it turned the corner and flowed along the ceiling toward the cold wall looking like a turbulent jet. When it reached the top of the cold wall, some of the

TABLE IV

CHARACTERISTICS OF FULL-SCALE AND SMALL-SCALE EXPERIMENTS[a]

	Full scale	Small scale
Dimensions		
Height, H	2.6 m	49 cm
Width, W	3.9 m	69 cm
Length, L	7.9 m	136 cm
Aspect ratios		
Height/length AR_L	0.33	0.36
Height/width AR_W	0.67	0.71
Prandtl number, Pr	0.7	0.8
Grashoff number, Gr_H	$1-5 \times 10^{10}$	$1-5 \times 10^{10}$
Typical hot-wall temperatures	25–40°C	25–50°C
Typical cold-wall temperatures	5–10°C	5–10°C
Emissivity of heating and cooling surfaces	low	low
Emissivity of floor and ceiling	high	high
Emissivity of side walls	high	low

[a] Experiments done by Olsen et al. [105].

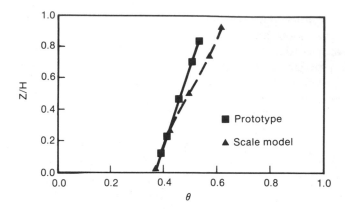

FIG. 44. Comparison of temperature versus height for the prototype and scale model midway between the hot and cold walls (from Olsen *et al.* [105]).

flow reversed direction, proceeding all the way back to the hot wall directly beneath the ceiling jet. A similar two-layer structure occurs near the floor, driven by the cold wall boundary layer. Although there were some differences in the magnitude of the core vertical temperature profiles (see Fig. 44), which were apparently a result of different thermal boundary conditions on the floor and ceiling in the small and large test cells, these differences did not affect the flow patterns or turbulent nature of the vertical boundary layers. The temperature measurements taken near the heated wall show qualitative agreement between the full and small scale.

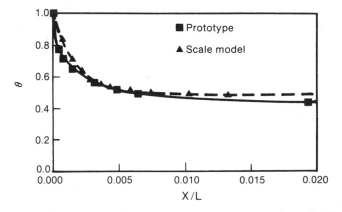

FIG. 45. Comparison of dimensionless temperature versus distance from the wall for the prototype and scale model hot-wall boundary layers (from Olsen *et al.* [105]).

There are a number of advantages associated with the use of liquids rather than gases in scale model studies. Fluids generally have higher thermal conductivities than gases, making it easier to approach adiabatic conditions on insulated walls. Also, the high density of liquids makes it relatively easy to suspend neutrally buoyant particles for use in flow visualization experiments or local velocity measurements. Direct application of the results from experiments that use fluids to buildings is based on the observation that, at high Rayleigh numbers, the Prandtl number effect on average heat transfer is largely accounted for by the Rayleigh number ($Ra_H = Gr_H \times Pr$).

Churchill and Chu [106] have successfully correlated experimental data for average natural convection heat transfer from a vertical plate in an unconfined medium for both laminar and turbulent regimes. Their correlation has the form

$$\overline{Nu}_H^{1/2} = 0.825 + 0.387\ Ra_H^{1/6}/\Psi(Pr) \tag{51}$$

with

$$\Psi(Pr) = [1 + 0.492/Pr^{9/16}]^{8/27} \tag{52}$$

The weak dependence of $\Psi(Pr)$ upon the Prandtl number tends to support the conclusion that average heat-transfer results are only weakly dependent upon the Prandtl number, provided that the Nusselt number is correlated as a function of the Rayleigh number. Numerical and experimental studies of natural convection in enclosures also exhibit only a weak dependence upon Prandtl number; however, experimental data for transitional and turbulent flows in enclosures are somewhat limited at the present time. Additional studies are required to fully determine the errors associated with the use of liquids to model transitional and turbulent natural convection in buildings.

3. *Heating and Cooling of Vertical and Horizontal Surfaces in Enclosures*

In solar buildings the hot and cold surfaces result from the presence of windows, thermal storage walls, auxiliary heating systems or solar illumination. In many cases of interest, both horizontal and vertical surfaces will be thermally active. A summary of natural convection studies for heat transfer to and/or from horizontal and vertical surfaces in single zone enclosures is shown in Table V. In all of these studies, two of the vertical walls on opposite sides of the enclosures were adiabatic. In our discussion we will use a shorthand notation consisting of four letters to specify the thermal boundary conditions on the remaining four walls of the enclosures, with H signifying an isothermal heated wall, C signifying an isother-

TABLE V

Summary of Enclosure Convection Studies with Vertical and Horizontal Heat Fluxes

Configuration	Parameters	Reference
	$Pr = 1.8 \times 10^4,\ 8.8 \times 10^4$ $4.0 \times 10^4 \leq Ra \leq 5.1 \times 10^4$ $A = 1, 3$ $0 \leq \Omega \leq 176.7$	Ostrach and Raghavan [107]
	$Pr = 8.9 \times 10^4$ $2.29 \times 10^4 \leq Ra \leq 5.99 \times 10^4$ $A = 1$ $0 \leq \Omega \leq 6$	Fu and Ostrach [108]
 	$Pr = 0.71$ $10^3 \leq Ra \leq 10^6$ $A = 1$ $-5 \leq \Omega \leq 5$	Shiralkar and Tien [109]
  	$Pr = 6.7$ $0.4 \times 10^{10} \leq Ra \leq 7 \times 10^{10}$ $A = 1$ $\Omega = 0, 1, \infty$	Kirkpatrick and Bohn [110, 111]

TABLE V   *(Continued)*

Configuration	Parameters	Reference
	$Pr = 0.7$ $Ra = 10^6$ $AR = 1$	Ozoe *et al.* [112]
	$Pr = 6.7$ $1 \times 10^{11} \leq Ra_H^* \leq 1 \times 10^{13}$ $AR = 1$ $0 \leq PWR \leq \infty$ $PWR = q_B/q_H$	Anderson *et al.* [113]
	$0.7 \leq Pr \leq 10$ $1 \times 10^6 \leq Ra_H^* \leq 1 \times 10^{13}$ $AR = 1$	Anderson and Lauriat [117]

mal cooled wall, Q signifying a constant flux heated wall and A signifying an adiabatic wall. The first letter in the sequence specifies the thermal boundary condition on the floor of the enclosure, followed by the thermal boundary condition on the right-hand-side wall, the ceiling, and finally left-hand-side wall of the enclosure. For example, the notation AHAC refers to the classical problem of an enclosure with heated and cooled vertical walls opposite to each other and adiabatic floor and ceiling surfaces.

Ostrach and Raghaven [107] and Fu and Ostrach [108] experimentally studied natural convection in enclosures with the configuration CHHC. They expected that the stable vertical temperature gradient caused by the cooled floor and heated ceiling would tend to damp out the natural convection flow in the enclosure. In both of these studies, the velocity distributions were measured by tracking the movement of particles suspended in large Prandtl number silicone oils. It was found that the imposition of a stable vertical temperature gradient caused a reduction in the velocity in the boundary layers next to the vertical walls and that small secondary circulation cells formed near the upper portion of the heated vertical wall and the lower portion of the cooled vertical wall.

Shiralkar and Tien [109] conducted numerical studies with stable and unstable vertical temperature gradients, corresponding to configurations HHCC and CHHC. Thermal instabilities such as Benard cells or thermal plumes were not observed when the enclosure was unstably heated. The temperature distribution in the core was found to be strongly dependent upon the heating configuration.

The controlling parameter that determines the relative strength of the vertical and horizontal temperature gradients is Ω, defined as the ratio of the two temperature differences $(T_T - T_B)$ and $(T_H - T_C)$, where $(T_T - T_B)$ is the difference between the temperature of the top surface T_T and the bottom surface T_B while $(T_H - T_C)$ is the difference between the thermally active sidewalls. Stable heating (CHHC) produced a motionless core with a high level of thermal stratification whereas unstable heating (HHCC) induced enough motion in the core to make it almost isothermal. Shiralkar and Tien found that the heat transfer from the vertical side walls increased with increasing Ω {= $(T_T - T_B)/(T_H - T_C)$} in enclosures with stable vertical temperature gradient because of the preheating or precooling caused by the floor and ceiling. This result indicates that the increase in $(T_H - T_B)$ or $(T_T - T_C)$, which occurs with increasing Ω, more than compensates for any corresponding velocity reduction caused by the imposition of a stable vertical temperature gradient.

Kirkpatrick and Bohn [110, 111] conducted a series of experiments in a water-filled cubical enclosure that was heated from below, while the thermal boundary conditions for two side walls and the top were varied. They tested the configurations HACA, HHCC, HHHC, and HCCC at Rayleigh numbers four orders of magnitude higher than in previous studies. They found that the thermal boundary condition at the top of the enclosure strongly influenced the temperature distribution and flow structure. When the vertical temperature gradient was reduced to zero by heating the ceiling (configuration HHHC) the fluid in the core was highly stratified. Unstable heating generated turbulent thermal plumes at the top and bottom surfaces and destroyed the thermal stratification in the core of the enclosure.

Ozoe et al. [112] used numerical calculations with a two-equation model of turbulence to study the building configuration HAAC. Only 57% of the cooled wall was thermally active, to simulate a wall with a cold window. The remainder of the cooled wall was insulated. The calculations of Ozoe et al. [112] were carried out for only one Rayleigh number ($Ra_L = 10^6$) and a Prandtl number of 0.7. They found the flow to be three dimensional with weak spiral flows near the side walls. A value of $Ra_L = 10^6$ placed their turbulent calculations intermediate to the onset of the turbulent regime at $Ra = 10^9$ in natural convection next to a vertical surface in an unconfined fluid, and at $Ra = 10^4$ for convection in an enclosure with a heated floor and a cooled ceiling. The calculations showed that the flow next to the floor consists of a horizontal boundary layer with no evidence of thermal instabilities.

Anderson *et al.* [113] experimentally examined the flow structure, heat transfer, thermal stratification, and temperature distributions in a closed cavity with the boundary condition QQAC. The ceiling and side walls of the cavity were insulated, the floor and one vertical wall were electrically heated, and the opposite vertical wall was cooled. They varied the level of heating provided to the floor and wall between $0 \leq PWR \leq \infty$. The parameter PWR is defined as the ratio of the energy per unit area convected from the floor divided by the energy per unit area convected from the vertical wall. The condition $PWR = 0$ corresponds to a closed cavity with differentially heated end walls. When $PWR = \infty$, the problem reduces to that of a cavity with a constant flux heated floor and a cooled vertical wall, similar to that studied by Ozoe *et al.* [112]. Anderson *et al.* [113] found that the level of the thermal stratification in the cavity is a strong function of the level of heating provided to the floor (see Fig. 46). The minimum level of

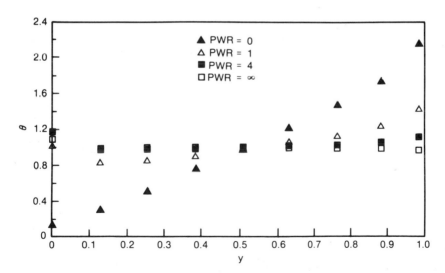

Fig. 46. Thermal stratification as a function of the relative levels of floor and wall heating in a direct gain zone (from Anderson *et al.* [113]).

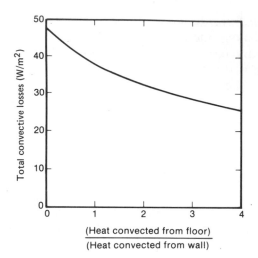

FIG. 47. Convective losses as a function of the relative levels of floor and wall heating in a direct gain zone (from Anderson et al. [113]).

thermal stratification occurred for pure floor heating (PWR = ∞). Anderson et al. [114] used these results to calculate the total convective losses from a direct gain zone over the range $0 \leq \text{PWR} \leq 4$. Their results are plotted in Fig. 47. The convective losses for PWR = 4 were found to be less than half of the convective losses for PWR = 0. They concluded that floor heating is more effective than wall heating for maintaining room temperature in a single building zone because floor heating minimizes the level of thermal stratification. Side wall heating causes a high level of thermal stratification, which increases convective losses to the cold surface. The experiments of Anderson et al. [113] covered the range $10^{11} \leq \text{Ra}_H^* \leq 10^{13}$ with Pr = 6.7. The flow next to the floor was found to be an extremely stable horizontal boundary layer, indicating that natural convection in enclosures with adiabatic ceiling behaves entirely differently from natural convection in enclosures with cooled ceilings. The following correlation is proposed for the temperature in the core of the enclosure as a function of the heat convected from the floor and vertical wall.

$$\frac{(T'_{\text{core}} - T_{\text{C}})K}{q_H H} = 40.74 \, \text{Ra}_H^*$$

$$- 0.304 \left\{ \frac{1}{2} + \frac{\text{PWR}}{\text{AR}} [0.454 + 0.001 \, \text{PWR}] \right\} \quad (53)$$

This correlation can be used to calculate the combination of floor and wall heating required to maintain a given air temperature in a single building zone with a vertical cold surface. Anderson *et al.* also provide correlations for surface temperatures, thermal stratification, and heat transfer.

The natural convection flow within an enclosure differs from an external natural convection flow in an unconfined medium because it recirculates and thus interacts with itself. It is difficult to model this interaction accurately by applying external flow results to internal flows. However, because results for internal flows did not exist until recently, the common practice has been to calculate building heat loads based upon external flow results. Bauman *et al.* [115] compared results for external and internal flows and found the heat transfer calculated from internal results to be 30–50% lower than results based upon external situations. Based upon these findings, Altmayer *et al.* [116] conducted a series of numerical experiments aimed at developing an improved set of correlations for use in rooms with isothermal heating and cooling of two vertical side walls. They succeeded, but the correlations are substantially more complicated to use then other methods.

Anderson and Lauriat [117] conducted a numerical study of natural convection in a closed cavity with an isothermal vertical wall and a heated floor. The heat transfer and flow patterns were calculated for cases where the floor was isothermal as well as for cases when the floor was a constant heat-flux surface (configurations HAAC and QAAC). A horizontal boundary layer was found to form adjacent to the heated floor, in agreement with numerical calculations by Ozoe *et al.* [112] and experimental observations by Anderson *et al.* [113]. The calculated structure of the horizontal boundary layer is shown in Fig. 48. A comparison of the heat-transfer results with correlations for horizontal and vertical surfaces in an unconfined medium, showed that the unconfined medium correlations overpredict heat transfer from the vertical wall by 13% and underpredict heat transfer from the floor by 40%. The primary reason for the difference between the unconfined medium results for external flow and enclosure results for internal flow is the temperature difference used to define the heat-transfer coefficient. In external flows this temperature difference is taken to be the temperature of the surface minus the temperature of the ambient fluid, whereas in an enclosure, the temperature difference that determines the heat transfer from a given surface is the temperature of the surface in question minus the temperature of the surface located in the upstream flow direction.

4. *Surface Roughness Effects*

Room surfaces can have uniformly distributed roughness elements, for example, a masonry wall, or they can have isolated projections due to

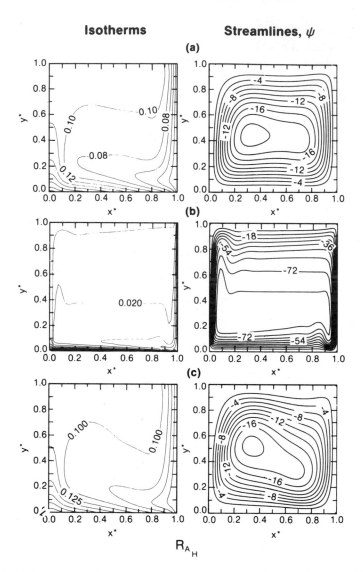

FIG. 48. Horizontal boundary layer structure in an enclosure (from Anderson *et al.* [114]).

windowsills, door soffits or ceiling beams. Anderson and Bohn [36] examined the effect on heat transfer of distributed roughness elements with the same length scale as the thermal boundary layer. They found that roughness was most effective for an isothermal wall, producing an average increase in total heat transfer of 10–15% and local increases of as much as 40%. The influence of the roughness elements upon the location of transition is shown in Fig. 49 for a constant-flux thermal boundary condition. The single solid line on Fig. 49 is the best fit curve demarking transition from laminar to turbulent flow in the absence of roughness. Transition in enclosures with heated and cooled vertical walls is delayed compared with transition for an isolated vertical surface (Fig. 49) [117a]. This delayed transition is caused by the strong thermal stratification in the core of the enclosure flow. The thermal stratification stabilizes the boundary layer by reducing the buoyancy force compared with an external flow in isothermal surroundings. Finite-size roughness elements in natural convection flows have been considered by Nansteel and Greif [118], ElSherbiny et al. [119], and Al-Arabi and El-Refaee [120]. Nansteel and Greif considered downward projections from the ceiling of an enclosure typical of door soffits. They found regions of intense turbulence downstream of the projections in a water-filled apparatus at $Ra_L = 10^{11}$. These turbulent regions did not exist if the projection extended across the entire width of the enclosure. ElSherbiny et al. [119] performed experiments in an enclosure with one V-corrugated and one flat surface. They found that the heat-transfer coefficient increased by up to 50% compared with an enclosure with two smooth surfaces. Al-Arabi and El-Rafaee [120] studied natural convection from an isolated V-corrugated plate. They also found this configuration provided higher heat transfer than a finned plate.

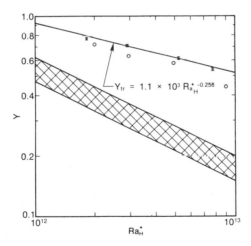

FIG. 49. Transition in an enclosure with a constant-flux heated surface. O, rough surface; +, smooth surface (from Anderson and Bohn [36]), The cross-hatched area is representative of the onset of transition for a heated plate in a semi-infinite medium as measured by Jaluria and Gebhart [117a].

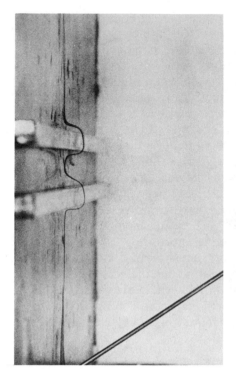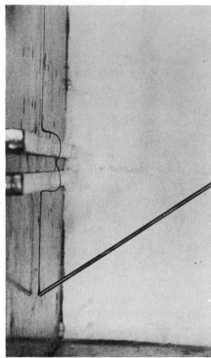

FIG. 50. Boundary-layer flow over horizontal roughness elements on a vertical surface (from Shakerin *et al.* [121]).

All of the studies mentioned above considered surface roughness in the context of heat-transfer enhancement. Surface roughness can also have the effect of blocking the natural convection flow and reducing heat transfer. Shakerin *et al.* [121] conducted a series of numerical and laboratory experiments with single and double roughness elements attached horizontally to a heated vertical surface in a rectangular enclosure. They found that if the spacing between the roughness elements was smaller than the height of the roughness element, then there was significant reduction in the ability of the natural convection flow to penetrate the space between the roughness elements. This effect is clearly shown in Fig. 50 for an experiment conducted in a water-filled enclosure. Dye was injected near the velocity maximum in the boundary layer next to the heated vertical wall. When the spacing between the roughness elements was reduced (Fig. 50b), the penetration depth of the dye was also reduced.

C. MULTIPLE BUILDING ZONES

Heat transfer between rooms separated by a partition in passive solar buildings occurs almost entirely by natural convection if the area of the flow aperture (e.g., doorway) is appreciably smaller than the overall cross-sectional area of the partition. Two approaches have been used to analyze interzonal convection in buildings. In one of these inviscid flow through a partition bounded by semi-infinite isothermal fluid reservoirs with different temperatures is assumed. This approach provides an accurate description of the flow through the partition under conditions of pure natural and forced/free convection when boundary layers are not important, but it predicts incorrect scaling for heat transfer when the flow is dominated by thermal boundary layers on the walls of the enclosure. The other approach recognizes the importance of the thermal boundary layers for interzone natural convection and thus heat-transfer results can be scaled correctly when thermal sources are present.

1. Bulk Density Driven Flow

Studies that have examined flow driven entirely by bulk density differences between zones include those of Emswiler [122], Brown and Solvason [123], Graf [124], Balcomb and Yamaguchi [125], and Kirkpatrick et al. [126]. Emswiler calculated the flow between zones with different density fluids for a partition with multiple openings by using Bernoulli's equation, but he did not treat the heat-transfer aspects of the problem. Brown and Solvason [123] conducted heat-transfer measurements in an air-filled enclosure that was divided into hot and cold regions by a single partition with a small variable-size opening. They developed an analytical expression for the heat transfer through the partition by the relation

$$\mathrm{Nu}_H = \frac{C}{3} \left(\frac{w}{W} \right) \left(\frac{l}{H} \right)^{3/2} (\mathrm{Ra}_H \, \mathrm{Pr})^{1/2} \qquad (54)$$

The constant C appearing in Eq. (54) is the discharge coefficient for the aperture and can have values $C \le 0.595$ [127]. The parameters l/H and w/W are the ratios of doorway height to total enclosure height and doorway width to total enclosure width, respectively. The temperature difference used in the definition of Nu_H and Ra_H in Eq. (54) is the temperature difference between the fluid in the hot and cold zones. Graf [124] examined invicid mixed forced and free convection by adding the pressure contribution from the forced flow to the pressure produced by the density differences between the fluid reservoirs. Examples of flow profiles resulting from the inviscid calculations are shown in Fig. 51. Balcomb and Yamaguchi [125] prepared a summary of velocity and temperature measurements taken in an occupied solar building. Kirkpatrick *et al.* [126] extended the model proposed by Brown and Solvason to include the effects of thermal stratification in the fluid reservoirs on either side of the partition. They measured thermal stratification levels in an unoccupied solar building and found that they could use their model to make accurate predictions of airflow and heat transfer that occurred as a result of the measured temperature differences between building zones.

2. *Boundary Layer Driven Flow*

In many building applications the heat transfer and fluid flow between zones are dominated by thin boundary layers that form next to heated and cooled surfaces. Studies of flow through two-dimensional partitions in the boundary layer region have been made by Janikowski *et al.* [128], Bejan

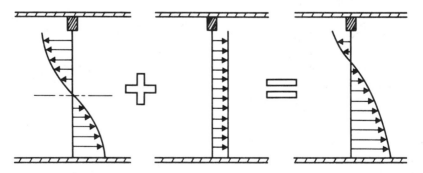

FIG. 51. Interzone flow profiles predicted by using Bernoulli's equation and assuming isothermal fluid reservoirs on either side of the partition; the skewed velocity profile is attributable to combined forcing pressure and density difference between the fluid reservoirs.

and Rossie [129], Nansteel and Greif [130], Bajorek and Lloyd [131], Chang et al. [132], Lin and Bejan [133], and Nansteel and Greif [118]. These studies were experimental, with the exception of Chang's, who used a finite-difference model for a geometry similar to that for the experimental work by Bajorek and Lloyd. Lin and Bejan, in addition to their experimental results, provided a perturbation solution valid in the limit Ra → 0. The only three-dimensional study is that by Nansteel and Greif [118] who considered a partition with a rectangular opening. A summary of the geometries and boundary conditions of these papers is shown in Table VI. Nansteel and Greif [130] and Lin and Bejan [133] demonstrated that the presence of a partition between zones tends to reduce the natural convection boundary-layer flow in subregions that are subjected to stable thermal boundary conditions. This effect reduces the wall area exposed to the primary boundary-layer flow and results in an overall reduction in the convective heat transfer between the hot and cold surfaces on either side of the partition. Nansteel and Greif [118] correlated their data to include this effect. Their correlation is

$$\mathrm{Nu}_H = 0.915(l/H)^{0.401} \, \mathrm{Ra}_H^{0.207} \tag{55}$$

The ranges of parameters for the above were

$$\tfrac{1}{4} \leq l/H \leq 1 \tag{56}$$

and

$$w/W = 0.093 \tag{57}$$

The temperature difference used in the evaluation of Nu_H and Ra_H in Eq. (55) is the temperature difference between the hot and cold end walls of the enclosure. Nansteel and Greif found that Eq. (55) can be used for width ratios w/W larger than 0.093 with a maximum error of 5–10%. The independence of heat transfer from doorway width in the boundary-layer regime is a result of the relative thinness of the boundary layer compared with the dimensions of the doorway.

If the area of the aperture is smaller than the area required for the passage of the boundary-layer flow, then the boundary-layer flow will have to accelerate to pass through the aperture. The additional driving force required to convect the flow through the aperture can only be provided by the creation of bulk density differences between the hot and cold zones of the enclosure. The flow area, A_{bl}, required by the boundary layers on heated and cooled surfaces can be calculated by summing the product of

TABLE VI

SUMMARY OF INTERZONE NATURAL CONVECTION STUDIES

Configuration	Type, parameters	Reference
	Air-filled $Gr_L = 1.1 \times 10^6$ $H/L_{enclosure} = 5$	Janikowski et al. [128]
	Water-filled $5 \times 10^6 \leq Ra_H \leq 5 \times 10^7$ $H/L_{duct} = \frac{1}{6}$	Bejan and Rossie [129]
	Water-filled $2 \times 10^{10} \leq Ra_L \leq 1 \times 10^{11}$ $H/L_{enclosure} = \frac{1}{2}$	Nansteel and Greif [130]
	Air-filled, CO_2-filled $10^5 \leq Gr_L \leq 3 \times 10^6$ $H/L_{enclosure} = 1$	Bajorek and Lloyd [131]
	Air-filled $10^3 \leq Gr_L \leq 10^8$ $H/L_{enclosure} = 1$	Chang et al. [132]
	Water-filled $10^9 \leq Ra_H \leq 10^{10}$ $H/L_{enclosure} = 0.31$	Lin and Bejan [133]
	Water-filled $10^{10} \leq Ra_L \leq 10^{11}$ $H/L_{enclosure} = \frac{1}{2}$	Nansteel and Greif [118]
	Water-filled $10^{11} \leq Ra_H \leq 10^{13}$ $H/L_{enclosure} = 1$	Scott et al. [135]

the thickness and width of each boundary layer in the enclosure, i.e.,

$$A_{bl} = \sum_{n=1}^{N} (\delta W)_n \tag{58}$$

According to the model described above, flow blockage will occur when

$$A_{bl}/lw = \sum_{n=1}^{N} (\delta W)_n/lw \sim 1 \tag{59}$$

For laminar flow it can be shown [134] that the natural convection boundary-layer thickness next to a vertical surface is scaled by the relationship

$$\delta/H \sim 1/Ra_H^{*1/5} \tag{60}$$

If we assume that the height, width, and average heat flux from each active surface are H, W, and q'', respectively, then the flow blockage criteria expressed by Eq. (59) can be rearranged into the simple form

$$\frac{l}{W} \frac{w}{W} \sim \frac{N}{Ra_H^{*1/5}} \tag{61}$$

The left-hand side of Eq. (61) is the ratio of the area of the aperture to the cross-sectional area of the enclosure and N is the number of active heat transfer surfaces in the enclosure. Equation (61) predicts that the onset of flow blockage is directly proportional to the number of active heat-transfer surfaces and is inversely proportional to the Rayleigh number that characterizes the natural convection flow.

A comparison between Eq. (54) and Eq. (55) demonstrates that the natural convection flow regime that governs the flow through the aperture (bulk density driven or boundary-layer driven) has a strong impact upon the geometric dependence of the heat-transfer coefficient. In the bulk density driven regime [Eq. (54)] the heat transfer between zones depends strongly upon both the aperture height ratio l/H and the aperture width ratio w/W. In the boundary-layer regime (Eq. 55) the heat transfer between zones depends weakly upon the aperture height ratio and appears to be independent of the aperture width ratio. Because of these differences, it is important to be able to predict when a multizone flow is in the bulk density driven or boundary layer driven regime. Scott, Anderson, and Figliola [135] conducted an experimental investigation to determine the onset of blockage of natural convection boundary-layer flow in a two-zone cavity with differentially heated end walls. They varied the width of the aperture between the zones while measuring the heat transfer and temperature distributions within the cavity. They found that flow blockage occurred when the area of the aperture was reduced below a critical value which was in qualitative agreement with Eq. (61). The temperature difference between the hot and cold zone reported by Scott *et al.* [135] is shown as a function of aperture width ratio in Fig. 52.

The data shown in Fig. 52 are for a constant zone-to-zone convective energy transport rate of 500 W. Also shown on Fig. 52 is the zone-to-zone temperature difference predicted by bulk density model of Brown and Solvason [123] [Eq. (54)]. As the width of the flow opening is reduced, the experimental data demonstrate that the zone-to-zone temperature difference required by the boundary-layer flow does not increase dramatically until $w/W \le 0.10$. A flow driven by bulk density differences, on the other hand, requires a steady increase in zone-to-zone temperature difference as the width of the flow aperture is reduced. This result demonstrates that boundary-layer flows more than potential to transport energy without requiring large zone-to-zone temperatures particularly in critical flow applications where flow aperture areas are limited.

3. *Trombe Walls*

The two-zone studies described above assumed that the zones on either side of the dividing partition have the same aspect ratios. In a Trombe wall the width of the direct gain zone is reduced until it becomes a vertical duct bounded by the window surface and the absorbing wall surface (Fig. 5b). The first studies of natural convection between vertical parallel plates were performed by Elenbaas [136] and Ostrach [137] for fully developed flows. Studies with specific applications to solar buildings have been done by

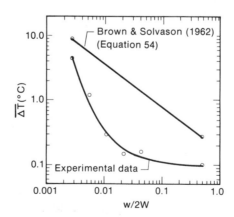

Fig. 52. Boundary-layer flow blockage in a two-zone enclosure (from Scott *et al.* [135]). $c = 0.6$.

Akbari and Borgers [138], Allen and Hayes [139], Tasdemiroglu *et al.* [140] and Ormiston, Raithby, and Hollands [141]. Bodoia and Osterle [142] considered the problem of developing flow between two isothermal plates with the same temperature. Miyatake and Fujii [143] and Miyatake *et al.* [144] calculated developing flow between two vertical plates when one plate was insulated and the other was an isothermal or constant flux surface. Aung *et al.* [145] conducted a numerical and experimental investigation of developing natural convection flow in a vertical duct with asymmetric side wall heating for both constant heat flux and constant temperature boundary conditions. Aung *et al.* [145] considered variable levels of the temperature or heat flux on the two side walls.

All of the studies referenced above that used numerical calculations assumed that the flow between the plates was parabolic and specified the velocity profile of the entering flow. Kettleborough [146] used an elliptic calculation method and found that regions of reverse flow could exist, particularly at high Rayleigh numbers. Sparrow *et al.* [147] observed flow reversals near the exit of a vertical channel with one insulated side wall and one isothermal side wall, but found that the average Nusselt number was unaffected by the presence of the recirculating zone. Sparrow *et al.* [148] considered natural convection combined with radiation in a vertical channel with one insulated wall and one isothermal wall. They found that the radiative transport between the walls increased the convective heat transfer by 50–70% for $1.1 \leq T_W/T_\infty \leq 1.25$, where T_∞ is the temperature of the fluid entering the channel.

The effect of channel width upon natural convection heat transfer between vertical parallel plates was studied experimentally by Sparrow and Azevedo [149]. They found that heat transfer was reduced dramatically if the width of the channel had the same order of magnitude as the boundary-layer thickness. The reduction in convective heat transfer that was measured in their experiment is plotted for different channel spacing in Fig. 53. Sparrow and Azevedo were able to reduce all of their data to a single curve by plotting the data as shown in Fig. 54. Their final correlation for heat transfer over the entire range of plate spacing $0.011 \leq L/H \leq 0.5$ and $3 \leq \mathrm{Ra}_L(L/H) \leq 10^8$ is

$$\overline{\mathrm{Nu}_L} = \left\{ \left(\frac{12}{(L/H\mathrm{Ra}_L)} \right)^2 + \left(\frac{1}{0.619(L/H\,\mathrm{Ra}_L)^{1/4}} \right)^2 \right\}^{-1/2} \tag{62}$$

Based upon their results, the channel width should satisfy the following inequality to avoid flow blockage effects

$$(L/H)\mathrm{Ra}_H^{1/4} \geq 5 \tag{63}$$

For a typical building application with $\mathrm{Ra}_H \sim 10^{10}$ and $H \sim 3$ m, the

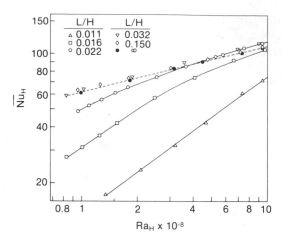

FIG. 53. Natural convection heat transfer as a function of channel with L (from Sparrow and Azevedo [149]).

channel width calculated from Eq. (62) is

$$L \geq 4.7 \quad \text{cm} \tag{64}$$

The equation assumes that there are no flow restrictions at the entrance and exit of the Trombe wall. In real applications that involve entrance and exit losses, the width of the duct at the entrance and exit should be increased beyond that recommended by Eqs. (63) and (64).

Ormiston *et al.* [141] completed a series of numerical calculations of

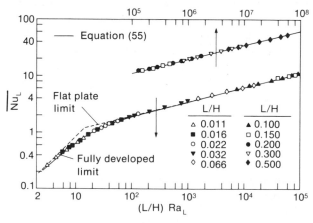

FIG. 54. Nusselt number correlation for natural convection in a parallel plate channel (from Sparrow *et al.* [149]).

natural convection for a Trombe wall channel that was coupled to a cold sink in a single-building zone. The problem was idealized by assuming that the cold sink was a perfect heat exchanger located in a trapezoidal volume near one end of the building zone. They found that their calculated heat-transfer results were 10% lower than the experimental results of Elenbaas [136] for parallel heated surfaces in an infinite medium. They attributed the difference to entrance and exit losses due to the turning of the flow at the top and bottom of the Trombe wall channel.

Tasdemiroglu *et al.* [140] conducted side-by-side tests of buildings, with and without a Trombe wall. They measured temperature distributions and incident solar radiation at intervals of 30 min and calculated the performance of the Trombe wall system. They found that the Trombe wall transmitted 15–35% of the incident solar radiation to the interior of the house.

IV. Summary and Conclusions

A. ACTIVE SOLAR COLLECTORS

There is a great deal of information available on natural convection phenomena under conditions and geometric configurations used in active solar collection systems. This information is adequate for calculating the heat loss by natural convection from flat-plate collectors at any orientation, but some of the scientific aspects of natural convection in small aspect ratio configurations used in convection suppression applications still elude complete explanations.

Available data on natural convection in line-focusing compound-parabolic configurations is adequate to estimate the heat loss in this type of collection system. Similarly, available information is adequate for estimating the heat loss in the annular space of line-focusing parabolic trough configurations, but most of the information is for uniform heat flux or uniform temperature boundary conditions. In real systems, the solar energy is reflected only onto a part of the receiver and there is relatively little information on the effect of nonuniform temperature or heat flux on the flow and heat loss in geometric configurations such as an annulus.

For point-focusing systems, there appears to be adequate information to calculate the heat loss from external-type central receivers. On the other hand our understanding of the mechanism and our ability to calculate natural convection losses from cavity-type central receivers is poor and additional work is required to clarify the situation. For parabolic dish receivers, relatively little experimental and analytic information exists. The situation is particularly complex because a receiver in such a system is in continuous motion, causing the orientation of the cavity with respect to the gravity vector to change during the day. In fact, only a single reference was found that deals specifically with natural convection loss from cavity receivers of point focusing parabolic dishes.

B. SOLAR BUILDINGS

Natural convection is a primary heat transport mechanism in solar buildings. Geometries of buildings are usually complex and, until recently, information on natural convection was essentially confined to heating and cooling of the floor and ceiling or of opposite vertical walls in single-zone rectangular enclosures. Only in recent years has research been conducted on heat transfer by natural convection in enclosures with complex geometries and nonuniform thermal boundary conditions such as those found in real building applications. As a result of this research, it is now possible to predict thermal stratification and heat-transfer rates as a function of thermal boundary conditions in rectangular and triangular enclosures, as well as in simple single-level, multizone configurations. There are several numerical codes available to predict natural convection flow in enclosures and their range of applicability and reliability for solar building energy simulation could be improved dramatically by validating them with some current research results. Because of the wide range of geometries and thermal boundary conditions that are encountered in building design, there are a number of research areas that require further consideration. Some of the most important of these areas are summarized below.

Most buildings operate with Rayleigh numbers large enough that at least a portion of the boundary layer over the interior and exterior walls is turbulent. A good deal of work has been done on the transition to turbulent flow in enclosures with large aspect ratios (like flat-plate collectors), but there is a need for additional analytical and experimental studies of transition as well as turbulent natural convection in enclosures with aspect ratio of order 1, the geometry found in rooms of residential and commercial buildings. In particular, there is a need to determine the effect of Prandtl number upon heat transfer in the transitional and laminar regimes.

Real buildings consist of several rooms at different levels. Consequently, natural convection occurs in multiple flow paths, both in the horizontal and vertical direction. Most heat-transfer studies of interzonal flows driven by natural convection have been limited to single-level geometries. There exists a real need to extend available information to three-dimensional flows in complex geometries, particularly those encountered in rooms at different levels, or shafts and connecting spaces as encountered in commercial buildings and astria.

Most laboratory investigations have been conducted in enclosures that have relatively smooth surfaces. In real buildings, however, there are furnishings such as draperies, blinds, and furniture, as well as obstructions such as door soffits that can interfere with normal boundary-layer flow. Our understanding of the effects of such real world amenities in buildings on the predictions based on laboratory experiments is limited. It is important to quantify these effects in order to determine the extent to which results from idealized experiments can be applied to the real world.

In many building applications the natural convection flows described in this paper interact with forced flow generated by auxiliary space conditioning systems. Very little is known about the combination of Rayleigh–Reynolds number ranges over which forced flow appreciably affects the results of natural convection predictions in enclosures.

We believe that building heat transfer information can be more conveniently and more cheaply obtained with water or Freon in small-scale models, than with air in full-scale models. However, there is a reluctance among architects to accept results obtained in model studies. Therefore, a carefully controlled experimental program aimed at convincing potential users of the validity and usefulness of correlations obtained from small scale models would be important.

A number of the areas described above are the subject of ongoing research efforts and substantial additional information should be available in the next five or ten years. It may take longer, however, to integrate this information into the architectural design methodology, and we recommend that a serious effort continue to be made to package the results of research in the thermal sciences in a form that can easily be used by architects to reduce energy consumption without sacrificing human comfort.

NOMENCLATURE

A_a	aperture area	m²		h	specific enthalpy	J/kg
A_r	receiver area	m²		$\bar{h}_c$	average convective heat-transfer coefficient	W/m² °C
AR	aspect ratio, H/L					
C	concentration ratio A_a/A_r			I_c	insolation on collector aperture	W/m²
c_p	specific heat at constant pressure	J/kg °C		I_b	beam insolation	W/m²
Gr_L	Grashof number $(g\beta/L^3\,\Delta T)/v^2$			I_d	diffuse insolation	W/m²
				k	thermal conductivity	W/m °C
Gr_H	Grashof number $(g\beta H^3\,\Delta T)/v^2$			L	spacing between enclosure walls or characteristic length as defined in text	m
g	gravitational acceleration	m/sec²				
H	enclosure height	m		l	doorway height	m

$\dot{m}$	mass flow rate	kg/sec	T_f	mean fluid temperature	°C
Nu_H	Nusselt number $\bar{h}_c H/k$			$(T_{\mathrm{in}} + T_{\mathrm{out}})/2$	
Nu_L	Nusselt number $\bar{h}_c L/k$		T_H	average receiver	°C
Pr	Prandtl number, $c_p \mu/k$			temperature	
PWR	$q_{\mathrm{sidewall}}/q_{\mathrm{floor}}$		T_s	surface temperature	°C
q_c	rate of heat transfer by	W/m²	ΔT	surface temperature	
	convection			difference	
q_u	rate of useful heat		$\overline{\Delta T}$	bulk fluid temperature	
	transfer	W/m²		difference	
R	thermal resistance	m² °C/W	U	total thermal loss	W/m² °C
Re_f	film Reynolds number			coefficient based on A_r	
	$V\delta/\nu$		V	velocity	m/sec
Ra_H	Rayleigh number		W	enclosure width	m
	$g\beta H^3 \, \Delta T/\nu\alpha$		w	doorway width	m
Ra_H^*	flux modified Rayleigh		x	distance from leading	m
	number $g\beta H^4 q_c \nu\alpha k$			edge or coordinate	
Re_L	Reynolds number $VL\rho/\mu$		y	vertical distance or	m
T_{amb}	ambient temperature	°C		coordinate	

Greek Symbols

α	thermal diffusivity	m²/sec	μ	viscosity	kg/m sec
β	coefficient of expansion	K^{-1}	ν	μ/ρ	m²/sec
δ	boundary-layer thickness	m	ρ	density	kg/m³
δ	film thickness		ϕ	collector tilt angle	degrees
θ	dimensionless		Ω	$(T_T - T_B)/(T_H - T_C)$	
	temperature,				
	$(T_{\mathrm{core}} - T_C)/(T'_{\mathrm{core}} - T_C)$				

Subscripts

B	bottom wall	o	outlet or outside
C	cold	T	top wall
H	hot	w	wall
i	inlet or inside		

Superscripts

'	midpoint	
—		average

ACKNOWLEDGMENTS

The authors would like to thank several individuals for reviews and helpful comments during the preparation of the manuscript. In particular, Ms. Mary Linskens, Dr. M. Abrams, and J. B. Wright have been very helpful. The work was partially supported by the U.S. Department of Energy (DOE) Solar Buildings Program, and the DOE Solar Thermal Program. Portions of this chapter were originally presented as lectures at a NATO Advanced

Study Institute (ASI) in 1984 and published in the Advanced Study Institute Book *Natural Convection,* (S. Kakac, W. Aung, and R. Viskanta, eds.), Hemisphere Publishing Corp., 1985. We express our appreciation to the publisher for the permission to use a portion of this lecture material in the development of the present chapter.

Support for the preparation of the final manuscript was provided by the DOE Solar Technical Information Program.

REFERENCES

1. C. F. Chen and D. H. Johnson, Double diffusive convection: A report of an engineering foundation conference. *J. Fluid Mech.* **138,** 405–416 (1984).
2. R. Viskanta, T. L. Bergman, and F. P. Incropera, Double diffusive natural convection. *In* "Natural Convection: Fundamentals and Applications " (S. Kakac, W. Aung, and R. Viskanta, eds.), Hemisphere, New York, 1985.
3. F. Kreith and J. F. Kreider, "Principles of Solar Engineering." McGraw-Hill, New York, 1978.
4. F. Kreith, R. Davenport, and J. Feustel, Status review and prospects for solar industrial process heat. *J. Sol. Energy Eng.* **105,** 385–400 (1983).
5. A. Rabl, "Active Solar Collectors and Their Applications." Oxford Univ. Press, London and New York, 1985.
6. F. Kreith and M. Bohn, "Principles of Heat Transfer," 4th ed. Harper & Row, New York, 1986.
7. J. D. Balcomb, Passive solar energy systems for buildings. *In* "Solar Energy Handbook" (J. F. Kreider and F. Kreith, eds.), pp. 16–27. McGraw-Hill, New York, 1981.
8. H. Buchberg, I. Catton, and D. K. Edwards, Natural convection in enclosed spaces — A review of application to solar energy collection. *J. Heat Transfer* **98,** 182–188 (1976).
9. I. Catton, Natural convection in enclosures. *Heat Transfer Int. Heat Transfer Conf., 6th, 1978* (1978).
10. S. Ostrach, Natural convection in enclosures." *Adv. Heat Transfer* **8** (1972).
11. S. Ostrach, Natural convection heat transfer in cavities and cells. *Heat Transfer, Proc. Int. Heat Transfer Conf., 7th, 1982* (1983).
12. J. E. Hart, Stability of the flow in a differentially heated inclined box. *J. Fluid Mech.* **47,** 547–576 (1971).
13. H. Ozoe, H. Sayama, and S. W. Churchill, Natural convection in an inclined rectangular channel at various aspect ratios and angles — Experimental measurements. *Int. J. Heat Mass Transfer* **18,** 1425–1431 (1975); see also Natural convection patterns in a long-inclined rectangular box heated from below. Part I. Three dimensional photography. *ibid.* **20,** 123–129 (1977).
14. D. W. Ruth, K. G. T. Hollands, and G. D. Raithby, On free convection experiments in inclined air layers heated from below. *J. Fluid Mech.* **96,** 461–479 (1980).
15. S. J. M. Linthorst, W. M. Schinkel, and C. J. Hoogendoorn, Flow structure with natural convection in inclined air-filled enclosures. *J. Heat Transfer* **103,** 535–537 (1981).
16. H. Inaba, Experimental study of natural convection in an inclined air layer. *Int. J. Heat Mass Transfer* **27,** 1127–1139 (1984).
17. R. J. Goldstein and Q.-J. Wang, An interferometric study of the natural convection in an inclined water layer. *Int. J. Heat Mass Transfer* **27,** 1445–1453 (1984).
18. H. Ozoe, K. Fujii, N. Lior, and S. W. Churchill, Long rolls generated by natural convection in an inclined, rectangular enclosure. *Int. J. Heat Mass Transfer* **26,** 1427–1438 (1983).
18a. W. M. M. Schinkel, National convection in inclined air-filled enclosures. Ph.D Thesis, Delft University of Technology, The Netherlands (1980).

19. S. M. ElSherbiny, G. D. Raithby, and K. G. T. Hollands, Heat transfer by natural convection across vertical and inclined air layers. *J. Heat Transfer* **104,** 96–102 (1982).

19a. J. G. A. DeGraaf and E. F. M. Van Der Held, The relation between the heat transfer and convection phenomena in enclosed plane air layers. *Appl. Sci. Res.* **3,** 393–409 (1953).

19b. E. R. G. Eckert and W. O. Carlson, Natural convection in an air layer enclosed between two vertical plates with different temperatures. *Int. J. Heat Mass Transfer* **2,** 106–120 (1961).

19c. W. M. M. Schinkel and C. J. Hoogendoorn, An Interferometric study of the local heat transfer by natural convection in inclined air-filled enclosures. *Heat Transfer, Int. Heat Transfer Conf., 6th, 1978,* Pap. NC-18 (1978).

19d. K. R. Randall, J. W. Mitchell, and M. M. El-Wakil, Natural convection heat transfer characteristics of flat-plate enclosures. *J. Heat Transfer* **101,** 120–125 (1979).

20. K. G. T. Hollands, L. Konicek, T. E. Unny, and G. D. Raithby, Free convection heat transfer across inclined air layers. *J. Heat Transfer* **98,** 189–193 (1976).

21. H. Buchberg and D. K. Edwards, Design considerations for solar collectors with cylindrical glass honeycombs. *Sol. Energy* **18,** 193–204 (1976).

22. H. Buchberg, O. A. Lalude, and D. K. Edwards, Performance characteristics of rectangular honeycomb solar-thermal converters. *Sol. Energy* **13,** 193–221 (1971).

23. K. G. T. Hollands, Honeycomb devices in flat-plate solar collectors. *Sol. Energy* **9,** 159 (1965).

24. K. L. D. Cane, K. G. T. Hollands, G. D. Raithby, and T. E. Unny, Free convection heat transfer across inclined honeycombed panels. *J. Heat Transfer* **99,** 86–91 (1977).

25. K. G. T. Hollands, G. D. Raithby, and T. E. Unny, "Studies on Methods of Reducing Heat Losses from Flat-Plate Collectors," Final Rep., ERDA Contract E(11-1)-2597. University of Waterloo, Ontario, Canada, 1976.

26. K. G. T. Hollands, G. D. Raithby, F. B. Russell, and R. G. Wilkinson, Coupled radiative and conductive heat transfer across honeycomb panels and through single cells. *Int. J. Heat Mass Transfer* **27,** 2119–2131 (1984).

27. C. J. Hoogendoorn, Natural convection suppression in solar collectors. *In* "Natural Convection: Fundamentals and Applications" (S. Kakac, W. Aung, and R. Viskanta, eds.). Hemisphere, New York, 1985.

28. K. G. T. Hollands and K. Iynkaran, Proposal for a compound-honeycomb collector. *Sol. Energy* **34,** 309–316 (1985).

29. J. N. Arnold, I. Catton, and D. K. Edwards, Experimental investigation of natural convection in inclined rectangular regions of differing aspect ratios. *J. Heat Transfer* **98,** 67–71 (1976).

30. J. N. Arnold, D. K. Edwards, and I. Cattan, Effects of tilt and horizontal aspect ratio on natural convection in rectangular honeycomb solar collectors. *J. Heat Transfer* **19,** 120–122 (1977).

31. P. A. Meyer, J. W. Mitchell, and M. M. El-Wakil, Natural convection heat transfer in moderate aspect ratio enclosures. *J. Heat Transfer* **101,** 655–659 (1979).

32. D. R. Smart, K. G. T. Hollands, and G. D. Raithby, Free convection heat transfer across rectangular-celled diathermanous honeycombs. *J. Heat Transfer* **102,** 75–80 (1980).

33. J. G. Symons and M. K. Peck, Natural convection heat transfer through inclined longitudinal slots. *J. Heat Transfer* **106,** 824–829 (1984).

34. J. G. Symons, Natural convection in inclined cavities with half and full partitions. *Proc. Australas. Conf. Heat Mass Transfer, 3rd,* pp. 69–76 (1985).

35. J. L. Balvanz and T. H. Kuehn, Effect of wall conduction and radiation on natural convection in a vertical slot with uniform heat generation on the heated wall. *HTD* [Publ.] *(Am. Soc. Mech. Eng.)* **8**, 55–62 (1980).
36. R. Anderson and M. Bohn, Heat transfer enhancement in natural convection enclosure flow. *HTD* [Publ.] *(Am. Soc. Mech. Eng.)* **39**, 29–38 (1985). Also *J. Heat Transfer* **108**, 330–337 (1986).
37. R. K. MacGregor and A. F. Emery, Free convection through vertical plane layers— Moderate and high prandtl number fluids. *J. Heat Transfer* **91**, 391–403 (1969).
38. W. M. M. Schinkel and C. J. Hoogendoorn, Natural convection in collector cavities with an isoflax absorber plate. *J. Sol. Energy Eng.* **105**, 19–22 (1983).
39. P. K. B. Chao, H. Ozoe, and S. W. Churchill, The effect of nonuniform surface temperature on laminar natural convection in a rectangular enclosure. *Chem. Eng. Commun.* **9**, 245–254 (1981).
40. G. S. H. Lock and R. S. Ko, Coupling through a wall between two free convective systems. *Int. J. Heat Mass Transfer* **16**, 2087–2096 (1973).
41. R. Anderson and A. Bejan, Natural convection on both sides of a vertical wall separating fluids at different temperatures. *J. Heat Transfer* **102**, 630–635 (1980).
42. R. Anderson and A. Bejan, Heat transfer through single and double vertical walls in natural convection: theory and experiment. *Int. J. Heat Mass Transfer* **24**, 1611–1620 (1981).
43. R. Viskanta and D. W. Lankford, Coupling of heat transfer between two natural convection systems separated by a vertical wall. *Int. J. Heat Mass Transfer* **24**, 1171–1177 (1981).
44. E. M. Sparrow and C. Prakash, Interaction between internal natural convection in an enclosure and an external natural convection boundary-layer flow. *Int. J. Heat Mass Transfer* **24**, 895–907 (1981).
45. F. Kreith, G. O. G. Lof, A. Rabl, and R. Winston, Solar collectors for low and intermediate temperature applications. *Prog. Eng. Cambridge Sci.* **6**, 1–34 (1980).
46. R. Winston, Principles of solar concentrators of a novel design. *Solar Energy* **16**, 89 (1974).
47. A. Rabl, Comparison of solar concentrators. *Sol. Energy* **18**, 93 (1976).
48. J. J. O'Gallagher, K. Snail, R. Winston, C. Peek, and J. D. Garrison, A new evacuated CPC collector tube. *Sol. Energy* **29**, 575–577 (1982).
49. U. Ortabasi and F. Fehlner, Cusp Mirror-heat pipe evacuated tubular solar thermal collector. *Sol. Energy* **24**, 477–489 (1980).
50. T. S. Lee, N. E. Wijeysundera, and K. S. Yeo, Free convection fluid motion and heat transfer in horizontal concentric and eccentric cylindrical collector systems. *In* "Solar Engineering 1984" (D. Y. Lowani, ed.), pp. 194–200. Am. Soc. Mech Eng., New York, 1984.
51. C. K. Hsieh, Thermal analysis of CPC collectors. *Sol. Energy* **27**, 19–29 (1981).
52. C. K. Hsieh and F. M. Mei, Empirical equations for calculation of CPC collector loss coefficient. *Sol. Energy* **30**, 487–470 (1970).
53. S. I. Abdel-Khalik, H.-W. Li, and K. R. Randall, Natural convection in compound parabolic concentrators—A finite element solution. *J. Heat Transfer* **100**, 199–204 (1978).
54. B. A. Meyer, J. W. Mitchell, and M. M. El-Wakil, Convective heat transfer in trough and C.P.C. collectors. *Proc. Annu. Meet., Am. Sect. ISES,* Vol. 3.1, pp. 437–440 (1980).
55. M. Collares Pereira, J. Dugue, A. Joyce, M. Delgado, G. Serrudo, and A. Rego-Teixeira, A 3X CPC-type concentrator with tubular receiver and tubular glass envelope to reduce

convective losses: Description and performance. *Proc. Int. Sol. Energy Cong., Sol. World Forum,* pp. 1718–1723 (1981).

56. K. N. Woo, Performance of the compound cylindrical concentrator. *Proc. Annu. Meet. Am. Soc. Eng. Sci.,* Vol. 6, pp. 465–470 (1983).

56a. A. Rabl, J. O'Gallagher, and R. Winston, Design and test of nonevacuated solar collectors with compound parabolic concentrators. *Sol. Energy* 25, 335–351 (1980).

56b. R. Patton, Design considerations for a stationary concentrating collector. *Conc. Sol. Collect., Proc. ERDA Conf., 1977,* pp. 3–37 (1977).

56c. D. E. Prapas, B. Norton, and S. D. Probert, Design of compound concentrating solar energy collectors. *J. Sol. Energy Eng.* (1985).

57. A. Rabl, P. Bendt, and W. W. Gaul, Optimization of parabolic trough solar collectors. *Sol. Energy* 29, 407–417 (1982).

58. T. H. Kuehn and R. J. Goldstein, An experimental and theoretical study of natural convection in the annulus between horizontal concentric cylinders. *J. Fluid Mech.* 74, 695–719 (1976).

59. T. H. Kuehn and R. J. Goldstein, Correlating equations for natural convection heat transfer between horizontal circular cylinders. *Int. J. Heat Mass Transfer* 19, 1127–1134 (1976).

60. T. H. Kuehn and R. J. Goldstein, An experimental study of natural convection heat transfer in concentric and eccentric horizontal cylinders. *J. Heat Transfer* 100, 535–540 (1978).

61. T. H. Kuehn and R. J. Goldstein, A parametric study of prandtl number and diameter ratio effects on natural convection heat transfer in horizontal cylindrical annuli. *J. Heat Transfer* 102, 768–770 (1980).

62. K. Butti and J. Perlin, "A Golden Thread." Van Nostrand-Reinhold, Princeton, New Jersey, 1980.

63. F. Kreith and R. T. Meyer, Large-scale use of solar energy with central receivers. *Am. Sci.* 71, 518–605 (1983).

64. M. Abrams, The status of research on convective losses from solar central receivers. **SAND 83–8224,** Sandia National Laboratory, Livermore, CA (1983).

65. D. L. Siebers, R. J. Moffat, and R. G. Schwind, Experimental mixed convection from a large, vertical plate in a horizontal flow. *Heat Transfer Proc. Int. Heat Transfer Conf., 7th 1982,* MC 13,1 (1983) (see also **SAND 83–8225**); Experimental, variable properties natural convection from a large, vertical flat surface. *Proc. ASME-JSME Therm. Eng. J. Conf., 1983,* Vol. 3, pp. 269–275 (1983).

66. A. M. Clausing, Advantages of a cryogenic environment for experimental investigations of convective heat transfer. *Int. J. Heat Mass Transfer* 25, 1255–1257 (1982).

67. A. M. Clausing, Convective losses from cavity solar receivers—Comparisons between analytical predictions and experimental results. *J. Sol. Energy Eng.* 105, 29–33 (1983).

68. B. Afshari and J. H. Ferziger, Computation of orthogonal mixed-convection heat transfer. *Proc.—ASME-JSME Therm. Eng. Jt. Conf., 1983,* Vol. 3, pp. 169–173 (1983).

69. E. Achenbach, The effect of surface roughness on the heat transfer from a circular cylinder to the cross flow of air. *Int. J. Heat Mass Transfer* 20, 359–369 (1977).

69a. K. Y. Wang and R. J. Copeland, Heat transfer in a solar radiation absorbing molten salt film flowing over an insulated substrate. Am. Soc. Mech. Eng. paper 84-WA/SOL-22 (1984).

70. P. K. McMordie, Convection losses from a cavity receiver. *J. Sol. Energy Eng.* 106, 98–100 (1984).

71. Mirenayat, Etude expérimentale du transfert de chaleur par convection naturalle dans

une Cavite isotherme ouverte. D. Eng. Thesis, University of Poitiers (1981); see also Experimental study of heat loss through national convection from an isothermal cubic open cavity. SAND 81-8014, Sandia National Laboratory, Livermore, CA 165–174 (1981).

72. V. Sernas and I. Kyriakidas, "Natural convection in an open cavity. *Heat Transfer, Proc. Int. Heat Transfer Conf., 7th, 1982,* Vol. 2, pp. 275–280 (1983).

73. Y. L. Chen and C. L. Tien, Laminar natural convection in shallow open cavities. *HTD [Publ.] (Am. Soc. Mech. Eng.)* **26,** 77–82 (1983).

74. C. F. Hess and R. H. Henze, Experimental investigation of natural convection losses from open cavities. *J. Heat Transfer* **106,** 333–338 (1984).

75. J. A. C. Humphrey, F. S. Sherman, and K. S. Chen, Experimental study of free and mixed convective flow of air in a heated cavity. **SAND84-8192,** Sandia National Laboratory, Livermore, CA (1985).

76. J. S. Kraabel, An experimental investigation of the natural convection from a side-facing cubical cavity. *Proc.—ASME-JSME Therm. Eng. Jt. Conf., 1983,* Vol. 1, pp. 299–306 (1983).

77. P. LeQuere, J. A. C. Humphrey, and F. S. Sherman, Numerical calculation of thermally driven two-dimensional unsteady laminar flow in cavities of rectangular cross section. *Numer. Heat Transfer* **4,** 249–283 (1981).

78. Y. L. Chen and C. L. Tien, A numerical study of two-dimensional natural convection in square open cavities. *Numer. Heat Transfer* **8,** 65–81 (1985).

79. J. A. C. Humphrey, F. S. Sherman, and W. M. To, "Numerical Simulation of Buoyant Turbulent Flow," Rep. FM-84-6. Dep. Mech. Eng., University of California, Berkeley, 1984.

80. R. F. Boehm, Review of thermal loss evaluations of solar central receivers," **SAND85-8019,** Sandia National Laboratory, Livermore, CA (1985): *J. Sol. Energy Eng.* (to be published).

81. R. Anderson, "Heat and Mass Transfer in Falling Film Receivers," SERI/PR-252-2822. Sol. Energy Res. Inst., Golden, Colorado, 1985.

82. R. T. Taussig, Aerowindows for Central solar receivers. *Proc. ASME Winter Annual Meeting* (1984).

83. J. A. C. Humphrey and E. W. Jacobs, Free-forced laminar flow convective heat transfer from a square cavity in a channel with variable junction. *Int. J. Heat Mass Transfer* **24,** 1584–1547 (1981).

84. A. R. Saydah, A. A. Koenig, R. H. Lambert, and D. A. Kugath, Final report on test of STEP, Shenandoah parabolic dish solar collector quadrant facility. **SAND82-7153,** Sandia National Laboratory, Albuquerque, NM (1983).

85. D. A. Kugath, G. Drenker, and A. A. Koenig, Design and development of a paraboloidal dish solar collector for intermediate temperature service. *Proc. ISES Silver Jubilee Congr.,* Vol. 1, p. 449–453 (1979).

86. A. A. Koenig and M. Marvin, "Convection Heat Loss Sensitivity in Open Cavity Solar Receivers," Final Rep. DOE Contract No. EG77-C-04-3985. Department of Energy, Oak Ridge, Tennessee, 1981.

87. J. A. Harris and T. G. Lenz, Thermal performance of solar concentrator/cavity receiver systems. *Sol. Energy* **34,** 135–142 (1985).

88. E. F. C. Somerscales and M. Kassemi, Electrochemical mass transfer studies in open cavities. *J. Appl. Electrochem.* **15,** 405–413 (1985).

89. J. D. Balcomb, R. W. Jones, R. D. McFarland, and W. O. Wray, "Passive Solar Heating Analysis: A Design Manual." ASHRAE, Atlanta, Georgia, 1984.

90. W. Jones and D. McFarland, "The Sunspace Primer: A Guide to Passive Solar Heating." Van Nostrand-Reinhold, Princeton, New Jersey, 1985.

84   Ren Anderson and Frank Kreith

91. A. E. Gill, The boundary-layer regime for convection in a rectangular cavity. *J. Fluid Mech.* **26**, 515–536 (1966).
92. A. Bejan, On the boundary layer region in a vertical enclosure filled with a porous medium. *Lett. Heat Mass Transfer* **6**, 93 (1979).
93. S. A. Korpela, Y. Lee, and J. E. Drummond, Heat transfer through a double-pane window. *J. Heat Transfer* **104**, 539–544 (1982).
94. A. Bejan, A synthesis of analytical results for natural convection heat transfer across rectangular enclosures. *Int. J. Heat Mass Transfer* **23**, 723–726 (1980).
95. R. D. Flack, T. T. Konopnicki, and J. H. Rooke, The measurement of natural convective heat transfer in triangular enclosures. *J. Heat Transfer* **101**, 648–654 (1979).
96. R. D. Flack, The experimental measurement of natural convection heat transfer in triangular enclosures heated or cooled from below. *J. Heat Transfer* **102**, 770–772 (1980).
97. D. Poulikakos and A. Bejan, Natural convection experiments in a triangular enclosure. *J. Heat Transfer* **105**, 652–655 (1983).
98. V. A. Akinsete and T. A. Coleman, Heat transfer by steady laminar free convection in triangular enclosures. *Int. J. Heat Mass Transfer* **25**, 991–998 (1982).
99. D. Poulikakos and A. Bejan, Fluid dynamics of an attic space. *J. Fluid Mech.* **131**, 251–269 (1983).
100. E. M. Sparrow and F. A. Azevedo, Lateral-edge effects on natural convection heat transfer from an isothermal vertical plate. *J. Heat Transfer* **107**, 977–979 (1985).
101. M. S. Bohn, A. T. Kirkpatrick, and D. A. Olsen, Experimental study of three-dimensional natural convection at high Rayleigh number. *J. Heat Transfer* **106**, 339–345 (1984).
102. M. S. Bohn and R. Anderson, "Temperature and Heat Flux Distributing in a Natural Convection Enclosure Flow," SERI/TR-252-2189. Sol. Energy Res. Inst., Golden, Colorado, 1984. Also *J. Heat Transfer* **108**, 471–475 (1986).
103. Y. Jaluria, Buoyancy-induced flow due to isolated thermal sources on a vertical surface. *J. Heat Transfer* **104**, 223–227 (1982).
104. K. Yamaguchi, Experimental Study of Natural Convection Heat Transfer through an Aperture in Passive Solar Heated Buildings, *Nat. Conf. Passive Solar, 9th, Columbus, Ohio,* 1984.
105. Olsen, D., Glicksman, L., and Yuan, X.-D., Natural convection modeling experiments of building interior spaces. *HTD* [Publ.] *(Am. Soc. Mech. Eng.)* **16/4** (1985). Also, personal communication, May 16, 1986.
106. S. W. Churchill and H. S. Chu, Correlating equations for laminar and turbulent-free convection from a vertical plate. *Int. J. Heat Mass Transfer* **18**, 1323–1329 (1975).
107. S. Ostrach and C. Raghaven, Effect of stabilizing thermal gradients on natural convection in rectangular enclosures. *J. Heat Transfer* **101**, 238–243 (1979).
108. B.-I. Fu and S. Ostrach, The effects of stabilizing thermal gradients on natural convection flows in a square enclosure. *HTD [Publ.] (Am. Soc. Mech. Eng.)* **16**, (1981).
109. G. S. Shiralkar and C. L. Tien, A numerical study of the effect of a vertical temperature difference imposed on a horizontal enclosure. *Numer. Heat Transfer* **5**, 185–197 (1982).
110. A. T. Kirkpatrick and M. S. Bohn, High Rayleigh number natural convection in an enclosure heated from below and from the sides. *ASME-AIChE Natl. Heat Transfer Conf.* (1983).
111. A. T. Kirkpatrick and M. Bohn, An experimental investigation of mixed cavity natural convection in the high Rayleigh number regime. To appear in *Int. J. Heat Mass Transfer* (1985).

112. H. Ozoe, A. Mouri, M. Hiramitsu, S. W. Churchill, and N. Lior, Numerical calculation of three-dimensional turbulent natural convection in a cubical enclosure using a two equation model. *HTD [Publ.] (Am. Soc. Mech. Eng.)* **32**, 25–32 (1984).

113. R. Anderson, E. M. Fisher, and M. Bohn, "Thermal Stratification in a Closed Cavity with Variable Heating of the Floor and One Vertical Wall," SERI/J-252-0146. Sol. Energy Res. Inst., Golden, Colorado, 1985.

114. R. Anderson, E. Fisher, and M. Bohn, "Thermal Stratification in Direct Gain Passive Heating Systems with Variable Heating of the Floor and One Vertical Wall," SERI/TP-252-2767. Sol. Energy Res. Inst., Golden, Colorado, 1985.

115. F. Bauman, A. Gadgil, R. Kammerud, E. Altmayer, and M. Nansteel, Convective heat transfer in buildings: Recent research results. *ASHRAE Trans.* **89**, Part 1A, 215–230 (1983).

116. E. F. Altmayer, A. J. Gadgil, F. S. Bauman, and R. L. Kammerud Correlations for convective heat transfer from room surfaces. *ASHRAE Trans.* **89**, Part 2A, 61–77 (1983).

117. R. Anderson and G. Lauriat, The horizontal natural convection boundary layer regime in a closed cavity. SERI/TP-252-2830, Sol. Energy Res. Inst., Golden, Colorado (1985). Also in *Proc. Int. Conf. Heat Transfer* **4**, 1453–1458 (1986).

117a. Y. Jaluria and B. Gebhart, On transition mechanisms in vertical natural convection flow. *J. Fluid Mech.* **66**, 309–339 (1974).

118. M. W. Nansteel and R. Greif, An investigation of natural convection in enclosures with two- and three-dimensional partitions. *Int. J. Heat Mass Transfer* **27**, 561–571 (1984).

119. S. M. ElSherbiny, K. G. T. Hollands, and G. D. Raithby, Free convection across inclined air layers with one surface V-corrugated. *J. Heat Transfer* **100**, 410–415 (1980).

120. M. Al-Arabi and M. M. El-Refaee, Heat transfer by natural convection from corrugated plates to air. *Int. J. Heat Mass Transfer* **21**, 357–359 (1978).

121. S. Shakerin, M. Bohn, and R. I. Loehrke, Convection in an enclosure with discrete roughness elements on a vertical heated wall. *Heat Transfer, Proc. Int. Heat Transfer Conf., 8th, 1986* (1986).

122. J. E. Emswiler, The neutral zone in ventilation. *Trans. Am. Soc. Heat Vent. Eng.* **32**, 59–74 (1926).

123. W. G. Brown and K. R. Solvason, Natural convection through rectangular openings in partitions 1. *Int. J. Heat Mass Transfer* **5**, 859–868 (1962).

124. A. Graf, Theoretische Betrachtung über den Luftaustausch zwischen zwei Räumen. *Schweiz. Bl. Heiz. Lueft.* **31**, 22–25 (1964).

125. J. D. Balcomb and K. Yamaguchi, Heat distribution by natural convection. *Natl. Passive Sol. Conf., 8th, 1983* (1983); see also J. D. Balcomb, "Heat Distribution by Natural Convection: Interim Report." Los Alamos Natl. Lab., New Mexico, 1985.

126. A. Kirkpatrick, D. Hill, and P. Burns, Interzonal natural and forced convection heat transfer in a passive solar building. *J. Sol. Energy Eng.* (to be published).

127. J. H. Lienhard, V and J. H. Lienhard, IV, Velocity coefficients for free jets from sharp-edged orifices. *J. Fluid Eng.* **106**, 13–17 (1984).

128. H. E. Janikowski, J. Ward, and S. D. Probert, Free convection in vertical, air-filled rectangular counties fitted with baffles. *Heat Transfer, Int. Heat Transfer Conf., 6th, 1978* (1978).

129. A. Bejan and A. N. Rossie, Natural convection in horizontal duct connecting two fluid reservoirs. *J. Heat Transfer* **103**, 108–113 (1981).

130. M. W. Nansteel and R. Greif, Natural convection in undivided and partially divided rectangular enclosures. *J. Heat Transfer* **103**, 623–629 (1981).

131. S. M. Bajorek and J. R. Lloyd, Experimental investigation of natural convection in partitioned enclosures. *J. Heat Transfer* **104**, 527–532 (1982).

132. L. C. Chang, J. R. Lloyd, and K. T. Yang, A finite difference study of natural convection in complex enclosures. *Heat Transfer, Proc. Int. Heat Transfer Conf., 7th, 1982* (1983).

133. N. N. Lin and A. Bejan, Natural convection in a partially divided enclosure. *Int. J. Heat Mass Transfer* **26**, 1867–1878 (1983).

134. A. Bejan, "Convection Heat Transfer," p. 116. Wiley, New York (1984).

135. D. Scott, R. Anderson, and R. Figliola, "Blockage of Natural Convection Boundary Layer Flow in a Multizone Enclosure," SERI/TP252-2847. Sol. Energy Res. Inst., Golden, Colorado, 1985. Also ASME paper **86-HT-390**.

136. W. Elenbaas, Heat dissipation of parallel plates by free convection. *Physica (Amsterdam)* **9**, 1 (1942).

137. S. Ostrach, "Laminar Natural Convection Flow and Heat Transfer of Fluids with and without Heat Sources in Channels with Constant Wall Temperature," NACA TN2863. (Nat. Advis. Comm. Aeronaut., Washington, D.C., 1952).

138. H. Akbari and T. R. Borgers, Free convective laminar flow within the Trombe wall channel. *Sol. Energy* **22**, 165–174 (1979).

139. T. Allen and J. Hayes, Measured performance of thermosiphon air panels. *ASES Natl. Passive Conf. Proc., 10th,* pp. 442–447 (1985).

140. E. Tasdemiroglu, F. Ramos Berjano, and D. Tinaut, The performance results of Trombe-wall passive systems under Aegean Sea climatic conditions. *Sol. Energy* **30**, 181–189 (1985),

141. S. J. Ormiston, G. D. Raithby, and K. G. T. Hollands, Numerical predictions of natural convection in a Trombe wall system. *Am. Soc. Mech. Eng. [Pap.]* **85-HT-36** (1985).

142. J. R. Bodoia and J. F. Osterle, The development of free convection between heated vertical plates. *J. Heat Transfer* **84**, 40–44 (1962).

143. O. Miyatake and T. Fujii, Free convection heat transfer between vertical parallel plates — One plate isothermally heated and the other thermally insulated. *Heat Transfer — Jpn. Res.* **1**, 30–38 (1972).

144. O. Miyatake, T. Fujii, M. Fujii, and H. Tanaka, Natural convection heat transfer between vertical parallel plates — One plate with a uniform heat flux and the other thermally insulated. *Heat Transfer — Jpn. Res.* **2**, 25–33 (1973).

145. W. Aung, L. S. Fletcher, and V. Sernas, Developing laminar free convection between vertical flat plate with asymmetric heating. *Int. J. Heat Mass Transfer* **15**, 2293–2308 (1972).

146. C. F. Kettleborough, Transient laminar free convection between heated vertical plates including entrance effects. *Int. J. Heat Mass Transfer* **15**, 883–896 (1972).

147. E. M. Sparrow, G. M. Chrysler, and L. F. Azevedo, Observed flow reversals and measured-predicted Nusselt numbers for natural convection in a one-sided heated vertical channel. *J. Heat Transfer* **106**, 325–332 (1984).

148. E. M. Sparrow, S. Shah, and C. Prakash, Natural convection in a vertical channel. I. Interacting convection and radiation. II. The vertical plate with and without shrouding, *Numer. Heat Transfer* **3**, 297–314 (1980).

149. E. M. Sparrow and L. F. A. Azevedo, Vertical channel natural convection spanning between the fully-developed limit and the single-plate boundary-layer limit. *Int. J. Heat Mass Transfer* **28**, 1847–1857 (1985).

Heat Transfer from Tubes in Crossflow

A. ŽUKAUSKAS

Academy of Sciences of the Lithuanian SSR, MTP-1 Vilnius 232600, USSR

I. Introduction

The power generation field, the chemical industry, and other technology-based industries increasingly employ heat exchangers involving tubes in crossflow. This has led to increased levels of research on the heat transfer and hydraulic behavior of single tubes and of different arrays of tubes in crossflow of a gas or a viscous fluid.

At the beginning of this century, rapid developments in the field of boiler equipment required more detailed information about the heat transfer performance of single tubes and banks of tubes in flows of gases than was available. Most of the early experiments were carried out in flows of air, as its physical properties scarcely differ from those of combustion gases.

Earlier investigations were concerned with the heat transfer of a single tube [1, 2], as well as with the intensity of the heat transfer of a tube in the inner rows of a bank [3, 4]. These were followed by detailed investigations of the influence on heat transfer of the geometrical parameters of banks [5–7]. The results led to the conclusion that the intensity of the heat transfer on a tube in a bank is higher than that of a single tube and depends on the arrangement of tubes in a bank.

At the same time, the development of a similarity theory improved the methods of generalization of test data. In the 1920s and 1930s, the calculation formulas, including dimensionless groups, became commonly used. Similarity theory, as applied to the process of convective heat transfer, not only suggested the necessary direction of experiments, but also supplied a common basis for the generalization and analysis of the experimental data obtained by different authors and yielded relations suitable for practical calculations.

According to the results of numerous investigations and the wide generalizations of Antuf'yev and Kozachenko [8], Grimison [9], and others [10, 11], Kuznetsov [12] and other authors suggested even more general relations for the calculations of heat transfer and hydraulic drag of banks of tubes in a crossflow of gas. These have been widely applied in the design of steam power boilers.

Further investigations of flow past a single tube and its heat transfer [13–16] facilitated the understanding of the physical phenomenon of heat transfer in banks of tubes in crossflow, the development of thermoanemometry, and the solution of other technical problems. Several books appeared in which extensive information was presented and summarized. Those by Antuf'yev and Beletsky [17], Kays and London [18], and Gregorig [19] are the first to be mentioned.

The works mentioned above dealt only with the heat transfer in flows of gases and only within a moderate range of Reynolds number. Recent accelerated development of science and technology has revealed a number of new problems in the field of heat transfer of tubes in crossflow. A need for reliable formulas for the calculation of the heat transfer of tubes in flows of gases at high Reynolds numbers became apparent. The fast growth of the chemical and power industries and the emergence of some new branches of engineering caused an increased interest in heat transfer in flows of viscous fluids at higher Prandtl and Reynolds numbers.

The shift to higher temperatures and forced heat fluxes requires more precise studies of local heat transfer and flow patterns, as well as the need to consider the influence of heat flux direction, temperature, and other factors on the intensity of heat transfer. More important became the choice of the reference temperature, to which physical properties, included in similarity criteria, are related.

In the past decade, studies of the heat transfer and hydraulic drag from tubes in flows of gases and viscous fluids at high Reynolds numbers[1] (Re) were performed at the Institute of Physical and Technical Problems of Energetics, Academy of Sciences of the Lithuanian SSR [20], at the Kernforschungsanlage, Jülich, Federal Republic of Germany, and at other research centers. We have reviewed our studies and the results of other authors in a previous publication [21].

But since that publication, a number of basically new studies have

[1] The properties in the dimensionless number, such as Re, Nu, Pr, Eu, are evaluated at the mean bulk temperature. In tube bundles with many rows, the mean bulk temperature is taken as the average of the inlet temperature and exit temperature of the tube bundle.

appeared. These studies cover the hydrodynamics and heat transfer of tubes in crossflow at very low Reynolds numbers, as well as in the super-critical flow regime. The range of Prandtl number (Pr) was also extended and different factors of local heat transfer were measured on tubes in crossflow. They all were included in this new survey.

In this article, we shall be concerned with important problems of heat transfer and the hydraulic drag of tubes as mentioned above, and in particular, with the heat transfer of single tubes, banks of tubes, and systems of tubes in crossflow. In connection with this, a separate section will be devoted to the influence of the physical properties of fluids on heat transfer. Extensive experimental data will be analyzed and will include investigations of banks of tubes of various arrangements and a single tube in crossflow in the range of Prandtl number from 0.7 to 500 and that of Reynolds number from 1 to 2×10^6.

Heat transfer is considerably influenced by the flow regime around the tube, while flow past banks of tubes is one of the most complicated problems of practical importance. Knowledge of these processes will enable us to carry out more extensive studies of heat transfer. Therefore, before discussing problems of heat transfer, we shall first consider descriptions of flow past a single tube and a tube in a bank.

II. Flow past a Single Tube

A. FLOW OF AN IDEAL FLUID

The theory of an ideal fluid is based on assumptions of a possibility of slip between the fluid and the wall and of the absence of internal friction. The velocity distribution in a flow of an ideal fluid past a tube is expressed by

$$u_x = u_0 (\sin \varphi)[1 + (R_0/R)^2] \tag{1}$$

where R_0 and R are distances from the tube axis to its surface and to the point considered, respectively, and u_0 is the velocity of the main flow. It is obvious from Eq. (1) that the tangential velocity u_x decreases as the distance from the surface increases, e.g., at $R = 2R_0$ it constitutes 1.25 of u_0 and reaches its maximum value on the tube surface:

$$u_x = 2u_0 \sin \varphi \tag{2}$$

Equation (2) suggests that the fluid velocity is zero in the front and rear stagnation points and reaches its maximum at $\varphi = 90°$. By use of Eq. (2) and Bernoulli's equation, the pressure distribution on the tube surface can be determined from

$$P = \frac{p_\varphi - p_0}{\rho u_0^2/2} = 1 - 4 \sin^2 \varphi \qquad (3)$$

where p_0 is the static pressure in the channel.

Pressure is inversely proportional to velocity, i.e., it is a maximum at stagnation points and a minimum in the main cross section. It follows from the symmetrical pressure distribution on the tube surface that the tube offers no resistance to the flow of an ideal fluid (d'Alembert paradox).

B. Flow of a Real Fluid

1. General Flow Pattern

Because of the viscosity of real fluids, on the front portion of the tube a laminar boundary layer is formed, the thickness of which increases downstream. According to the ratio of inertial and viscous forces in the flow, characterized by its Reynolds number, several types of flow regimes can be distinguished.

At Re < 1, inertial forces are negligibly small as compared to frictional forces, and if the tube is streamlined, the laminar boundary layer separates from the surface at the rear stagnation point.

With increasing Reynolds number, the relative influence of inertial forces increases, and at Re > 5, the laminar boundary layer separates from the tube surface without reaching the rear stagnation point. Behind the tube, there appears a symmetrical pair of stable vortices which form a circulation region confined by flow lines. With a further increase in Reynolds number, the vortices become extended downstream, and at Re > 40, the stability of the movement in the circulation region is lost, and the vortices are periodically shed from the rear of the tube.

The phenomenon of boundary-layer separation is due to internal friction within the boundary layer and is closely connected with the pressure and velocity distribution around the tube. As is known, a certain amount of energy is consumed in overcoming the internal friction in the boundary layer and thus is dissipated. With a velocity decrease and pressure increase over the rear portion of the tube ($dp/dx > 0$), the remainder of the energy does not suffice to overcome the increased pressure. Fluid particles of the boundary layer that have low velocity because of friction become even slower, until they eventually stop and start moving in the opposite direction. Fluid sheets of opposite movement begin to curl and give rise to vortices that shed from the tube. The boundary-layer separation point is at approximately $\varphi = 80°$. This flow pattern is observed up to the critical Reynolds number of 2×10^5.

With a further increase of the Reynolds number, the boundary layer gradually becomes turbulent and additionally receives kinetic energy from the main flow through turbulent fluctuations. As a result, the turbulent part of the boundary layer is shifted downstream, and its separation is shifted to $\varphi = 140°$ (Fig. 1), as shown by the measurements of Achenbach in air [22] and our measurements in water [23].

According to Achenbach [22], at $Re > 2 \times 10^6$, in the so-called supercritical flow regime, the point of separation is again shifted upstream to $\varphi = 115°$. But our measurements at $Re = 2 \times 10^6$ did not validate this observation; the boundary-layer separation was still at $\varphi = 140°$.

At present there is no generally accepted approach to the laminar–turbulent boundary-layer transition. Two concepts are under discussion. According to the first concept, in the critical range of Re, a laminar boundary layer separates on the front part of the cylinder and forms a separation buble, which then reattaches to the surface and again separates at $\varphi = 140°$. According to the second concept, which is conventional in various treatments of heat transfer studies, in the critical range of Re a laminar boundary layer loses its stability and enters a transitional flow, followed by a turbulent boundary layer and by its separation, as before. We are ardent supporters of the first concept of the critical flow pattern, since it is validated by measurements of shear stress and of pressure distribution, as discussed below. The second concept of the flow pattern applies to the supercritical flow regime.

An analysis of our results suggests that the separation bubble occupies a $\delta\varphi = 10°$ part of the perimeter. Consequently, the laminar separation and the laminar–turbulent transition are very close to each other. Figure 2 illustrates the dynamic behavior of the laminar–turbulent transition for a range of Re and for different main flow turbulence levels in water and air

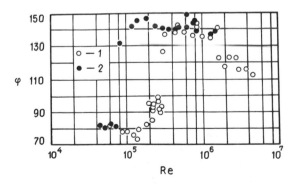

FIG. 1. Position of the boundary-layer separation point on a tube in dependence on Re. (1) From Achenbach [22]; (2) from Daujotas *et al.* [23].

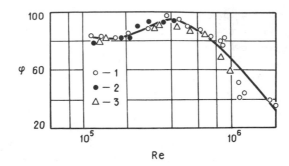

FIG. 2. Position of the outset of laminar–turbulent transition in dependence on Re at Tu = 1% [23]. (1) $k_q = 0.28$ and (2) $k_q = 0.17$ in water; (3) $k_q = 0.25$ in air.

[23]. We see that in the range of Re from 2×10^5 to 3×10^5 the initiation of the laminar–turbulent transition is at φ from 80 to 85°. But, with a further increase of Re up to 5×10^5 and at Tu = 1%, this point is shifted downstream to $\varphi = 95°$. This shift is caused by the separation bubble. A still further increase of Re causes the laminar–turbulent transition to shift to $\varphi = 35°$, when the laminar boundary layer becomes turbulent directly on the surface, and no separation bubble is formed.

The laminar–turbulent transition in the boundary layer is related to its loss of stability. The loss of stability is mainly related to the boundary-layer thickness and to the shape of the velocity profile, which is directly connected to the pressure gradient. Calculations [24] located the points of the loss of stability on different bodies. It is interesting to note that on a circular cylinder the point of the loss of stability undergoes an insignificant shift with an increase of Re, and only at Re = 10^7 does it move from $\varphi = 90°$ to $\varphi = 75°$. But other measurements [22, 23] suggest that, in real turbulent flows, the laminar—turbulent boundary-layer transition should not be related to the loss of stability, because on a cylinder in crossflow at Re $\approx 2 \times 10^6$, the transition occurs as early upstream as at $\varphi = 30°$.

2. Flow in the Wake

As noted above, at Re > 40 a vortex path is formed behind the tube. It appears as the result of the loss of stability in the circulation region, formed previously by a pair of vortices. With Reynolds number increasing further, a regular path of staggered vortices is observed in the wake. According to the measurements of Roshko [25], velocity fluctuations in the wake have a distinct periodicity, noted even at a considerable distance behind the tube. Interaction between vortices and fluid in the wake does not disturb the flow in the range of Reynolds number up to 150. With a further increase in

Reynolds number, irregular velocity fluctuations in the wake are observed. These give rise to small vortices which are formed at a certain distance from the tube and are destroyed much earlier than the ones formed from the laminar layer.

At Re > 300, vortices formed behind the tube consume only a certain part of the energy of disturbances of the separated boundary layer. The rest of the energy dissipates in the wake in the form of small turbulent vortices. This flow pattern persists up to the critical regime. In the supercritical regime, regular fluctuations are superimposed on the turbulent ones.

Thus in a flow with regular vortex shedding there is a constant and intensive exchange of substance and momentum between the circulation region and the undisturbed flow. At Re < 40, there is only molecular exchange since the vortices are not shed.

Periodicity of vortex shedding is characterized by the Strouhal number, which establishes the relation between vortex shedding frequency, cylinder diameter, and main flow velocity:

$$Sh = fD/u_0$$

It is obvious from Fig. 3 that, with increasing Reynolds number, the Strouhal number increases up to 0.2 but is practically constant in the range of Re from 10^3 to 2×10^5. In the critical regime, the Strouhal number increases greatly up to 0.46. In the supercritical regime, it decreases to 0.25, as reported by Roshko [25].

Flow analysis reveals that, in the predominant range of Reynolds num-

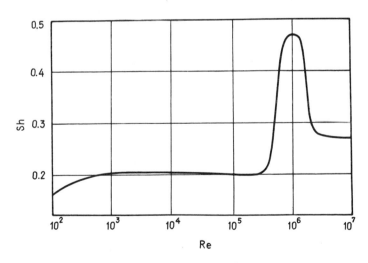

FIG. 3. Dependence of Sh for a tube on Re.

ber, vortices are shed intermittently from the two sides of the tube and symmetrically to the wake axis. Periodical and asymmetrical variations in pressure distribution on the surface give rise to forces of intermittent direction, which lead to crosswise vibration of the tube. This has to be taken into account in the design of heat exchangers.

3. *Velocity Distribution around the Tube*

The variations in the hydrodynamic conditions in the flow around the tube are illustrated by the distribution of pressure and local velocity.

Figure 4, compiled from experimental data of various authors [25–27], shows that up to $\varphi = 50°$ the distribution of pressure as well as of local velocity around the tube outside the boundary layer does not depend on Reynolds number and correlates well with calculations according to Eq. (3). The influence of Reynolds number begins at $\varphi > 50°$. The open circles in the figure represent the points of boundary-layer separation. The location of these points is characterized by the angle φ at which the pressure downstream is stable and, in curve 3, the angle φ at which a bubble is formed.

From measurements of pressure on the surface and by the use of Bernoulli's equation, the velocity distributions on the outer edge of the boundary layer can be determined. The velocity distribution in a real fluid differs

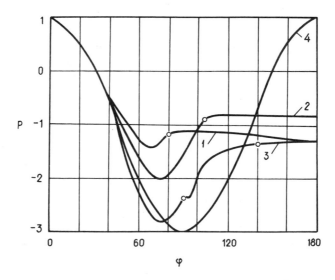

FIG. 4. Distribution of the pressure coefficient $P = (p_{x=\varphi} - p_0)/(\rho u_0^2/2)$ around a tube at different Re. (1) From Fage *et al.* [27] at Re = 1.1×10^5, (2) from Roshko [25] at Re = 8.4×10^6, (3) from Žukauskas and Žiugžda [26] at Re = 4.5×10^5, and (4) potential flow.

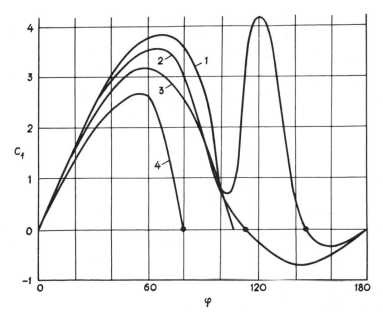

FIG. 5. Distribution of the skin friction coefficient $c_f = 100\tau/(\rho u_0^2)$ around a tube at different Re. (1) At $Re = 8.5 \times 10^5$ and (2) at $Re = 3.6 \times 10^6$ (from Achenbach [22]), (3) the calculation curve, and (4) at $Re = 1.1 \times 10^5$ (from Fage and Folkner [27]).

considerably from that in an ideal fluid because of the displacement of the main flow by the boundary layer and the complicated vortical flow in the wake.

A regular velocity distribution outside the boundary layer is characteristic only for the front part of a tube and is expressed by

$$u_x/u_0 = 3.631(x/D) - 2.171(x/D)^3 - 1.514(x/D)^5 \qquad (4)$$

4. Drag of the Tube

The total drag D_f of the tube is equal to the sum of the forces of friction and pressure. At very low Reynolds numbers, the tube is streamlined. According to theoretical calculations and measurements, frictional forces prevail in the drag.

With an increase in Reynolds number, the relative influence of inertial forces increases and that of frictional forces becomes negligibly small. In the range of an intensive vortical flow past a tube, the frictional drag constitutes just a small percentage of the total drag. Nevertheless, the flow pattern may be judged from the frictional distribution around the tube. Data of different authors (Fig. 5) for air flow suggest an increase of friction from zero to a maximum at $\varphi = 60°$ and a subsequent decrease to a minimum. The solid circles represent boundary-layer separation points, where the friction is equal to zero. The first minimum of the curve at $Re = 8.5 \times 10^5$ corresponds to the boundary-layer transition from laminar to turbulent flow.

Figure 6 shows the variation of frictional drag of a heated tube in air flow as measured by Žiugžda and Ruseckas [28]. A definite influence of heating on frictional drag can be noted.

From the position of the separation points on the tube, the width of the circulation region and consequently the pressure drag may be determined. Pressure drag is caused mainly by boundary-layer separation and depends on the width of the circulation region and the frequency of vortex shedding. The maximum width of the vortex region in the subcritical flow regime corresponds to the maximum pressure drag coefficient and minimum Strouhal number, and in the critical flow regime, the minimum width of the vortex region corresponds to the minimum pressure drag coefficient and maximum Strouhal number.

The general flow pattern past a tube is reflected in the variation of drag coefficient C_D with Reynolds number. It is clear from Fig. 7 that the

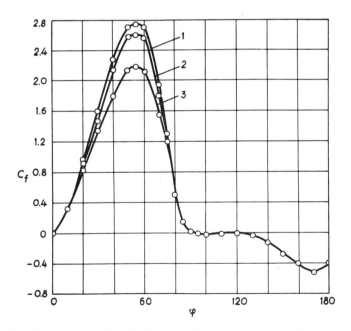

FIG. 6. Distribution of the skin friction coefficient around a tube in the presence of heat transfer at $Re = 5.22 \times 10^4$. (1) $\Delta t \simeq 50°C$, (2) $\Delta t \simeq 30°C$, and (3) $\Delta t = 0$.

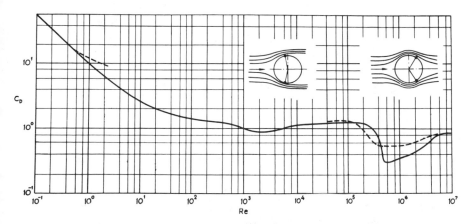

FIG. 7. The dependence of drag coefficient of a circular cylinder on Re.

dependence of C_D on Re in the range of the subcritical regime is negligible and at Re $> 2 \times 10^5$ the drag coefficient decreases sharply. This is because the boundary-layer separation point is moved downstream, as mentioned above, and with the convergence of the wake, the drag decreases considerably. The dotted line at low Reynolds numbers corresponds to calculations of potential flow past a tube.

Turbulence in the main flow exerts a certain influence on the flow pattern around immersed bodies. In the subcritical regime, turbulence has no effect on the drag coefficient, but in the range of the critical regime, the variation of drag coefficient, indicated by a dotted line in Fig. 7, depends on the turbulence. The origin of the critical flow regime is displaced to lower Reynolds numbers with an increase of turbulence.

A rough surface of the immersed tube has a similar influence on the drag coefficient and flow pattern. As the relative roughness of the surface increases, the Reynolds number at which the critical flow regime is established decreases. On the other hand, the presence of roughness increases the drag coefficient C_D in the critical flow regime.

C. FLOW IN A RESTRICTED CHANNEL

In practice, circular tubes are usually placed in flows restricted by walls and with considerable blockage of the flow cross section (Fig. 8). In the case of a circular tube in a channel, blockage is expressed by the ratio D/H. As the blockage ratio increases, the velocity in the boundary layer around the outside of the tube and the pressure and velocity distribution are changed accordingly.

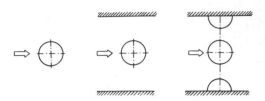

FIG. 8. Different flow patterns over a tube.

In Fig. 9, the velocity distribution at a distance of 4.2 mm from the tube surface is represented as measured by Akilba'yev *et al.* [29] at a main flow velocity of 20 m/sec and for blockage ratios of 0.39 and 0.63. In the separation region, particles of fluid near the surface move in the opposite direction, as seen from Fig. 9. As was stated by the same author, increasing the blockage ratio from 0 to 0.8 causes the minimum pressure point to be displaced from $\varphi = 70$ to $90°$, and the separation point moves downstream to $\varphi = 100°$.

The change in blockage ratio alters flow patterns in the wake. Periodical vortex shedding in the rear is observed at low blockage ratios. At higher

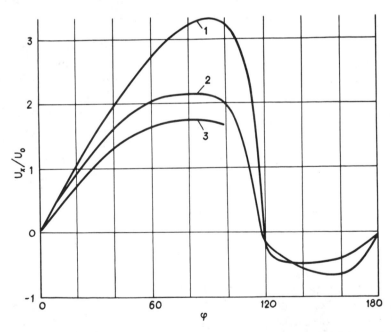

FIG. 9. Velocity distribution on a tube. (1) At $D/H = 0.63$, (2) at $D/H = 0.39$ (from Akilba'yev *et al.* [29]), and (3) at $D/H = 0.16$ (from Perkins and Leppert [30]).

ratios exceeding 0.6, unperiodical vortex flow behind the tube [29] is established. As a consequence, the drag coefficient C_D increases sharply.

To take into account the blockage ratio, the mean velocity in the minimum free cross section of the channel is usually used as a reference in technical calculations of heat transfer and drag:

$$u = [1/(1 - D/H)]u_0 \qquad (5a)$$

where u_0 is the velocity of an unrestricted flow.

This choice of reference velocity, however, allows no comparison between drag and heat transfer in systems of tubes in various arrangements, since different velocity distributions on the surface are not taken into account. Calculated reference velocities may be the same on a tube in an unrestricted flow, a tube in a channel, and one in the first row of a bank, velocities in the front portion being different. For this reason, the introduction of an average velocity seems reasonable, i.e., the average value of the velocity evaluated by integrating over the cylinder surface,

$$u = \frac{1}{l} \int_0^l u(x)\, dx \qquad (5b)$$

or

$$u = u_0 \left(1 - \frac{\pi}{4}\frac{D}{H}\right)^{-1} \qquad (5c)$$

At high blockage ratios, the flow pattern is basically changed, and the above-mentioned reference velocities are no longer applicable in determining drag and heat transfer coefficients. Thus in the calculation of drag and heat transfer at $D/H > 0.8$, various semiempirical equations [29, 30] are recommended for the evaluation of a reference velocity.

III. Flow past a Tube in a Bank

A. GENERAL FLOW PATTERN

The flow pattern around a tube in a bank is influenced by the surrounding tubes. In a contraction between adjacent tubes of a transverse flow, the pressure gradient changes even more. This causes a corresponding change of velocity distribution in the boundary layer and of the flow pattern in the rear.

The flow pattern around a tube in a bank is determined by the arrangement and geometrical parameters of the bank. Banks of in-line and a staggered arrangement of tubes are most common (Fig. 10), and they are

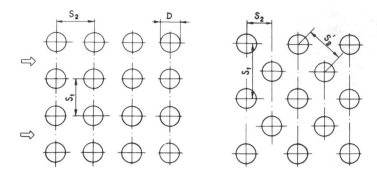

FIG. 10. Arrangement of tubes in a bank.

usually defined by the relative transverse $a = s_1/D$ and longitudinal $b = s_2/D$ center-to-center distances, called transverse and longitudinal pitches, respectively. In banks of both arrangements, flow around a tube in the first row is similar to that around a single tube, but the flow pattern in subsequent rows is different.

In staggered banks, the flow is comparable with flow in a curved channel of periodically converging and diverging cross section. Thus the velocity distributions around tubes in different rows in a staggered bank have a similar character. Flow in an in-line bank is sometimes comparable with that in a straight channel. The velocity distribution in the minimum cross section of an inner row is mainly determined by the pitch ratio. Two extreme cases of the relative longitudinal pitch equal to unity and to infinity can be noted here. Flow, in the first case, is closely similar to a flow in a channel, and in the other case, it is a flow through a single transverse row, with the velocity profile of the flow behind the preceding row being straightened. In intermediate cases, inner rows are located in the circulation regions of the preceding rows, and the flow preceding one of the inner rows is vortical with a nonuniform velocity distribution.

At low Reynolds numbers, the flow in a bank is laminar with large-scale vortices in the circulation regions, their effect on the boundary layer of the front portion of a subsequent tube being eliminated by viscous forces and a negative pressure gradient. The flow in the boundary layer is laminar. Such a flow pattern existing at $Re < 10^3$ may be described as predominantly laminar.

With increasing Reynolds number, the flow pattern in a bank undergoes considerable variations. The flow between tubes becomes vortical with a higher degree of turbulence. Although the front portion of an inner tube is influenced by the vortical flow, a laminar boundary layer persists on it. The pattern of flow, with a laminar boundary layer on the tube being under the influence of a turbulent flow and with an intensive vortical flow in the rear, may be described as mixed.

The intensity of turbulence and its production in a bank depend on the bank arrangement and Reynolds number. In banks of large longitudinal pitches, the transition from laminar to turbulent flow in the bank is gradual and depends on the increase in Reynolds number. The flow in the bank consists initially of large-scale vortices, the size of which decrease with increasing velocity. Thus in the range $10^3 < \text{Re} < 10^4$, there may exist a flow pattern intermediate between a predominantly laminar and a mixed flow.

In staggered banks with small longitudinal pitches at $\text{Re} > 10^3$, small vortices appear suddenly and flow in the bank becomes turbulent instantly.

The mixed-flow regime covers a wide range of Reynolds numbers and alters its character only at $\text{Re} > 2 \times 10^5$. Then flow in the bank becomes highly turbulent. The total drag of the bank varies like that of a single tube in the critical flow regime.

Thus we may distinguish three flow regimes in banks with respect to the Reynolds numbers: a predominantly laminar flow regime at $\text{Re} < 10^3$, mixed or subcritical flow regime at $5 \times 10^2 < \text{Re} < 2 \times 10^5$, and a predominantly turbulent or critical flow regime at $\text{Re} < 2 \times 10^5$.

B. Velocity Distribution

Pressure and velocity distributions around a tube in a longitudinal row of a bank differ substantially from that around a single tube. Figure 11 illustrates the variation of pressure coefficient around tubes in the first and the fourth rows of in-line and staggered banks for $a = 2.0$ and $b = 2.0$ at $\text{Re} = 10,800$ [31]. The pressure coefficient for a tube in a bank is given by

$$P = 1 - (p_{\varphi=0} - p_{\varphi})/(\rho u^2/2) \qquad (6)$$

where u is the mean velocity in the minimum free cross section.

Equation (3) does not apply to the determination of pressure coefficient variation for a tube in a bank, because the static pressure varies considerably along a bank.

It is obvious that flow around tubes in the first row is similar to that around a single tube in the subcritical flow regime. But on any of the inner tubes of a staggered bank, the pressure in front of the separation point is higher than that on a single tube. In inner rows of an in-line bank, the maximum pressure point is at about $\varphi = 40°$, where the impact of the main stream on the tube surface occurs. Thus there are two impact points and two maximum pressure points on any of the inner tubes of an in-line bank. It has been noticed by Bressler [31] that, for in-line banks of small

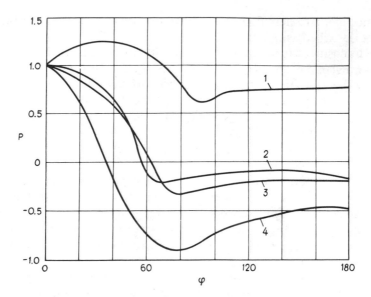

FIG. 11. Distribution of the pressure coefficient $P = 1 - [(p_{\varphi=0} - p_\varphi)/(\rho u^2/2)]$ on a tube in a bank [31]. (1) Fourth row and (2) first row in an in-line bank; (3) First row and (4) fourth row in a staggered bank.

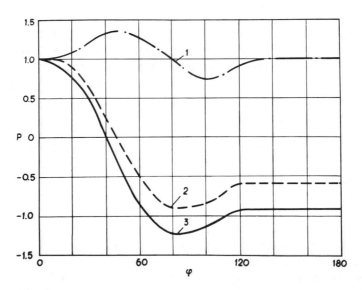

FIG. 12. The dependence of the pressure coefficient on the longitudinal tube spacing [32]. (1) $L/D = 3$, (2) $L/D = 6$, and (3) $L/D = 9$.

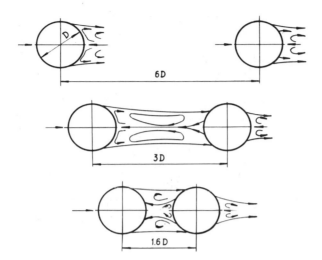

Fig. 13. Changes in the flow pattern in a line of tubes.

pitch, the pressure at one of the points is higher. From the third row on, the reciprocal positions of the minor and major pressure points interchange from row to row.

For in-line banks, the position of the impact point is dependent on the longitudinal pitch and Reynolds number. An investigation [32] of the pattern in the wake behind the tube at Re = 13,000 is interesting in this respect. The distance between the two tubes was changed in the range $1.6 < L/D < 9$. Figure 12 shows the qualitative change in the pressure distribution along the perimeter of the second tube for $L/D \leq 3$, substantially differing from that of a single tube in an infinite flow. For $L/D = 6$, the point of attack and impact point coincide at $\varphi = 0$, and for $L/D = 1.6$, the impact point is removed to $\varphi = 75°$. The general flow pattern based on the results of Kostič [32] is illustrated in Fig. 13.

The position of the impact point is also influenced by the Reynolds number. Investigation [20] of flow around a tube in the inner rows of an in-line bank, 2.0×1.3, suggests that, for low Reynolds numbers, the impact point is close to the front stagnation point, but with an increase of Re to 4000, it is removed to $\varphi = 55°$, and later at Re > 4000, again returns to the front. Figures 13 and 14 imply that for in-line banks, laminar boundary layers develop from the two impact points and separate from the tube at approximately $\varphi = 145°$. Sometimes there is a laminar–turbulent transition in the boundary layer.

In the case of flow around a tube in staggered banks, as in the case of a single tube, the flow is divided at the front stagnation point and a laminar

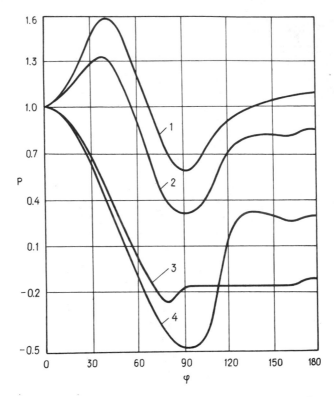

FIG. 14. Distribution of the pressure coefficient on a tube in an in-line bank of 2.0 × 1.25. Inner row, (1) at Re = 10^5; (2) at Re = 10^6, first row; (3) at Re = 10^5; and (4) at Re = 10^6 [35].

boundary layer develops on the front portion of an inner tube. Figure 15 [33, 34] suggests a certain influence of Reynolds number on pressure distribution of an inner tube in the staggered bank. The position of the boundary-layer separation point in this case differs from that of a single tube because the transition from a laminar to a turbulent boundary layer moves this point to $\varphi = 150°$. At high Reynolds numbers, the position of the separation point (Fig. 16) is somewhat different.

In studies of the velocity distribution in banks, the influence of the transverse pitch should be mentioned. With a decrease of the transverse pitch, the velocity in the free cross section increases rapidly. The calculations of mean velocity in the minimum free cross section, as a function of the transverse pitch, may be based on the constant rate of volume

$$u = u_0[a/(a - 1)] \tag{7}$$

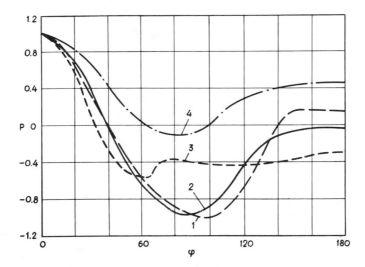

FIG. 15. Distribution of the pressure coefficient on a tube in a staggered bank in dependence on Re. (1) Second row and (2) fourth row at Re $= 1.5 \times 10^6$ (from Achenbach [33]), (3) second row and (4) fourth row at Re $= 2.7 \times 10^4$ (from Kostič and Oka [34]).

The mean velocity in any free cross section is then calculated by

$$u_\varphi = u_0[a/(a - \sin \varphi)] \tag{8}$$

Theoretical calculations of velocity distributions, as well as their determination by the method of electrical analogy for a potential flow through a row, yield velocity profiles in the minimum free cross section which have a more or less pronounced concave shape in the central part, their values depending on the transverse pitch.

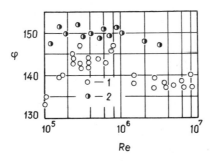

FIG. 16. The point of boundary-layer separation on a tube in a staggered bank in dependence on Re. (1) From Achenbach [33] and (2) from Žukauskas [35].

In the flow of a real fluid through a row, under the influence of viscous forces, boundary layers are formed on the surface. The velocity near the surface decreases, the velocity profiles are straightened, and the concave shapes in the curves diminish accordingly. Plots of pressure distribution imply that velocity distributions around a tube in a longitudinal row differ substantially from that of a single tube. Changes of the longitudinal pitch cause considerable changes in velocity distributions on the front and rear portions of an inner tube.

C. Drag of a Tube in a Bank

The distribution of surface friction on a tube in a bank may be judged by Fig. 17, which presents the experimental results from Achenbach [33] and Žukauskas [35] in staggered banks of 1.5×1.5 and 2.0×1.4 at different Re. Open circles which correspond to the theoretical analysis [35] coincide with the experimental curve. At the impact point and at the separation, $c_f = 0$. In relation to Re the maximum values of c_f are at $50° < \varphi < 80°$. The friction drag of a tube in a bank constitutes only about 5% of the total drag, and in the critical flow regime, it decreases to about 0.5%. Surface roughness causes an increase of friction drag.

The pressure drag of a tube in a bank is determined mainly by the longitudinal pitch. In an in-line bank for $b < 3$ under the influence of the preceding tube, the pressure drag decreases suddenly. The hydraulic drag of banks of tubes will be generally considered in Section VIII.

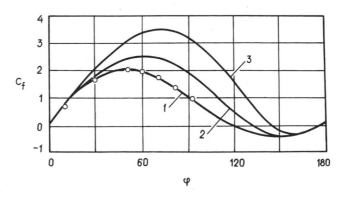

FIG. 17. Distribution of the skin friction coefficient $c_f = (\tau/\rho u^2)\sqrt{Re}$ in staggered banks. Open circles, analysis up to separation (from Žukauskas [35]). (1) Bank of 1.5×1.5 at $Re = 10^3$ (from Žukauskas [35]), (2) bank of 1.5×1.5 at $Re = 1.3 \times 10^5$ (from Žukauskas [35]), and (3) bank of 2.0×1.4 at $Re = 1.4 \times 10^6$ (from Achenbach [33]).

IV. Influence of Fluid Properties on Heat Transfer

The heat transfer of a single tube and a tube in a bank is determined mainly by flow velocity, physical properties of the fluid, heat flux density and its direction, and the arrangement of the tubes. The dimensionless relation is

$$\text{Nu} = f\left(\text{Re}, \text{Pr}, \frac{\mu_f}{\mu_w}, \frac{\lambda_f}{\lambda_w}, \frac{c_{pf}}{c_{pw}}, \frac{\rho_f}{\rho_w}, \frac{s_1}{D}, \frac{s_2}{D}\right) \tag{9}$$

For the generalization of experimental data the following power equation based on the functional relation (9) is commonly used:

$$\text{Nu} = c\text{Re}^m\text{Pr}^n \tag{10}$$

For gases of equal atomicity, i.e., for which the Prandtl numbers are equal and constant, Eq. (10) becomes

$$\text{Nu} = c\text{Re}^m \tag{11}$$

A. INFLUENCE OF PRANDTL NUMBER

Most of the fluids commonly used in practice have Prandtl numbers ranging from 1 to 1000. This means that in calculations of heat transfer a wrong choice of power index of Prandtl number may lead to considerable errors. A power index of Prandtl number equal to $0.31-0.33$ for tubes in crossflow is still accepted by some authors. This value of n is suggested by theoretical investigations of heat transfer in a laminar boundary layer on a plate. However, later calculations and experimental measurements [35] have revealed the dependence of the power index of the Prandtl number on the flow regime in the boundary layer. For a laminar boundary layer on a plate $n = 0.33$, but for a turbulent boundary layer it amounts to 0.43.

The above suggests that for mean transfer from a tube the value of n may be somewhere between 0.33 and 0.43.

Investigations of heat transfer in laminar and turbulent boundary layers suggest certain changes in the value of the power index with large variations of Prandtl number. Numerical solutions of heat transfer in the region of the front stagnation point, performed by Makarevičius [36] and this author, gave $n = 0.37$ for $\text{Pr} < 10$ and $n = 0.35$ for $\text{Pr} > 10$.

Our detailed studies of mean heat transfer from a tube in crossflows of transformer oil, water, and air [37] at $T_w = $ constant yielded an approximate value of the power index of the Prandtl number between 0.37 and 0.38. In later investigations of local heat transfer from a tube in flows of various fluids [38] at $q_w = $ constant, it was determined that n varies along the tube perimeter, reaching the value of 0.39 in the rear. The mean value of n is equal to 0.365. The mean value of n for a circular tube, 0.37, is acceptable for $Pr < 10$ and somewhat lower for $Pr > 10$.

Our investigations of heat transfer [20] in 27 banks of tubes of different arrangements in flow of various fluids in the range of Pr from 0.7 to 500 suggest (Fig. 18) that, for the mean heat transfer of all sorts of banks, the power index of the Prandtl number has the value 0.36. In practical calculations of the mean heat transfer of banks of tubes, $n = 0.36$ is sufficiently accurate.

B. CHOICE OF REFERENCE TEMPERATURE

In the process of heat transfer, the fluid temperature varies, which causes variations of its physical properties. Thus evaluation of the influence of the fluid physical properties on heat transfer is closely connected with accounting for the temperature variations in the boundary layer, in other words, with the choice of the so-called reference temperature, according to which the physical properties are evaluated.

The influence of the variations of physical properties may be established by two methods. By the first method, the physical properties are referred to the main flow temperature, and an additional parameter is introduced in Eq. (10) to account for the properties variation. The second method is to choose a certain value of temperature between that of the flow and the wall, which enables evaluation of the influence of the physical properties on heat transfer. In this case, the relations for the heat transfer calculation remain the same as for constant physical properties.

Analysis of relations for heat transfer calculations in the flow of gases,

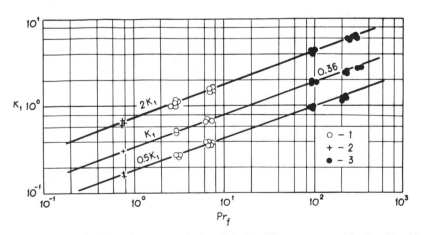

FIG. 18. Determination of the power index of Pr for different staggered banks, $K_1 = Nu_f\ Re_f^{-0.60}$. (1) Water, (2) air, and (3) transformer oil.

proposed by different authors, shows that the main cause of discrepancies is due to different choices of the reference temperature.

Detailed experimental investigations of heat transfer from a single tube [15] and banks of tubes [39] in the crossflow of gases suggest that the physical properties should be evaluated at the main flow temperature T_f. This generalizes the results of heat transfer in gas flow with sufficient accuracy, and no additional parameters for the evaluation of temperature difference are necessary. In investigations of heat transfer of banks of tubes in air at high Reynolds numbers [40], the properties are also referred to the main flow temperature T_f.

A certain mean temperature of the boundary layer has sometimes been proposed as the reference temperature by some authors [9]. We prefer to take the main flow temperature as the reference in the range of moderate temperatures. This method is simple in practice and sufficiently accurate for practical purposes.

In flows of viscous fluids, the intensity of heat transfer markedly depends on physical property variation in the boundary layer with heat flux direction and temperature difference. Experimental results referred to the main flow temperature are higher for heating than for cooling, the discrepancy increasing with the increase of temperature difference.

Mikhe'yev [41] has proposed to account for the influence of sharp changes in the physical properties of fluids in the boundary layer near the surface by the ratio Pr_f/Pr_w to the power 0.25. The ratio μ_f/μ_w is often used also. It should be noted that for viscous fluids, like water and oil, it is mainly viscosity that changes with temperature, and therefore $Pr_f/Pr_w \approx \mu_f/\mu_w$. However, calculations of heat transfer in laminar boundary layers on a plate in flows of various fluids [36] lead to the conclusion that the influence of other physical properties constitutes up to 7% of the total influence.

Our investigations [37] confirm that variations of viscosity and other properties in the boundary layer on a circular tube may well be accounted for by the ratio Pr_f/Pr_w, with a corresponding power index. In this case, the results are referred to the main flow temperature.

Computer-aided calculations [36] suggest that for a laminar boundary layer on a plate the power index of Pr_f/Pr_w is 0.25 for heating and 0.19 for cooling. It was also found to depend to a certain extent on the value of Pr_w, increasing slightly with high values of the latter.

Our investigations [42] on the changes of the power index at Pr_f/Pr_w for the case of a turbulent boundary layer on a plate give $n = 0.25$ for heating and $n = 0.17$ for cooling (Fig. 19). From calculations [35], a small decrease of n is observed in turbulent boundary layers at high Pr. Thus the power index of Pr_f/Pr_w is lower for cooling than for heating. In any case, for most practical purposes, $n = 0.25$ is sufficiently accurate for both cases.

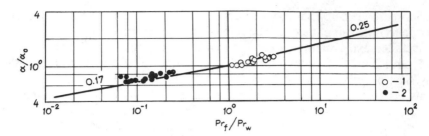

FIG. 19. Local heat transfer of a plate in a turbulent flow versus heat flux direction. (1) Transformer oil and (2) glycerine.

Some other parameters have also been proposed to account for radical changes of physical properties. However, we still prefer the ratio Pr_f/Pr_w as the most simple and convenient to use.

In Fig. 20, the changes of intensity of heat transfer in staggered banks as a function of Pr_f/Pr_w are shown from our experiment [20]. When physical properties are referred to the main flow temperature, the ratio $(Pr_f/Pr_w)^{0.25}$ accounts satisfactorily for changes of the physical properties both for heating and cooling. For gases with constant Prandtl number, $Pr_f/Pr_w = 1$.

Therefore, for the calculation of heat transfer from a tube in the cross-flow of viscous fluids, the following relation will be used:

$$Nu_f = cRe_f^m Pr_f^n (Pr_f/Pr_w)^{0.25} \qquad (12)$$

V. Heat Transfer of a Single Tube

A. LOCAL HEAT TRANSFER

1. *Theoretical Calculations of the Local Heat Transfer*

The analysis presented in Section II suggests that flow past a single tube is a rather complicated process. The same applies to its heat transfer. At various Reynolds numbers, the boundary layer on the front portion of a tube is laminar. Therefore, theoretical methods for the calculation of heat transfer may be applied. The rear portion is in the region of a complicated vortical flow. Here the theoretical calculation of heat transfer is almost impossible, though some attempts have been made in recent years. Furthermore, the heat transfer in the front stagnation point and in the boundary-layer separation region has some peculiar features.

First, the calculations of heat transfer on the front portion of a tube by means of an approximate integral method were performed by Kruzhilin [43]. This method has been used for the calculation of heat transfer from a plate and in a modified form was applied to the evaluation of heat transfer from the front portion of a tube. The following relation for local heat transfer was obtained:

$$\mathrm{Nu}_x = [2/F(\varphi)]\mathrm{Re}^{0.5}\mathrm{Pr}^{0.33} \tag{13}$$

where $F(\varphi)$ is a hydrodynamical factor, accounting for the changes in the boundary layer around the tube and the velocity distribution outside the boundary layer. The relation applies only for $\mathrm{Pr} \geq 2.6$.

Other theoretical methods have been applied to the heat transfer calculations of wedge-shaped bodies, having a velocity variation outside the boundary layer which is expressed by the exponential relation

$$u_x = cx^{m_1} \tag{14}$$

These methods can be used only for some definite portions of a circular tube. Thus for the region near the front stagnation point the value of m_1 in Eq. (14) is unity and for the separation region $m_1 = -0.0804$. In the first case, the velocity outside the boundary layer changes linearly along the perimeter.

Boundary-layer equations can be applied to the heat transfer from wedge-shaped bodies:

$$\rho u \frac{\partial u}{\partial x} + \rho v \frac{\partial u}{\partial y} = \mu \frac{\partial^2 u}{\partial y^2} - \frac{dp}{dx}$$

$$\frac{\partial u}{\partial x} + \frac{\partial v}{\partial y} = 0 \tag{15}$$

$$\rho c_p u \frac{\partial T}{\partial x} + \rho c_p v \frac{\partial T}{\partial y} = \lambda \frac{\partial^2 T}{\partial y^2}$$

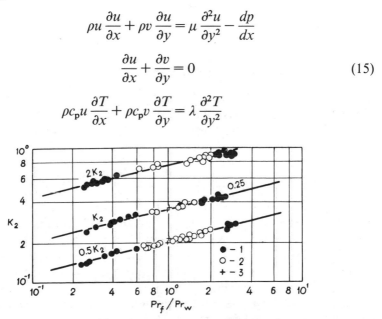

FIG. 20. Determination of the power index of $\mathrm{Pr}_f/\mathrm{Pr}_w$ for different staggered banks, $K_2 = \mathrm{Nu}_f \, \mathrm{Re}_f^{-0.60} \, \mathrm{Pr}_f^{-0.36}$. (1) Transformer oil, (2) water, and (3) air.

together with Eq. (14) and with boundary conditions

$$\text{for} \quad y = 0, \quad u = v = 0, \quad T = T_{\text{w}}$$
$$\text{for} \quad y = \infty, \quad u = u_x, \quad T = T_{\text{f}}$$

The flow function

$$\psi = f(\eta) \left(\frac{2}{m_1 + 1} uvx \right)^{1/2}$$

and the following new variables are also introduced:

$$\theta = \frac{T - T_{\text{w}}}{T_{\text{f}} - T_{\text{w}}}, \qquad \eta = y \sqrt{\frac{m_1 + 1}{2} \frac{u}{vx}} \tag{16}$$

Equations (15) have been solved for the above conditions by a number of authors. Heat transfer coefficients calculated for different wedge profiles for the range of Prandtl number from 0.7 to 10 were presented by Eckert [44] in the form

$$\text{Nu}_x / \sqrt{\text{Re}_x} = 0.56A / \sqrt{2 - \beta} \tag{17}$$

The generalization of calculated data [44] yields the following expression for A:

$$A = (\beta + 0.2)^{0.11} \text{Pr}^{0.333 + 0.067\beta - 0.026\beta^2} \tag{18}$$

where

$$\beta = 2m_1 / (m_1 + 1) \tag{19}$$

By substituting for β in Eq. (17), the local heat transfer can be calculated in regions near the separation point assuming the above values of m_1.

For flow past curvilinear bodies, the power index in Eq. (14) may be determined from

$$m_1 = \frac{x}{u_x} \frac{du_x}{dx} \tag{20}$$

where u_x is the local velocity outside the boundary layer.

For a circular tube in crossflow, the velocity distribution on the front portion is determined from Eq. (4), according to Hiemenz [45]. The power index determined from Eq. (20) accounts for the velocity distribution around the tube decreasing downstream from unity in the front stagnation point to zero and below zero.

For heat transfer at the front stagnation point, the following relation is derived from Eqs. (17) and (18) for Pr < 10:

$$\text{Nu}_x = 0.57 \text{Re}_x^{0.5} \text{Pr}^{0.37} \tag{21}$$

Equation (18) suggests that the power index of the Prandtl number decreases downstream along the perimeter, reaching the value of 0.3 in the separation region.

Merk [46] proposed a simplified method for heat transfer calculations from the front portion of a tube for Pr > 10. He used only the first term in the polynomial of the influence of Pr and β, neglecting the rest as infinitesimally small at high Prandtl numbers. Therefore, the relation for heat transfer calculation at the front stagnation point is

$$Nu_x = 0.569Re_x^{0.5}Pr^{0.33} \tag{22}$$

A numerical solution similar to that of Eckert's [44] performed by Makarevičius and the author [35, 36] in the range of Pr from 10 to 100 yielded the following relation for heat transfer in the region near the front stagnation point:

$$Nu_x = 0.57Re_x^{0.5}Pr^{0.35} \tag{23}$$

As a result of the exact solution of the boundary-layer equations, Frössling [47] proposed an asymptotic relation for the calculation of local heat transfer from the front portion of a tube in air:

$$Nu_x = [0.9450 - 0.7696(x/D)^2 - 0.3478(x/D)^4]Re_x^{0.5} \tag{24}$$

The relation was obtained by using a velocity distribution outside the boundary layer analogous to that in Eq. (4).

The relations presented above refer only to the heat transfer of the front portion of a tube. Theoretical calculations by Krall and Eckert [48] apply to the whole perimeter of a tube at low Reynolds numbers. At first, the lines of flow in the front and rear regions were calculated, and the heat transfer was determined from the temperature gradient at the tube surface, as a result of numerical solutions of the boundary-layer equations in cylindrical coordinates by the method of finite differences. From 800 to 1000 iterations were performed. Solutions were obtained for the two boundary conditions on the wall, i.e., constant surface temperature and constant heat flux. The results of the calculations at two different values of Reynolds number are presented in Fig. 21. The point of minimum heat transfer is clearly removed downstream and is located in the region between 125 and 145°, which is considered characteristic of low Reynolds number flow. The influence of boundary conditions on the heat transfer is also obvious. For a constant heat flux, the heat transfer is more intensive than for a constant surface temperature, the average difference of the mean heat transfer being in the range of 15 to 20%, depending on the Reynolds number. It is interesting to note that in the front portion up to $\varphi = 30°$ the heat transfer is independent of the boundary conditions on the wall. Dif-

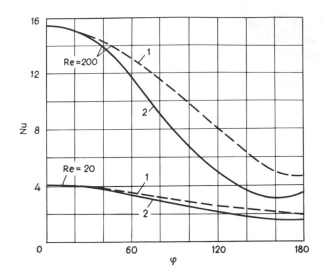

FIG. 21. Theoretical calculations of the local heat transfer around a tube (from Eckert and Soengen [49]). (1) At q_w = const. and (2) at t_w = const.

ferent theoretical calculations of the heat transfer of the front portion of a tube in air lead to similar results (Fig. 22).

Some recent publications have described attempts to calculate the heat transfer in the rear portion of a circular tube at higher values of Reynolds numbers. However, these approximate methods are based on simplified

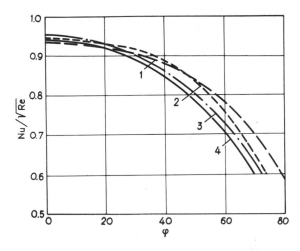

FIG. 22. Comparison of heat transfer calculations on the front part of a circular cylinder. (1) From Kruzhilin [43], (2) from Merk [46], (3) from Eckert [44], and (4) from Frössling [47].

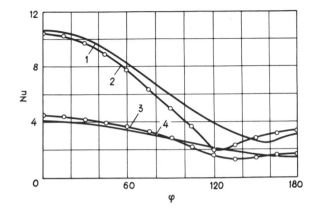

FIG. 23. Local heat transfer of a tube at low Re. (1) Analysis from Krall and Eckert [48], at Re = 100; (2) experiments at Re = 120 and (3) at Re = 23 (from Eckert and Soengen [49]); and (4) analysis at Re = 20.

flow patterns and do not take into account the numerous factors of separation. Therefore, experimental data are preferred in practice for this case.

2. *Heat Transfer in Gas Flow*

At low Reynolds number, as long as the vortical path is formed, theoretical results of local heat transfer [48] correlate well with experimental data [49] (Fig. 23). The heat transfer in the rear portion of a tube is at a minimum.

For higher values of Reynolds number, the heat transfer is a minimum at the separation point and increases downstream (Fig. 24). Nevertheless, the variation of heat transfer is greatly under the influence of the Reynolds number. At low Reynolds numbers, the heat transfer in the front portion is higher than in the rear (Fig. 23). With an increase of Reynolds number, the heat transfer in the rear portion increases, and for Re > 5 × 10^4 it is higher than in the front portion. For Re > 2 × 10^5 the variation of the heat transfer coefficient acquires a new character under the influence of the transition from laminar to turbulent flow in the boundary layer. In Fig. 24 [50], curve 4, a sharp increase of the heat transfer is noted at $\varphi = 90°$. According to our concept, in the critical flow regime the separation of a laminar boundary layer is accompanied by the formation of a separation bubble, which is followed by the development of a turbulent boundary layer. The formation of a bubble after the separation facilitates the access of cold fluid to the surface, and the heat transfer increases sharply. In this case the heat transfer reaches its maximum at $\varphi = 110°$ and later decreases with the development of a turbulent boundary layer. The turbulent bound-

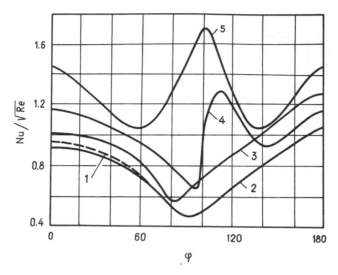

FIG. 24. Local heat transfer of a cylinder at different Re. (1) Analysis after Frössling [47], (2) at Re = 2.1 × 10⁴ (from Kruzhilin and Schwab [14]), (3) at Re = 9.9 × 10⁴, (4) at Re = 2.1 × 10⁵ (after Giedt [50]), and (5) at Re = 7.7 × 10⁵ and Tu = 1.2% (from Žukauskas and Žiugžda [26]).

ary layer separates at $\varphi = 140°$, where a second heat minimum is observed. Heat transfer increases again in the rear, in the recirculation region.

With a further increase of Re the supercritical flow regime is established. It is noted for a direct laminar–turbulent transition in the boundary layer and for an upstream shift of the transition point with an increase of Re. Our measurements [26] (Fig. 24, curve 5) suggest a laminar—turbulent boundary-layer transition at $\varphi = 60°$ for Re = 7.7 × 10⁵.

It should be noted that heat transfer is also influenced by the turbulence of the main flow. The calculations refer to the laminar main flow. Therefore, a coincidence of calculated results with experimental data obtained in the flow with low turbulence is possible. Experimental points for a highly turbulent main flow are considerably higher than the theoretical curves in Fig. 24.

Recent investigations [26, 51–53] suggest that the local heat transfer of a circular tube increases considerably with the turbulence of the main flow. Figure 25 presents the results of heat transfer measurements around a tube [26, 51] for different levels of turbulence of the main flow. A regular increase of heat transfer with higher turbulence levels is obvious. It is interesting to note that, even for Re = 39,000 with an increase of turbulence to Tu = 11.5%, two minima appear in the heat transfer, the first of which is due to the transition from laminar to turbulent flow. Thus, with an increase in turbulence of the main flow, at moderate Reynolds numbers the critical flow regime is established.

According to our results [26] for the supercritical flow regime, with an increase of the main flow turbulence, the first heat transfer minimum is shifted upstream, and the part of the surface covered by a laminar boundary layer decreases.

Experiments suggest a nonuniform influence of turbulence along the perimeter, which can be described by the following factor:

$$\varepsilon_\varphi = Nu_{Tu \neq 0}/Nu_{Tu=0} \tag{25}$$

As seen in Fig. 26, in the subcritical flow regime the dependence of heat transfer on turbulence at different Re has a similar character. For higher values of Re and Tu, because of the transition to turbulent flow in the boundary layer, the influence of turbulence is most appreciable at $\varphi = 110°$. It may be concluded from the results that the critical flow regime is established at

$$Re\ Tu \geq 150{,}000$$

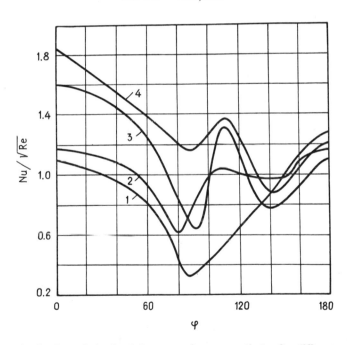

FIG. 25. Distribution of the local heat transfer on a cylinder for different turbulence intensities of the flow. (1) Tu = 0.9% at Re = 1.1 × 10⁵ (from Zapp [51]), (2) Tu = 1.2% at Re = 1.4 × 10⁵ (from Žukauskas and Žiugžda [26]), (3) Tu = 11.5% at Re = 3.9 × 10⁴ (from Zapp [51]), and (4) Tu = 15% at Re = 1.4 × 10⁵ (from Žukauskas and Žiugžda [26]).

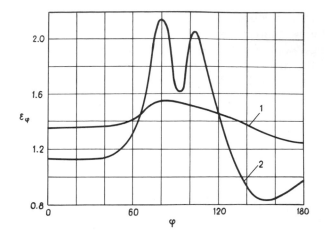

FIG. 26. Variation of ε_φ around a tube. (1) At Re $= 1.56 \times 10^4$ and Tu $= 12\%$ (from Dyban and Epick [53]) and (2) at Re $= 6.7 \times 10^4$ and Tu $= 2.5\%$ (from Katinas *et al.* [38]).

where Tu is expressed as a percentage. The influence of turbulence on heat transfer is least in the rear critical point.

Dyban and Epick [53] proposed the relation

$$\varepsilon_\varphi = 1 + 0.01(\text{Re Tu})^{0.5} \tag{26}$$

for the evaluation of the influence of turbulence on heat transfer from the front portion up to $\varphi = 60°$, which is thought to overestimate the heat transfer rate. This would mean that, for Re $= 60,000$, with the turbulence level increasing from 0.2 to 4%, the heat transfer increases by 30% in the front of the tube.

3. Heat Transfer in the Flow of Viscous Fluids

The local heat transfer from a circular tube in air flow, transformer oil, and water at $q_w = $ const. is thoroughly investigated in Katinas *et al.* [38]. It is obvious from Fig. 27 that for approximately equal Reynolds numbers, the variations of local heat transfer coefficient have a similar character in all of the fluids examined. The ratio of the heat transfer intensities in the front and in the rear is also the same for different fluids, in spite of the variation of Pr from 0.7 to 95. This implies that in the range of Re examined, the physical properties of the fluid have no effect on the character of heat transfer. The variations of the local heat transfer in different flow regimes show that with an increase in Reynolds number and turbulence level the curve of heat transfer in the rear portion of the tube undergoes similar changes in flows of viscous fluids and in air. Curve 4, representing variations of local heat transfer in water for Re = 130,000 and Tu = 2%, has two minima, the first one relating to the laminar–turbulent transition in the boundary layer and the second one to the separation of the turbulent boundary layer.

Our joint studies with Daujotas and Žiugžda [23] in a wide range of critical and supercritical Re numbers revealed that the beginning of the laminar–turbulent boundary-layer transition and the formation of a turbulent thermal boundary layer were related to Re (Fig. 28) and to the main flow turbulence Tu (Fig. 25). With an increase of either Re or Tu, the thermal turbulent boundary layer is sharply shifted upstream to $\varphi = 30–35°$ and occupies a large part of the perimeter on the front of the cylinder. This phenomenon is responsible for a large increase of the heat transfer.

Results in Fig. 28 (Re = 8.6×10^5) suggest a complex behavior of the heat transfer in the critical flow regime. Heat transfer on the front part is going on under a laminar boundary layer, same as in the subcritical flow regime, and decreases with the development of the laminar boundary layer.

In the critical flow regime the first heat transfer minimum at $\varphi = 80–90°$ corresponds to the separation of the laminar boundary layer and the

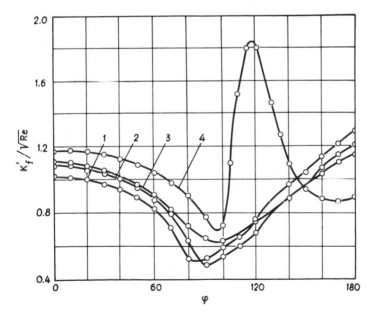

FIG. 27. Variation of the heat transfer around a tube in different fluids [38], $K_f^* = Nu_f Pr_f^{-0.37} (Pr_f/Pr_w)^{-0.25}$. (3) At Re = 3.9×10^4 in air flow, (2) at Re = 2.5×10^4 and (4) at Re = 1.3×10^5 in water flow, and (1) at Re = 2×10^4 in transformer flow.

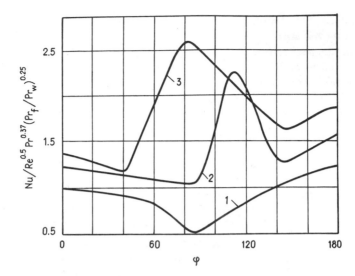

FIG. 28. Effect of Re on the local heat transfer of a cylinder in a flow of water at Tu = 1.5%
[23]. (1) At Re = 5.5×10^4, (2) at Re = 8.6×10^5, and (3) at Re = 2.03×10^6.

formation of the separation bubble, which is followed by the development
of a turbulent boundary layer. In the supercritical flow regime (Re =
2.03×10^6), the first heat transfer minimum corresponds to the beginning
of the laminar–turbulent transition in the boundary layer itself. In both
cases, the second heat transfer minimum at $\varphi \approx 140°$ corresponds to the
separation of a newly developed turbulent boundary layer.

With an increase of Tu and Re, the separation bubble disappears, the
laminar boundary layer directly turns turbulent, and the transition point is
shifted upstream. From analysis of experimental results, a relatively small
level of turbulence (Tu $\approx$ 1%) enhances the onset of the critical flow regime. The results suggest a local effect of turbulence, same as in the
subcritical flow regime. The effect is most pronounced in the front stagnation point and on the front part of a cylinder. In the medium part, the
effect decreases with a growth of Tu and Re (Figs. 25 and 28). In the rear
part, a regular effect of turbulence on the heat transfer cannot be observed.
The effect of external turbulence is weak in the vortical and highly turbulent flow in the rear. Our results for the supercritical flow regime are
supported by the studies of Achenbach in air [54].

The variations in the flow pattern around the tube lead to changes of the
power indexes in Eq. (12). According to Katinas et al. [38], the average
power index of the Reynolds numbers is 0.5 and 0.73 in the front and in
the rear portions, respectively. The power index of the Prandtl number for

Pr > 10 is equal to approximately 0.34 and 0.39 in the front and in the rear, respectively.

In Fig. 29, the experimental points of local heat transfer at the front portion of a circular tube are located on the same curve for flows of air, water, and transformer oil and may well be generalized by a single relation:

$$Nu_{fx} = 0.65 \, Re_{fx}^{0.5} \, Pr_f^{0.34} (Pr_f/Pr_w)^{0.25} \qquad (27)$$

In this case, experimental data were referred to the local velocity from Eq. (4).

The local heat transfer is higher for q_w = const. than for T_w = const. So, in heat transfer calculations and in the analysis of experimental results, a consideration of surface temperature variation is necessary. Different temperature distributions on the wall lead to different values of the heat transfer coefficient. Therefore, applying the relations of heat transfer for a tube with a constant surface temperature to the case of a variable surface temperature will be only approximate, and in the case of a considerable surface temperature gradient, the relation may be misleading. The influence of a temperature gradient is analogous to that of a pressure gradient. The surface temperature gradient determines the temperature distribution in the thermal boundary layer and its thickness. The thermal boundary-layer thickness varies under the influence of a surface temperature gradient, causing corresponding variations of the heat transfer coefficient, which is inversely proportional to the thickness of this layer.

The derivation of a general relation for the local heat transfer of the rear portion of a tube is connected with the difficulties of determining the local velocity. This problem is very complicated and requires a more detailed analysis, which will be the object of our future investigations. The mean

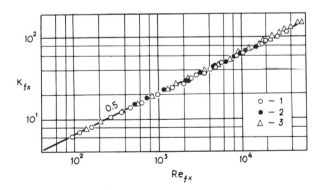

FIG. 29. Local heat transfer of the front part of a cylinder in different fluids [38], $K_{fx} = Nu_{fx}Pr_f^{-0.34} (Pr_f/Pr_w)^{-0.25}$. (1) Transformer oil, (2) water, and (3) air.

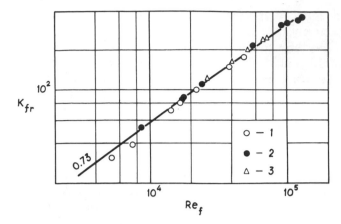

FIG. 30. Mean heat transfer of the rear part of a cylinder, $K_{fr} = Nu_f Pr_f^{-0.4} (Pr_f/Pr_w)^{-0.25}$.

heat transfer in the rear portion of a circular tube in flows of liquids for $q_w = $ const. is presented in Fig. 30. Calculations of the mean heat transfer from a tube are referred to the velocity in the minimum free cross section according to Eq. (5a).

B. Mean Heat Transfer

If the coefficient of local heat transfer along the perimeter has been determined, the mean heat transfer of the tube as a whole can be derived. The choice of the averaging method of calculation is of some importance. According to the first of the two known methods, the mean heat transfer coefficient is determined by integrating the local heat transfer relation

$$\alpha = \frac{1}{l} \int_0^l \alpha(x) \, dx \tag{28}$$

Dividing an integral mean heat flux by an integral mean temperature difference,

$$\alpha = \frac{(1/l) \int_0^l q_w(x) \, dx}{(1/l) \int_0^l \Delta T(x) \, dx} \tag{29}$$

is the second averaging method of the mean heat transfer coefficient.

Calculations by these two different methods may lead to discrepancies in the results. The results are identical for isothermal surfaces, but with surface temperature variations, discrepancies between the two methods do occur. For example, the mean heat transfer coefficient, determined from Eq. (28), is considerably higher for $q_w = $ const. than for $T_w = $ const., while the mean heat transfer determined by the second method from Eq. (29) is approximately equal for both cases.

In theoretical calculations, the first method is more convenient. In experiments, however, the mean heat transfer is usually found by the second method. In the case of heat transfer from wires at low Reynolds numbers, the mean temperature is determined from their electrical resistance.

1. Range of Low Reynolds Numbers

At low Reynolds numbers, heat transfer of a circular cylinder is influenced by free convection. At high values of Re, the effect of free convection is negligible as compared to that of forced convection. Different ways of evaluating the effect of free convection were reviewed by Morgan [55].

The total heat transfer may be determined by adding up the components of free and forced convection [56], where the relation

$$(Nu - 0.35)\sqrt{1 - \left(\frac{0.24Gr^{1/8} + 0.41Gr^{1/4}}{Nu - 0.35}\right)^2} = 0.5Re^{0.5} \qquad (30)$$

is used for calculations at low Reynolds numbers. For the case of a weak influence of free convection, it simplifies to

$$Nu = 0.35 + 0.5Re^{0.5} \qquad (31)$$

For calculations of heat transfer from wires at $Re < 0.5$, when they are streamlined, the following relation is proposed [57]:

$$Nu(T_m/T_f)^{-0.17} = 1/(1.18 - 1.1 \log Re) \qquad (32)$$

where T_m is the arithmetical mean value of the wire temperature and T_f is the main flow temperature.

In connection with the developments in thermoanemometry, a number of works have been devoted to the heat transfer of thin wires which are used as sensors. Specific problems of heat transfer from thin wires lie outside the scope of the present study. Publications [58, 59] in this field can be recommended.

Figure 31 presents the data of various authors on heat transfer in air at low Reynolds numbers. A certain scatter of the points is obvious, but the slope of the curves corresponds to the power index of Reynolds number $m = 0.40$ at $Re < 40$ and $m = 0.50$ at $Re > 40$.

The process of heat transfer in liquids at low Reynolds numbers is similar (Fig. 32) [60]. The influence of the physical properties of the liquid in this case is accounted for by the Prandtl number to the power 0.37. But

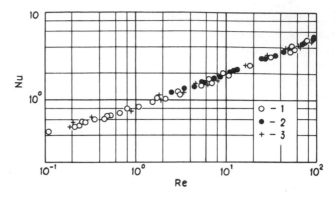

FIG. 31. Heat transfer of a cylinder in air flow at Re $< 10^2$. (1) From Hilbert [13], (2) from King [1], and (3) from Collis and Williams [57].

at low Reynolds numbers, the power index $n = 0.33$ may be assumed with sufficient accuracy.

2. Range of High Reynolds Numbers

The preliminary results of the correlation of our experimental data on mean heat transfer in flows of air, water, and transformer oil at various heat flux directions are presented in Fig. 33. All the experiments were performed under identical conditions (T_w = constant, Tu $< 1\%$) in the same experimental apparatus.

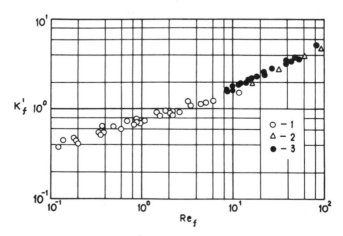

FIG. 32. Heat transfer of a cylinder in fluid flows at Re $< 10^2$, $K_f' = \mathrm{Nu_f Pr_f^{-0.37}}$ ($\mathrm{Pr_f}$/ $\mathrm{Pr_w})^{-0.25}$. (1) From Piret et al. [60]; (2) from Davis [16], in water; and (3) from Žukauskas [37], in transformer oil.

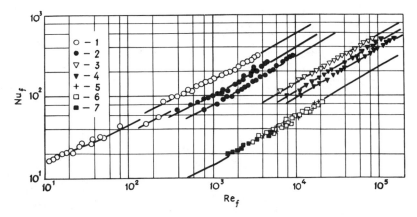

FIG. 33. The effect of the type of fluid on the heat transfer of a tube (from Žukauskas [37]). (1) Heating and (2) cooling of transformer oil, (3) heating and (4) cooling of water, (5) air, (6) air (from Mikhe'yev [15]), and (7) air (from Hughes [2]).

It is obvious that the experimental points for the flow of transformer oil and water are higher for heating than for cooling. Thus the relative positions of the heat transfer data are determined by fluid type, fluid temperature, and heat flux direction. Our data in air flow correlate well with the results of other authors [2, 15] presented here. Through the experimental points two straight lines of different slopes may be drawn, corresponding to different values of the power index of Reynolds number. At Re $< 10^3$, $m = 0.48$ and in the range of Reynolds numbers from 10^3 to 2×10^5, $m = 0.61$ both for liquids and air.

In Fig. 34, the final results according to Eq. (12) are presented, together with the results [38] of investigations of mean heat transfer from a circular tube in crossflow of air, water, and transformer oil for $q_w = $ const. The power index of the Prandtl number, based on an analysis of the experimental results $n = 0.37$, is chosen. As is seen from the figure, all experimental data correlate well and definitely determine the process. Averaged data on heat transfer according to Eq. (29) at $q_w = $ const. agree well with mean heat transfer at $T_w = $ const.

Figure 35 presents our experiments [23] on the mean heat transfer of circular cylinders in water at Re from 4×10^4 to 2×10^6 and the boundary condition $q = $ const. In the critical flow regime at Re $\simeq 2 \times 10^5$ the mean heat transfer increases considerably. Results for Re $> 3.5 \times 10^5$ are fitted by a single curve with the tangent of the inclination angle of 0.8. Thus in the supercritical flow regime, the heat transfer increases and follows a new law in which the power index of Re is 0.8. It is interesting to note that the same value of m was found for a turbulent flow over a plate. At Re from

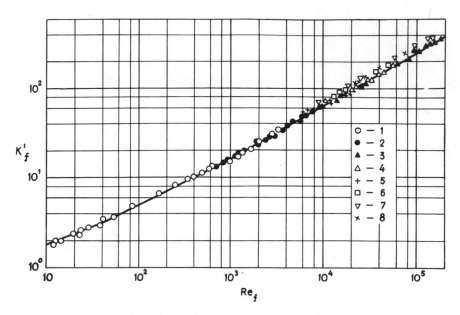

FIG. 34. Heat transfer of a tube at different conditions. (1–5) t_w = const. (from Žukauskas [37]) legend as in Fig. 33, (6) transformer oil, (7) water, and (8) air, q_w = const. (from Katinas *et al.* [38]).

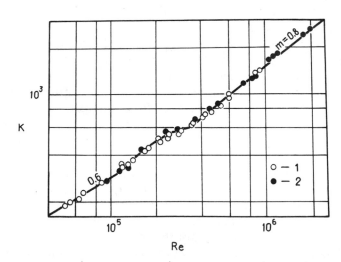

FIG. 35. Mean heat transfer of a cylinder at high Re (from Daujotas *et al.* [23]). (1) Cylinder of D = 30.7 mm and (2) cylinder of D = 50 mm. $K = K_f'$.

1.5×10^5 to 3.5×10^5, the results deviate from the general heat transfer law. This is a feature of the critical flow regime with a higher frequency of vortex shedding and the formation of a separation bubble. Similar mean heat transfer results were obtained in our studies [26] and by Achenbach [54] in air.

3. Relations for Heat Transfer Calculations

The solid line in Fig. 36 represents heat transfer data of a circular tube in crossflow of viscous fluids and gases, when the turbulence of the main flow is less than 1%. The curve has been derived from our experimental data in the range of Re from 1 to 1×10^7 [23, 26, 37, 38]. Data of other authors presented for comparison correlate well with our relation. Thus the following equation is recommended for practical calculations:

$$\text{Nu}_f = c\text{Re}_f^m\text{Pr}_f^{0.37}(\text{Pr}_f/\text{Pr}_w)^{0.25} \tag{33}$$

The values of c and m at various Reynolds numbers are given in Table I.

For gases, Eq. (33) is simplified, e.g., for air $\text{Pr} = 0.7 = \text{const.}$ and $\text{Pr}_f^{0.37}$ $(\text{Pr}_f/\text{Pr}_w)^{0.25} = 0.88$.

For $\text{Re} \le 1$, the heat transfer in flows of gases may be calculated from

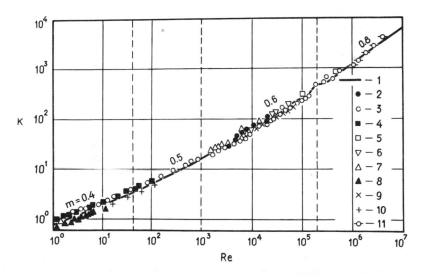

FIG. 36. Comparison of heat transfer of a tube. (1) From Žukauskas and Žiugžda [26], (2) from Mikhe'yev [15], (3) from Hilpert [13], (4) from Collis and Williams [57], (5) from Schmidt and Wenner [62], (6) from Ilyin [63], (7) from Hughes [2], (8) from Piret et al. [60], (9) from Fand [61], (10) from Davis [16], and (11) from Achenbach [54]. $K = K_f'$.

TABLE I

VALUES OF CONSTANTS IN Eq. (33)

Re	c	m
1–40	0.76	0.4
$40–1 \times 10^3$	0.52	0.5
$1 \times 10^3–2 \times 10^5$	0.26	0.6
$2 \times 10^5–1 \times 10^7$	0.023	0.8

Eqs. (30)–(32). In the range of low Reynolds numbers (Re $< 10^3$) when laminar boundary layers prevail, the value of the power index of Pr may be assumed $n = 0.33$, but in the supercritical flow regime with a predominantly turbulent boundary layer $n = 0.4$.

The relation of the type

$$Nu = c_1 Re^{0.5} + c_2 Re^{0.7} \tag{34}$$

is proposed for the whole range of Reynolds numbers [30], where the first term stands for the heat transfer through the laminar boundary layer on the front portion of a tube, and the second one stands for the heat transfer in the rear portion. It has, however, been shown that the power index of the Prandtl number is not constant along the perimeter. Thus for viscous liquids Eq. (34) becomes more complex.

On the basis of the known heat transfer regularities in separate parts of bodies in which laminar or turbulent boundary layers prevail, Gnielinski [64] recommended the following relation for the heat transfer of circular tubes and other bodies:

$$Nu_L = c + \sqrt{Nu_{L,lam}^2 + Nu_{L,turb}^2} \tag{35}$$

where L represents the perimeter length, $Nu_{L,lam}$ represents the Nusselt number from the equation for the heat transfer of a laminar boundary layer over a plate, $Nu_{L,turb}$ represents the Nusselt number from the equation for the heat transfer in a turbulent boundary layer over a plate, and $c = 0.3$ for a cylinder.

Calculations according to Eqs. (34) and (35) are more complex and time consuming.

The above-mentioned relations for the heat transfer of a cylinder in crossflow were compiled at moderate turbulences, Tu $\leq 1\%$, and are applicable for such conditions. In most practical applications and in heat exchangers, we deal with much higher turbulences. A number of investigations [26, 35, 52, 53 a.o.] suggest that an increase in the main flow turbulence causes higher heat transfer from a cylinder. In the calculations of the heat transfer, the effects of different factors, and first of all of the main flow turbulence, must be evaluated.

In our studies [26, 35] considerable increases of heat transfer were observed at higher turbulences. It is clear from Fig. 37 that in air an increase of turbulence to 15% in the subcritical range of Re_f leads to a 40% increase of mean heat transfer. With further growth of Re_f, the effect of Tu on the mean heat transfer increases, with the exception of a small interval of Re_f at the beginning of the critical range. In the supercritical flow regime with a further increase of Re to 1×10^6, a large increase of the mean heat transfer occurs. The value of m approaches 0.8, and an increase of Tu from 1.2 to 15% gives a 50% increase of the heat transfer.

The experimental results are in good agreement after the introduction of $Tu = (\sqrt{u'^2}/u) \times 100\%$ in the power of 0.15. At Tu $>$ 1%, the mean heat transfer can be approximated by Eq. (33), assuming that it is multiplied by the value of Tu raised to the power of 0.15. An exact evaluation of Tu is only possible if the turbulence macroscale is taken into account [26].

C. HEAT TRANSFER OF A TUBE IN A RESTRICTED CHANNEL

The influence of channel blockage on the flow pattern was analyzed in Section II. Heat transfer also changes under the variation of free space

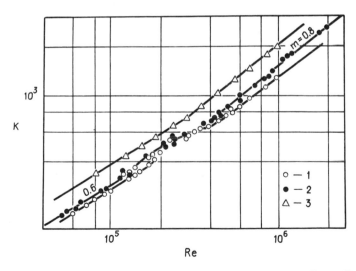

FIG. 37. Effect of turbulence on mean heat transfer of a cylinder in different flows (from Žukauskas and Žiugžda [26]). (1) Tu = 1.2% in air, (2) Tu = 4% in water, and (3) Tu = 15% in air. $K = K_f'$.

between the tube and the wall of the channel. According to theoretical calculations, the heat transfer on the front portion of a tube increases with an increase in the blockage ratio. This is confirmed by Fig. 38, representing calculations [29] by the method of Merk [46] using the potential flow velocity distribution around a tube.

Measurements [26, 29] of local heat transfer of a circular tube in a channel (Fig. 39) suggest that, for $D/H > 0.6$, the distribution of heat transfer coefficient in the rear also has a different character. In the region near $\varphi = 120°$, the second minimum of heat transfer appears. Curves 1, 2, and 4 [29] in the figure correspond to Re = 50,000 and are referred to the velocity of the main flow of air, and curve 3 [26] corresponds to Re = 1.35×10^6. Heat transfer of a cylinder in a restricted channel depends on the ratio between the cylinder diameter and the channel height, i.e., $\psi = D/H$.

The influence of the blockage ratio on the mean heat transfer is taken into account when choosing the reference velocity. In calculations of the mean velocity for the case $D/H < 0.8$, Eq. (5c) may be applied. But for higher blockage ratios, its influence on the heat transfer is different in the front and in the rear. Therefore, in Akilba'yev et al. [29], Eq. (34) is

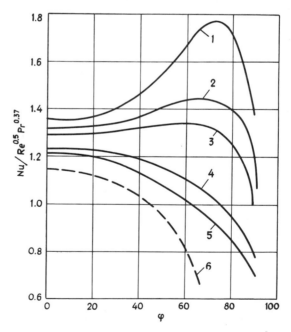

FIG. 38. Theoretical calculations of the heat transfer of a cylinder for different D/H [29]. (1) 0.83, (2) 0.75, (3) 0.71, (4) 0.52, (5) 0.39, and (6) 0.

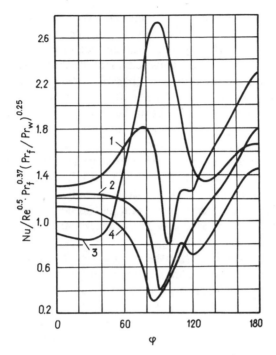

FIG. 39. Local heat transfer of a cylinder for different D/H. (1) 0.83 and (2) 0.63 at Re = 50,000 (from Akilba'yev et al. [29]), (3) 0.68 at Re = 1.35 × 10⁶ and Tu = 1% (from Žukauskas and Žiugžda [26]), and (4) 0.39 at Re = 50,000 (from Akilba'yev et al. [29]).

suggested for heat transfer calculations at $D/H > 0.8$, with separate empirical corrections for the front and the rear of a tube.

Our studies [26, 35] in the range of Re from 10³ to 10⁶ suggest that an increase of ψ at constant Re and Pr is accompanied by an increase of heat transfer which depends on the blockage factor. Experimental results fit nicely to Eq. (33), which is recommended for calculations, but with the following corrections:

$$Re_f = u^*D/v$$

where $u^* = (1 - \psi^2)u$ and u is defined in Eq. (5a).

VI. Heat Transfer of a Tube in a Bank

A. LOCAL HEAT TRANSFER

Heat transfer peculiarities of a tube in a bank are similar to those of a single tube. The variation of heat transfer around a tube in a bank is determined by the flow pattern, which depends greatly on the arrangement of the tubes in the bank. Thus in banks of in-line arrangement, two impact points and consequently two heat transfer maxima are observed. On the other hand, in banks of a staggered arrangement, the process of heat transfer is to some extent similar to that of a single tube.

As mentioned above, a tube in one of the inner rows is influenced by a highly turbulent flow, and the boundary layer near the impact point is purely laminar only at low Reynolds numbers.

Heat transfer variations in inner rows of banks of in-line and staggered arrangements are compared with that of a single tube in Fig. 40. In banks of both arrangements, a higher turbulence intensity in the flow causes an increase in the heat transfer at the front as well as the rear portions of a tube. Nevertheless, the maximum value of the heat transfer in the case of an in-line bank is observed at $\varphi = 50°$ because the impact of the stream on the surface occurs at this point.

Let us consider the heat transfer in different rows of a staggered bank (Fig. 41). The heat transfer from a tube in the first row is similar to that of a single tube. As the fluid passes through the first row, it is disturbed. This causes an increase in heat transfer in subsequent rows. This is observed up to the third row, downstream of which the heat transfer becomes stable and equal to the value of the latter. This applies to the front and rear portions of the tube [65].

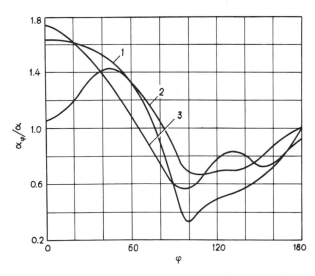

FIG. 40. Variation of the heat transfer around a tube. (1) Single tube, (2) tube in an in-line bank, and (3) tube in a staggered bank.

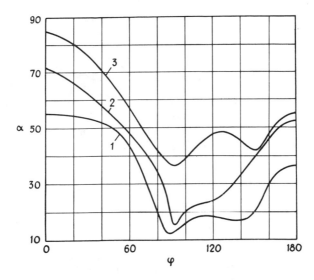

FIG. 41. Local heat transfer of a tube in a staggered bank of 2.0×2.0 at $Re = 1.4 \times 10^4$ (from Mikhaylov [65]). (1) First row, (2) second row, and (3) inner row.

In banks with a staggered arrangement, a change of longitudinal and transverse pitches from 1.3 to 2.0 has hardly any effect on the character of the heat transfer (Fig. 42) [65–69]. An increase in heat transfer is observed at about $\varphi = 120°$ in the banks examined, which corresponds to the laminar–turbulent transition in the boundary layer. Boundary-layer separation occurs at $\varphi = 150°$.

The variation of heat transfer around a tube in a staggered bank is almost independent of Reynolds number in the subcritical flow regime. In closely spaced staggered banks, the heat transfer of the inner tubes decreases in the front up to $\varphi = 50°$ and later increases again. This is explained by large pressure gradients and subsequent velocity increase.

Figure 43 presents heat transfer variations in an in-line bank [65]. The heat transfer is stable from the fourth row on. Tubes of the second and subsequent rows are "shaded" by the preceding ones with the character of heat transfer changing accordingly. Because of the two impact points, expressed by sharp maxima of heat transfer, the laminar boundary layer on an inner tube begins at $\varphi = 40–60°$, instead of the front stagnation point. With subsequent growth of the boundary layer in both directions from the impact points, the heat transfer decreases.

It is interesting to note that the heat transfer at the front stagnation point of the inner tubes increases from row to row and approaches the maximum heat transfer at the impact point. This is connected with the increase of

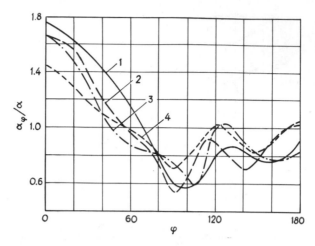

FIG. 42. Local heat transfer of tubes in different staggered banks at Re $< 7 \times 10^4$ (1) 2.0×2.0 (from Mikhaylov [65] and Kazakevich [66]), (2) 1.5×1.5 (from Mayinger and Schad [67]), (3) 1.3×1.13 (from Bortoli *et al.* [68]), and (4) 1.5×2.0 (from Winding and Cheney [69]).

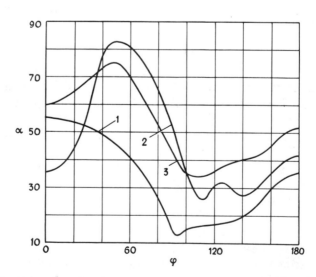

FIG. 43. Local heat transfer of a tube in an in-line bank of 2.0×2.0 at Re $= 1.4 \times 10^4$ (from Mikhaylov [65]). (1) First row, (2) second row, and (3) inner row.

flow turbulence in the bank. This also leads to more intensive heat transfer in the rear. Heat transfer in the rear constitutes a significant part of the total heat transfer, which means that the flow in the rear is not stagnant.

With an increase of Reynolds number, the flow in the bank becomes more turbulent, and the distribution of local heat transfer coefficients on tubes in inner rows levels out. This is confirmed by Fig. 44, comparing the results of various authors, where tube arrangement has little effect on the character of heat transfer.

On an inner tube in a staggered bank with the onset of the supercritical flow regime (Re $\approx 3 \times 10^5$) and with a further increase of Re, the first heat transfer minimum, which corresponds to the laminar—turbulent boundary-layer transition, begins to move upstream (Fig. 45). At $Re_f = 6.85 \times 10^5$ the transition occurs at $\varphi = 100°$; at $Re_f = 2.59 \times 10^5$ it is moved to $\varphi = 80°$; and at $Re_f = 9.98 \times 10^5$ it is at $\varphi \approx 25°$. The second heat transfer minimum, which corresponds to the turbulent boundary-layer separation, hardly moves away from $\varphi \approx 150°$ with an increase of Re_f. Analogous results were found for water flows [35, 70].

On an inner tube inside an in-line bank with the onset of the critical flow regime (Re $\approx 2 \times 10^5$), the point of laminar–turbulent transition also moves upstream, but its shift is less pronounced than in a staggered bank. In most in-line banks of conventional arrangement, the impact point lies not at $\varphi = 0$ but at $\varphi = 40-50°$; therefore the transition point only reaches this position, instead of $\varphi = 0$ as in staggered banks. But sometimes at higher Reynolds numbers, turbulent boundary layers begin to develop in the recirculation regions in front of the tubes.

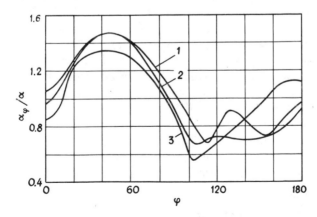

FIG. 44. Local heat transfer of a tube in different in-line banks at Re $= 3 \times 10^4$. (1) 1.5 × 1.5 (from Mayinger and Schad [67]), (2) 1.6 × 2.0 (from Kazakevich [66]), and (3) 1.5 × 1.1 (from Mayinger and Schad [67]).

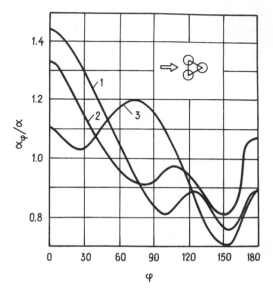

Fig. 45. Local heat transfer of a tube in an inner row of a staggered bank in a flow of air at different Re. (From Poškas *et al.* [70]). (1) At Re = 6.85 × 10⁴, (2) at Re = 2.59 × 10⁵, and (3) at Re = 9.98 × 10⁵.

No effect of the fluid physical properties on the dynamics of the heat transfer processes has been noted, and the distribution of the local heat transfer coefficient is similar in different fluids. This was proved in our numerous studies [20, 35] on the local heat transfer inside staggered and in-line banks in flows of air, water, and different oils. The relative rate of the local heat transfer in liquids and gases is somewhat dependent on Re and on the transverse and longitudinal pitches.

We attempted a determination of the value of m, the power index of Re, in the generalization of the experimental results from the local values of heat transfer and velocity. In the front region from the impact point to the separation point, we found $m = 0.5$.

Over the rest of the perimeter, the value of m varies from 0.5 to 0.8, which implies a change of the flow pattern at the wall. We conclude that the mean heat transfer of a tube in a bank depends on the interchange of different flow patterns on its surface, which determines the values of m.

B. Mean Heat Transfer

For practical purposes, the power index of the Prandtl number $n = 0.36$ is sufficiently accurate for all sorts of banks (see Section IV). Thus the following relation may be applied for heat transfer from banks of tubes:

$$Nu_f = cRe_f^m Pr_f^{0.36}(Pr_f/Pr_w)^{0.25} \tag{36}$$

The results are referred to the tube diameter and to the flow velocity in the minimum free cross section.

The final results for the mean heat transfer in the form of

$$K_f = Nu_f Pr_f^{-0.36}(Pr_f/Pr_w)^{-0.25} = f(Re_f) \tag{37}$$

are presented in the subsequent figures.

1. *Heat Transfer Variation along a Bank*

Experiments suggest that the heat transfer from a tube is determined by its position in the bank. In most cases, the heat transfer from tubes in the first row is considerably lower than in inner rows (Figs. 46 and 47).

In the range of low Re heat transfer from a tube in the first row is similar to that from a single tube and from tubes in inner rows. In specific in-line geometries, the heat transfer may be even higher from the front row. An increase of flow turbulence in a bank at higher Reynolds numbers leads to an increase of heat transfer of the inner tubes, as compared to the first row. The rows of tubes in a bank in fact act as a turbulence grid. In most banks, the heat transfer becomes stable from the third or fourth row in the

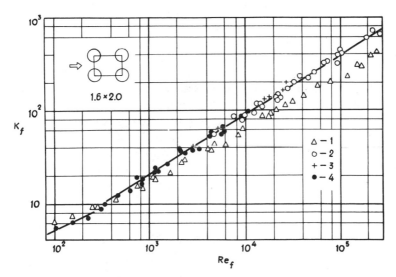

FIG. 46. Heat transfer of a tube in an in-line bank of 1.6 × 2.0. (1) First row; (2) inner row, water; (3) inner row, air; and (4) inner row, transformer oil.

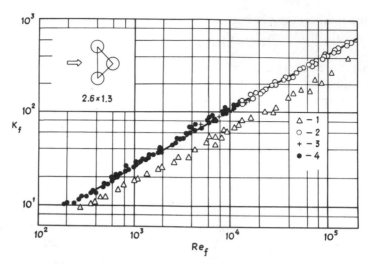

FIG. 47. Heat transfer of a tube in a staggered bank of 2.6 × 1.3. (1) First row; (2) inner row, water; (3) inner row, air; and (4) inner row, transformer oil.

mixed-flow regime. A comparison of heat transfer between the first row and inner rows in a fully developed flow reveals the influence of turbulence intensity on heat transfer.

The heat transfer of inner tubes generally increases as the longitudinal pitch decreases. This correlates well with known investigations of the heat transfer of a tube placed at various distances from the turbulence grid [20].

Due to turbulence the heat transfer from the inner tubes, depending on longitudinal pitch, is 30–100% higher than from the leading row. Thus in most cases, the heat transfer intensity of the inner rows is determined by the level of turbulence, which increases with a decrease of the distance from a turbulence grid, i.e., from the preceding rows.

The development of flow turbulence in a bank is different for different values of Reynolds number. In a staggered bank, the heat transfer of a tube in the second row is lower than that of an inner tube for $Re < 10^4$, and equal for $Re > 10^4$. The same applies to banks with in-line arrangements. The heat transfer of a tube in the second row is in most cases 10–30% lower than that of inner tubes.

2. Heat Transfer in a Range of Low Reynolds Numbers

As noted in Section III, at low Reynolds numbers, laminar flow patterns prevail with large vortices in the circulation zones. This is also reflected in the character of the heat transfer. The influence of free convection is observed in some cases of predominantly laminar flow. However, this is not usually taken into account in calculation formulas. Experimental data for the heat transfer of banks of tubes for low Reynolds numbers are presented in several publications [71–75].

Figures 48 and 49 present heat transfer from tubes in staggered and in-line banks with relative pitches 2.0 × 2.0. At Re < 10^3 heat transfer of a tube in the first row of a 2.0 × 2.0 in-line bank is 25% higher than that of inner tubes. In this bank any inner tube is in the wake of the preceding one, and its heat transfer is lower because of lower velocity in the recirculation region. This is the "shading" effect.

In contrast to in-line banks, in a staggered bank the heat transfer of a tube in the front row is lower than in an inner tube. The difference amounts to 7% for a 2.0 × 2.0 bank at values of Re_f up to 30, but it increases to 35% with a further increase of Re_f. This is a nice illustration of the different flow patterns in staggered and in-line banks.

In the predominantly laminar flow regime, the value of m, the power index of Re_f increases sharply with an increase of Re_f. At $Re_f < 10^2$ we have $m = 0.33–0.4$; later $m = 0.5$, but in the mixed-flow regime it becomes 0.6 and even 0.63 for both in-line and staggered banks.

A comparison of data on the heat transfer in the inner rows reveals a close agreement of data from different sources at $Re_f > 10^2$ and larger discrepancies at $Re_f < 10^2$.

Over the whole range of the predominantly laminar flow, the heat transfer from the inner tubes is lower for in-line banks than for staggered ones. From the available experimental results for the predominantly laminar flow, the following relations were found for the mean heat transfer of tubes in staggered banks: at $1.6 \leq Re_f \leq 40$

$$Nu_f = 1.04 Re_f^{0.4} Pr_f^{0.36} (Pr_f/Pr_w)^{0.25} \tag{38}$$

at $40 \leq Re_f \leq 10^3$

$$Nu_f = 0.71 Re_f^{0.5} Pr_f^{0.36} (Pr_f/Pr_w)^{0.25} \tag{39}$$

for in-line banks at $16 \leq Re_f \leq 100$

$$Nu_f = 0.9 Re_f^{0.4} Pr_f^{0.36} (Pr_f/Pr_w)^{0.25} \tag{40}$$

at $100 < Re_f < 10^3$

$$Nu_f = 0.52 Re_f^{0.5} Pr_f^{0.36} (Pr_f/Pr_w)^{0.25} \tag{41}$$

3. Heat Transfer in the Mixed- (Subcritical) Flow Regime

This regime covers Re from 10^3 to 2×10^5. Transition of the predominantly laminar to the mixed flow takes place at different Reynolds num-

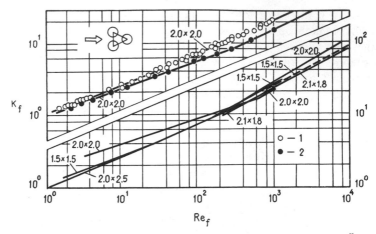

FIG. 48. Mean heat transfer of a tube in staggered banks at different Re_f (from Žukauskas [35]), $K_f = Nu_f Pr_f^{-0.36} (Pr_f/Pr_w)^{-0.25}$. (1) Inner row and (2) first row. The solid curve of a 2.0×2.0 bank below (from Isachenko [74]).

bers, depending on the tube arrangements (Figs. 48 and 49). For $Re > 10^3$ laminar boundary layer forms on the front of an inner tube, but the main portion of it is influenced by vortical flow. The character of the heat transfer is determined by the flow regime in the boundary layer. Thus the power index m of the Reynolds number varies from 0.55 to 0.73 for banks of different arrangements.

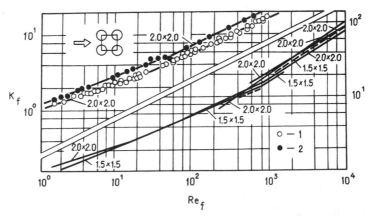

FIG. 49. Mean heat transfer of a tube in in-line banks at different Re_f (from Žukauskas [35]), $K_f = Nu_f Pr_f^{-0.36} (Pr_f/Pr_w)^{-0.25}$. (1) Inner row and (2) first row. The solid curve of a 2.0×2.0 bank blow (from Isachenko [74]); dotted curve (from Bergelin et al. [72]).

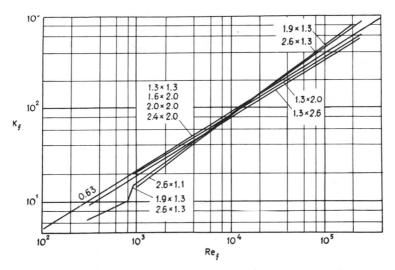

FIG. 50. Comparison of the heat transfer of different in-line banks (from Žukauskas et al. [20]).

Figure 50 shows a comparison of the heat transfer from 15 various banks of in-line arrangements [20]. It suggests an increase of m with a constant longitudinal and a decreasing transverse pitch. In fact, the value of m is influenced by changes in the ratio of longitudinal and transverse pitches (Fig. 51).

In the mixed-flow regime $m = 0.63$ is acceptable for most banks of in-line arrangements. The mean heat transfer of an inner tube is calculated from

$$\mathrm{Nu}_f = 0.27\mathrm{Re}_f^{0.63}\mathrm{Pr}_f^{0.36}(\mathrm{Pr}_f/\mathrm{Pr}_w)^{0.25} \tag{42}$$

In banks with $a/b < 0.7$, experimental heat transfer measurements are much lower than those calculated according to Eq. (42). Banks of this type are considered ineffective as heat exchangers.

Final experimental results of the heat transfer of different staggered banks in flows of viscous fluids [20] are presented in Fig. 52. The power index of Re is equal to 0.60 for all banks.

The effect of tube pitch is clear. Heat transfer increases with a decrease in the longitudinal pitch and, to a lesser extent, with an increase of the transverse pitch. The variation of c may be evaluated by the geometrical parameter a/b to the power 0.2 for $a/b < 2$. For $a/b > 2$, $c = 0.40$. In such banks the minimum free section is diagonal with respect to the main flow. Thus changes in c involve certain changes in the conditions of flow through a bank. The generalized formulas for heat transfer of inner tubes in various

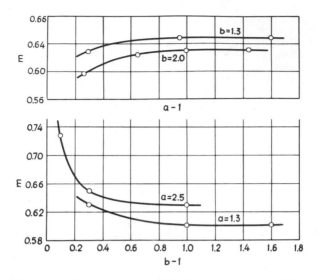

FIG. 51. The value of the power index of Re for in-line banks of different pitches.

staggered banks are for $a/b < 2$

$$Nu_f = 0.35(a/b)^{0.2}Re_f^{0.60}Pr_f^{0.36}(Pr_f/Pr_w)^{0.25} \qquad (43)$$

and for $a/b > 2$

$$Nu_f = 0.40Re_f^{0.60}Pr_f^{0.36}(Pr_f/Pr_w)^{0.25} \qquad (44)$$

4. Heat Transfer in the Critical Flow Regime

At about $Re_f = 2 \times 10^5$ the supercritical flow regime is established, and a more intensive heat transfer is observed. The value of the power index of Re_f increases to 0.8 or more. Detailed studies of the mean heat transfer in this range of Re_f were conducted at the Institute of Physical and Technical Problems of Energetics, Academy of Sciences of the Lithuanian SSR, in flows of air and water [70, 76, 77].

Figure 53 presents data on the mean heat transfer from tubes in the first row and in the inner rows of a 1.5×1.25 in-line bank. In the inner rows heat transfer is much higher. At $Re_f = 2 \times 10^5$ the value of m in the heat transfer relation is 0.6–0.65, but at high Re_f it increases to 0.77–0.8 and indicates a different law of the heat transfer. The increase of the power index of Re_f is mainly related to the upstream shift of the laminar–turbulent boundary-layer transition. This was shown in the analysis of the local heat transfer. As a result, over a larger part of the tube surface heat

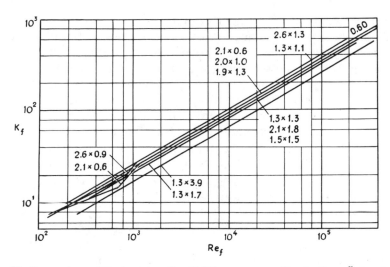

FIG. 52. Comparison of the heat transfer of different staggered banks (from Žukauskas *et al.* [20]).

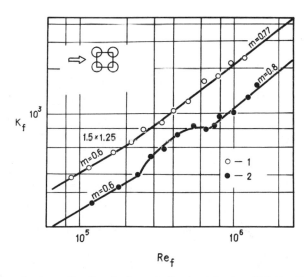

FIG. 53. Mean heat transfer of a tube in an in-line bank in air at high Re_f (from Žukauskas [35]). (2) First row and (1) inner row.

exchange occurs in a turbulent boundary layer. The analysis suggests a different character of the heat transfer augmentation of a tube in the first row as compared to that of an inner row. On a tube in the first row a deviation from the power law heat transfer is observed at Re_f from 2.5 × 10^5 to 7 × 10^5 and is similar to the situation on a single tube in a flow of moderate turbulence. This deviation is related to the formation of a separation bubble and of a sharp increase of the heat transfer in the rear part of the tube. We have already noted that this type of deviation is not observed on single cylinders in flows of high turbulence. The situation on inner rows is similar for in-line banks (Fig. 53) and staggered banks. At $Re_f > 2 × 10^5$ the value of m is higher for staggered banks, where it reaches 0.8. Analogous results were found in in-line and in staggered banks in the supercritical flow regime of water [35].

Figures 54 and 55 give a comparison of the heat transfer in inner rows of different in-line and staggered banks at high Re_f [35]. At $Re_f > 2 × 10^5$ the results correlate well with the earlier results on the heat transfer from banks. In the supercritical flow regime, the effect of pitch on the heat transfer is similar to that in the mixed-flow regime. Heat transfer in staggered banks is most intensive, but the difference among separate banks never exceeds 25%.

Similar results were obtained for in-line banks. Most intensive heat transfer is observed in banks of large longitudinal and transverse pitch. For the heat transfer in the inner rows of in-line banks, we suggest

$$Nu_f = 0.033 Re_f^{0.8} Pr_f^{0.4} (Pr_f/Pr_w)^{0.25} \qquad (45)$$

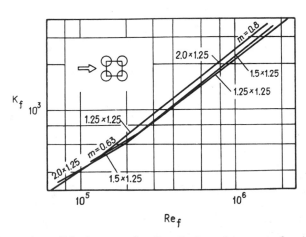

Fig. 54. Comparison of the heat transfer of a tube in an inner row of an in-line bank at high Re_f (from Žukauskas [35]).

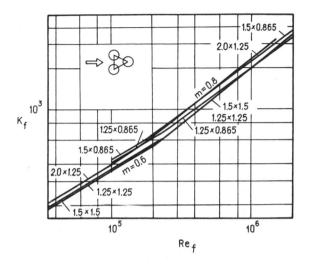

FIG. 55. Comparison of the heat transfer of a tube in an inner row of a staggered bank at high Re_f (from Žukauskas [35]).

and for staggered banks

$$Nu_f = 0.031(a/b)^{0.2}Re_f^{0.8}Pr_f^{0.4}(Pr_f/Pr_w)^{0.25} \qquad (46)$$

and for gaseous flows when $Pr_f = 0.7$

$$Nu_f = 0.027(a/b)^{0.2}Re_f^{0.8} \qquad (47)$$

It is evident from Figs. 52–55 that, in the case studied, the supercritical flow regime is established at $Re_f > 2 \times 10^5$, where the value of m increases to 0.6–0.8. This was supported by Hammecke et al. [78]. In some banks the power index m is over 0.8 and is close to its value for a turbulent boundary layer over a plate. Inasmuch as the front part of the tube is under a laminar boundary layer, such an increase of m suggests a higher augmentation of the heat transfer in the rear part as compared to that of a plate under a turbulent boundary layer. The rate of increase of the heat transfer in the rear part of a tube is close to that at a laminar–turbulent transition on a plate [79]. The investigation suggests that the onset of the critical flow regime is governed not only by the bank arrangement and Reynolds number but also by the surface roughness, temperature difference, and fluid physical properties.

5. Heat Transfer of Closely Spaced Banks

In the previous cases of heat transfer of banks of tubes, the results were referred to the velocity in the minimum free cross section from Eq. (7), i.e.,

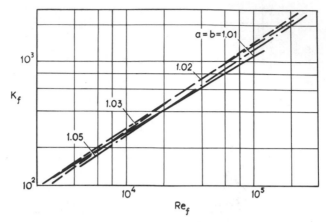

FIG. 56. Heat transfer of closely spaced in-line banks (from Žukauskas *et al.* [20]).

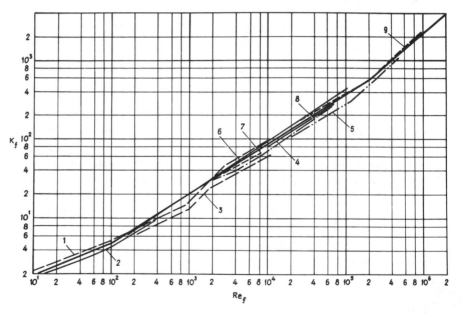

FIG. 57. Heat transfer of in-line banks. (1) 1.25 × 1.25 and (2) 1.5 × 1.5 (from Bergelin *et al.* [72]), (3) 1.25 × 1.25 (from Kays and London [18]), (4) 1.45 × 1.45 (from Kuznetsov and Turilin [39]), (5) 1.3 × 1.5 (from Lyapin [80]), (6) 2.0 × 2.0 (from Isachenko [74]), (7) 1.9 × 1.9 (from Grimison [9]), (8) 2.4 × 2.4 (from Kuznetsov and Turilin [39]), and (9) 2.1 × 1.4 (from Hammecke *et al.* [78]).

to the maximum velocity. In fact, the heat transfer is determined not by the maximum velocity but by the average velocity, integrated over the perimeter of the tube. In wider spaced banks, the average velocity scarcely differs from the maximum, and the acceptance of the latter is justified by the simpler calculation. But when the free cross section is small, the maximum velocity exerts its influence only on a small portion of the surface.

Investigations of heat transfer [20] in staggered and in-line banks, with the clearance between tubes from 0.5 to 4 mm, revealed that the heat transfer curves for separate banks referred to the maximum velocity and differ from each other. Therefore, for closely spaced banks the results should be referred to as the average velocity given by Eq. (8). Such results for various closely spaced banks correlate satisfactorily (Fig. 56).

6. Comparison of the Results

Our Eqs. (38)–(47) concerning the heat transfer in different banks with various fluids actually describe the results of investigations of 79 banks of different arrangements at Re from 1 to 2×10^6 and Pr from 0.7 to 5000.

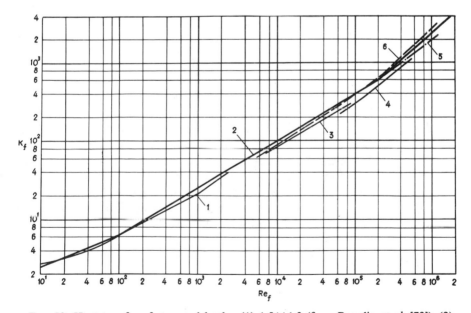

FIG. 58. Heat transfer of staggered banks. (1) 1.5×1.3 (from Bergelin et al. [72]); (2) 1.5×1.5 and 2.0×2.0 (from Grimison [9] and Bergelin et al. [73]); (3) 2.0×2.0 (from Antuf'yev and Beletsky [17], Kuznetsov and Turilin [39], and Kazakevich [66]); (4) 1.3×1.5 (from Lyapin [80]); (5) 1.6×1.4 (from Dwyer and Sheeman [81]); and (6) 2.1×1.4 (from Hammecke et al. [78]).

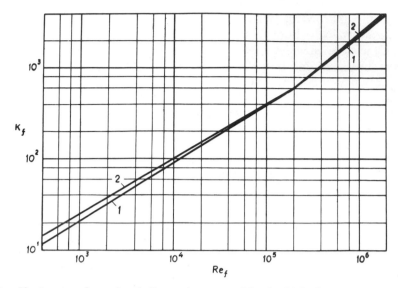

FIG. 59. A comparison of an in-line and a staggered banks. (1) In-line and (2) staggered bank of $a/b > 2$.

These investigations have been performed at the Institute of Physical and Technical Problems of Energetics, Lithuanian Academy of Sciences. Our results for heat transfer, denoted by the continuous line, are compared with the results of other authors for in-line and staggered banks in Figs. 57 and 58, respectively. The results correlate well over the whole range of Re examined.

The results of heat transfer in staggered banks are compared with those in in-line banks (Fig. 59), using the experimental data of square in-line banks [20]. For the case of the staggered arrangement, this bank was turned by 45°. Figure 59 shows that at low Reynolds numbers the heat transfer of in-line banks is considerably lower than that of staggered banks. With an increase of Reynolds number, the heat transfer of in-line banks increases at a higher rate, and at high Re it approaches the heat transfer intensity of staggered banks. However, it should be mentioned that the efficiency of banks depends not only on heat transfer intensity but also on hydraulic drag.

VII. Hydraulic Drag of Banks

A. DRAG CALCULATION METHODS

Hydraulic drag is one of the most important features of heat exchangers and is characterized by the total pressure drop in flow across banks of tubes. As is known, the total pressure drop across a bank is a function of flow velocity, bank arrangement, and the physical properties of the fluid. The pressure drop of a bank in viscous flow of constant density is expressed by the following functional relation:

$$\Delta p = f(u, s_1, s_2, D, z, \mu, \rho) \tag{48}$$

The dimensionless form of this relation will be

$$\text{Eu} = \varphi(\text{Re}, s_1/D, s_2/D, z) \tag{49}$$

or the exponential form

$$\text{Eu} = k\text{Re}^r z \tag{50}$$

In generalizations of experimental results and in calculations according to Eq. (50), the choice of reference velocity and number of rows to which pressure drop is related is of considerable importance. The use of the average velocity, calculated along the perimeter from the front stagnation point to $\varphi = 90°$, as reference velocity is most suitable. This allows the comparison of the hydraulic drag of various types of banks.

In banks with $a > 1.25$, the average value of the velocity is close to the maximum, and here the latter is more convenient. In closely spaced banks with $a < 1.25$, the average value of the velocity may be used as reference.

For simplicity the maximum velocity was used as a reference in the summary of experimental data. This procedure reflects the actual drag with sufficient accuracy, except for closely spaced banks.

Experiment suggests that the pressure drop across banks is proportional to the number of rows and is determined by the tube arrangement. With a decreasing number of rows, the entrance and exit conditions in the bank contribute more to the total loss of energy. This must be taken into account in calculations for banks with a small number of rows. However, in experiments, banks with a large number of rows are usually used. The results of hydraulic drag are represented by the Eu number referred to one row or to a bank of ten rows.

In the treatment of experimental results with the aid of Eq. (50), the choice of a reference temperature, a reference velocity, and a number of rows to which the pressure drop is referred must be considered. Physical properties in the similarity numbers are referred to the main flow (i.e., mean bulk) temperature. For nonisothermal conditions the same relation may be applied, but with the mean flow temperature in the bank as a reference temperature for the physical properties. At high values of Re the influence of variation of fluid physical properties, depending on the temperature of pressure drop, is insignificant because surface shear contributes little to the total drag.

At low Re in viscous flows the effect of variable fluid physical properties is evaluated by

$$Eu_f = Eu(\mu_w/\mu_f)^p \tag{51}$$

where Eu_f is the Euler number for heating or cooling, Eu is the Euler number for isothermal conditions, and μ_w and μ_f are the dynamic viscosity at wall temperature and flow temperature, respectively.

The value of the power index of μ_w/μ_f for heating is given by

$$p = -1.8 \times 10^{-3}Re + 0.28 \tag{52}$$

and for cooling

$$p = -2.6 \times 10^{-3}Re + 0.43 \tag{53}$$

B. MEAN DRAG OF BANKS

Let us consider now the drag of banks related to the tube arrangement and Reynolds number. Measurements of the pressure drop across 79 different banks in flows of air and liquids [20, 35] suggest that drag is mainly determined by the transverse pitch a and increases with a decrease of the latter. This applies both to staggered and in-line banks. With an increase of the longitudinal pitch b, the larger space between two neighboring rows causes the formation of vortices which in many cases affects the drag of the bank. The effect is related to the bank geometry and to the flow regime. The dependence of hydraulic drag versus Re suggests different laws in the predominantly laminar flow and in the mixed flow.

1. Predominantly Laminar Flow

Viscous forces prevail in the drag of in-line banks at $Re < 10^3$, and the power index of Re is equal to -0.5. In this range the drag of staggered banks is determined mainly by the value of the minimum free cross section and increases with a decrease of the latter.

Fig. 60 presents generalized results of the drag of several staggered banks, related to one row according to Eq. (50). The variation of drag of in-line banks is similar. The relative drag of in-line banks is, however, somewhat lower than that of staggered banks.

2. Mixed Flow

Data on hydraulic drag (Fig. 60) show that the transition of the predominantly laminar to the mixed flow occurs at $Re = 10^3$ in staggered banks and at lower Re in banks in which the minimum free cross section s_2' is

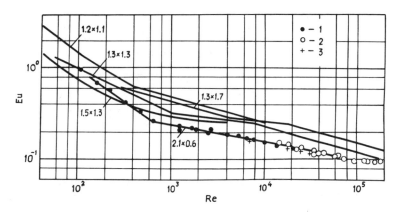

FIG. 60. Hydraulic drag of staggered banks. (1) Transformer oil, (2) water, and (3) air.

found along the diagonal (Fig. 10). After the transition and with a further increase of Re from 10^4 to 2×10^5, curves in the figure fall more gradually.

An increase of the longitudinal pitch leads to an increase of the drag, the latter being influenced mainly by the space between the tubes in the bank, in which the vortex flow is forming. The flow pattern in staggered banks with $a > 2$ is to some extent similar to that of in-line banks. A decrease of the longitudinal pitch involves a decrease of the free cross section, and the effect of tube arrangement on drag in such banks is reflected only in the value of the pitch along a diagonal. This applies for banks with $a < 2$.

In most in-line banks, the transition of the predominantly laminar to the mixed flow also takes place at Re = 10^3. In banks of large longitudinal pitch ($b > 1.70$), Eu becomes essentially independent of Reynolds number. In this case, the drag of a bank is determined solely by the transverse pitch. In banks with $b < 1.5$, the pressure drop coefficient depends on Re and is influenced by the longitudinal pitch.

3. Critical Flow

Let us first consider the magnitude of the flow drag in the transition to the critical regime as a function of the number of rows in a bank.

Experiments on staggered banks in air flows [76] suggest, as might be expected, that the drag of a single row is analogous to that of a single tube. It decreases at the critical Reynolds numbers and increases a little with a further increase of Re (Fig. 61). The pressure drop coefficient of a staggered bank with a large number of rows also decreases with an increase of Re, and at Re > 2×10^5 it becomes independent of Reynolds number, so a self-preserved flow is established. As can be seen, the character of the drag

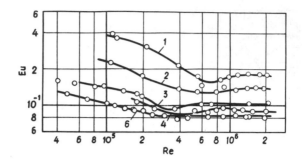

FIG. 61. Hydraulic drag of a staggered bank of 1.5 × 1.04 referred to as one row. The figures stand for the number of rows in the bank.

of a bank differs slightly from that of a single row, due to the effect of the turbulence generated by preceding rows.

In Fig. 62 the drag characteristics of staggered banks with many rows are presented. Transition to the critical regime is noted at Re > 2 × 10⁵ in all the banks.

The drag of banks is influenced also by the amount of free space between the tubes in a bank. Closely spaced banks with a staggered arrangement can be imagined as a number of obstacles, periodically narrowing and widening the channel, which leads to the disturbance of the flow. Thus for banks with $a/b < 1.7$, an automodel character of the pressure drop variation takes place with Eu = const. and independent of Re. In wider spaced banks with $a/b < 1.7$, the flow across banks can be stable only at high Reynolds numbers.

Figure 63 presents data on the pressure drop of in-line banks with a crossflow of air. It is obvious that in the closely spaced bank, 1.3 × 1.3, and banks with a large longitudinal pitch, the automodel character of the pressure drop variation is established at Re = 2 × 10⁵. In banks of a large transverse and small longitudinal pitch, 2.5 × 1.3, the automodel character

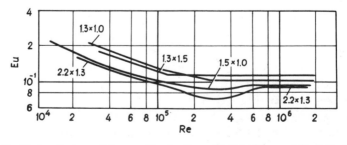

FIG. 62. Hydraulic drag of staggered banks (from Stasiulevičius and Samoška [76]).

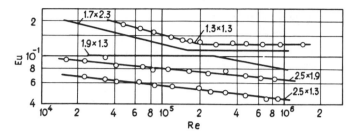

FIG. 63. Hydraulic drag of in-line banks (from Stasiulevičius and Samoška [76]).

of the pressure drop variation never occurs over the range of Re examined. Apparently, banks of this type are not capable of making the flow sufficiently turbulent, and if the automodel character of the pressure drop is established, it occurs at even higher Reynolds numbers.

In predominantly turbulent flow, the pressure drop across banks of in-line arrangement with large transverse and small longitudinal pitches is less. The largest pressure drop is observed, of course, across banks with closely spaced tubes.

C. Proposals for Calculation of Hydraulic Drag

As suggested by the analysis of the experimental results, for simplicity of calculation a graphical interpretation of the data is most convenient. General graphs have been compiled from our results described previously, including the results of other authors on the pressure drop across banks in flows of gases and liquids. A satisfactory correlation has been achieved.

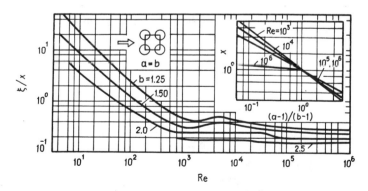

FIG. 64. Pressure drop coefficient of in-line banks versus the relative longitudinal pitch b as the reference.

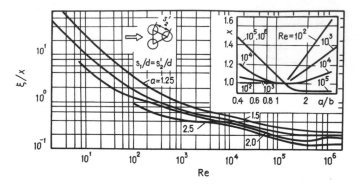

FIG. 65. Pressure drop coefficient of staggered banks versus the relative transverse pitch a as the reference.

The general graphs of in-line banks are based on the drag of banks of square arrangements, with the reference distance being the longitudinal pitch. Graphic corrections have been introduced for other banks to account for different pitches and Reynolds numbers.

The graphs of staggered banks are based on the equilateral–triangular arrangement with corrections for the evaluation of different arrangements and Re.

The graphs of the pressure drop coefficient as a function of Re for in-line and staggered banks are presented in Figs. 64 and 65 ascribed for one row of a bank. Comparison of the pressure drop across staggered and in-line banks suggests that, in the transition region from the predominantly laminar to the mixed flow, the banks of in-line arrangement exhibit a considerably lower pressure drop coefficient which must be ascribed to the structure of the turbulent flow in the rear portions of tubes due to different pitches.

With increasing Reynolds number, the flow in a bank becomes increasingly turbulent, and the pressure drop across in-line and staggered banks becomes equal.

VIII. Calculation of Banks of Tubes in Crossflow

The preceding sections were devoted to the heat transfer of tubes in banks and the hydraulic drag of banks. The main factors exerting an influence on the heat transfer process were analyzed. Calculation formulas were proposed, reflecting the general characteristics of the heat transfer of tubes in crossflow. The derivation of various equations and charts for various cases of tubes in crossflow are beyond the scope of this article.

Equations and charts for all practical cases can be easily derived from the graphs presented here. Thus Fig. 56 may be used for heat transfer calculations in closely spaced banks.

For gases, the formulas are simplified; say, for air, $Pr = 0.7 = const.$ and $Pr_f^{0.36}(Pr_f/Pr_w)^{0.25} = 0.88$. Multiplication of the constants in the formulas by 0.88 gives simplified formulas for banks in air at moderate temperature differences. Thus for $Re > 10^3$ the heat transfer of a tube of an in-line bank is determined by the relation

$$Nu_f = 0.27 \times 0.88 Re_f^{0.63} = 0.24 Re_f^{0.63} \qquad (54)$$

The author is not concerned with the processes of convective heat transfer at high temperatures of gases and in the presence of chemical reactions in the boundary layer, and the formulas cannot be applied for these conditions.

In Section VI, formulas were proposed for the calculations of heat transfer of the inner tubes or banks of many rows. If the number of rows is small, the lower heat transfer intensity of the first rows should be taken into account. In banks of less than 20 rows, the difference between the heat transfer of the first row and the inner rows is evaluated by the factor c_z, relating to the corresponding Nusselt numbers

$$Nu_z = c_z Nu_{z \geq 20} \qquad (55)$$

The values of c_z as a function of number of rows are presented in Fig. 66.

The formulas apply only if the flow is perpendicular to the tube axis. At attack angles $\psi < 90°$, the heat transfer decreases and is determined by the multiplication of the heat transfer coefficient by the correction factor

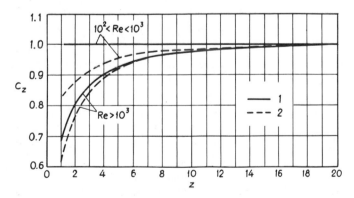

FIG. 66. Correction for the number of rows in the heat transfer calculations of banks. (1) In-line and (2) staggered.

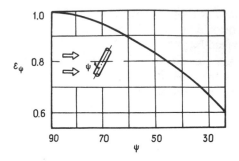

FIG. 67. The relation of heat transfer of banks versus the angle of attack.

depending on the angle

$$\varepsilon_\psi = \alpha_\psi / \alpha_{\psi=90°}.$$

The values of the correction factor for different angles of attack are given in Fig. 67. At $\psi = 0$, we are dealing with the case of longitudinal flow past a bank of tubes.

The optimal arrangement of tubes is one of the main problems in the construction of heat exchangers. The choice in each case is connected with the amount of investment and operation cost, which makes the knowledge of the power characteristics of heat exchangers so important.

The efficiency of heat exchangers from the energy point of view is characterized by the ratio of the amount of heat transferred through a definite surface to the energy consumed to overcome the hydraulic drag. This problem can be solved using the data for heat transfer and hydraulic drag.

An efficiency comparison of in-line and staggered banks reveals that for Re from 5×10^2 to 5×10^4 the in-line banks are more efficient. In spite of the fact that the heat transfer of in-line banks is lower in this range of Re, their efficiency is increased by the lower hydraulic drag. At higher Reynolds numbers, the efficiency of banks of different types becomes comparable, being mainly determined by the pitch.

The efficiency of the process of heat transfer in a tube bank reflects its efficiency from an energy point of view. A decrease of velocity leads to higher efficiency, and from an energy point of view, heat transfer seems more effective at low velocities. But in this case, the heated surface increases correspondingly, and only a complex solution of capital investment and operating costs leads to the optimal results for each specific case.

The problems of optimal arrangements and the calculations of heat exchangers have been analyzed in a number of special publications [17–19, 35, 82–85].

NOMENCLATURE

a	relative transverse pitch, s_1/D	r	power index of Re, Eq. (50)
b	relative longitudinal pitch, s_2/D	s_1	transverse pitch of bank of tubes (m)
C_D	drag coefficient, $D_f/(\rho u_0^2/DL/2)$	s_2	longitudinal pitch of bank of tubes (m)
c	constant		
c_p	specific heat at constant pressure (J/kg C)	s_2'	diagonal pitch of staggered bank (m)
D	diameter of tube (m)	T	temperature (°C)
D_f	total drag of tube (N)	u	fluid flow velocity (m/sec)
F	heat transfer surface (m^2)	u_0	free-stream velocity (m/sec)
f	frequency (Hz)	u_x	local velocity outside boundary layer (m/sec)
g	acceleration of gravity (m/sec²)	v	normal component of velocity (m/sec)
H	height (m)		
K_f	complex dimensionless terms, $Nu_f Pr_f^{-0.36}(Pr_f/Pr_w)^{-0.25}$	x	distance measured from front stagnation point (m)
L	length (m)	y	distance measured normal to wall (m)
m	power index of Re		
m_1	power index, Eq. (14)	z	number of tube rows in bank
n	power index of Pr	Gr	Grashof number, $(gL^3/\nu^2)\beta\,\Delta t$
P	pressure coefficient, Eqs. (3) and (6)	Nu	Nusselt number, $\alpha D/\lambda$
p	pressure (N/m²)	Nu(x)	local Nusselt number
p_0	static pressure (N/m²)	Pr	Prandtl number, $c_p\mu/\lambda$
Δp	pressure drop (N/m²)	Re	Reynolds number, uD/ν
q	specific heat flux (W/m²)	Eu	Euler number, $\Delta p/\rho u^2$
$q(x)$	local specific heat flux (W/m²)	Sh	Strouhal number, fD/u
R	radius of tube (m)	Tu	turbulence intensity, $\sqrt{\overline{u'^2}}/u$

Greek Symbols

α	heat transfer coefficient (W/m²K)	ν	kinematic viscosity (m²/sec)
β	coefficient of expansion (°C⁻¹)	ρ	density (kg/m³)
ζ	pressure drop coefficient, $2\,\Delta p/\rho u^2 z$	τ	shear stress (N/m²)
λ	thermal conductivity (W/mK)	φ	angle measured from front stagnation point (°)
μ	dynamic viscosity (Nsec/m²)		

Subscripts

f, 0	conditions of the free stream	w	conditions on the wall
m	mean value	x	local conditions

REFERENCES

1. L. V. King, *Philos. Trans. R. Soc. London, Ser. A* **214**, 374 (1914).
2. J. A. Hughes, *Philos. Mag.* [6] **31**, 118 (1916).
3. H. Rietschel, *Mitt. Pruefunganst. Heiz. Luft. Koenig. Tech.* Hochsch., Berlin **3** (1910).
4. W. H. Carrier and F. L. Busey, *Trans. ASME* **33**, 1055 (1911).
5. H. Thoma, "Hochleistungskessel," Springer-Verlag, Berlin and New York, 1921.
6. H. Reiher, *Forschungsarbeiten* **269**, 1 (1925).

7. W. L. Lohrisch, *Forschungsarbeiten* **322,** 1 (1929).
8. V. M. Antuf'yev and L. S. Kozachenko, *Sov. Kotloturbostroyeniye* **5,** 24 (1937).
9. E. D. Grimison, *Trans. ASME* **59,** 583 (1937).
10. O. L. Pierson, *Trans. ASME* **59,** 563 (1937).
11. E. S. Huge, *Trans. ASME* **59,** 573 (1936).
12. N. V. Kuznetsov and V. A. Lokshin, *Teplo Sila* **10,** 19 (1937).
13. L. Hilpert, *Forsch. Geb. Ingenieurwes.* **4,** 215 (1933).
14. G. N. Kruzhilin and V. A. Shwab, *Zh. Tekh. Fiz.* **5,** 703 (1935).
15. M. A. Mikhe'yev, *Zh. Tekh. Fiz.* **13,** 311 (1943).
16. A. H. Davis, *Philos. Mag.* [6] **47,** 972 (1924).
17. V. M. Antuf'yev and G. S. Beletsky, "Teploperedacha i aerodinamicheskiye soprotivleniya trubchatikh poverkhnostey v poperechnom potoke." Mashgiz, Moscow, 1948.
18. W. M. Kays and A. L. London, "Compact Heat Exchangers." McGraw-Hill, New York, 1958 (2nd ed., 1964).
19. R. Gregorig, "Wärmeaustausch und Wärmeaustauscher," Verlag II, p. 944. R. Sauelländer, Frankfurt am Main, 1973.
20. A. Žukauskas, V. Makarevičius, and A. Šlančiauskas "Teplootdacha puchkov trub v poperechnom potoke zhidkosti" (Heat transfer in Banks of Tubes in Crossflow of Fluid), p. 210. Mintis, Vilnius, 1968.
21. A. Žukauskas, *Adv. Heat Transfer* **8,** 93 (1972).
22. E. Achenbach, *J. Fluid Mech.* **34,** 625 (1968).
23. P.M. Daujotas, J. Žiugžda, and A. A. Žukauskas, *Int. Chem. Eng.* **16,** 476 (1976).
24. H. Schlichting, *in* "Encyclopedia of Physics," Vol. 8, p. 351. Springer-Verlag, Berlin and New York, 1959.
25. A. Roshko, *Natl. Advis. Comm. Aeronaut., Rep.* **1191** (1954); *J. Fluid Mech.* **10,** 345 (1961).
26. A. Žukauskas and J. Žiugžda, "Heat Transfer of a Cylinder in Crossflow," p. 208. Hemisphere Publ. Corp. Washington, D.C., 1985.
27. A. Fage and V. M. Folkner, *Aerosp. Res. Counc. (London), Rep. Mem.* **1369** (1931).
28. V. Žiugžda and T. Ruseckas, *Mokslas Tech.* **3,** 61 (1971).
29. Zh. S. Akilba'yev, S. I. Isata'yev, P. A. Krashtalev, and N. V. Masle'yeva, *Probl. Teplo-Prikl. Teplofiz.* **3,** 180 (1966).
30. H. C. Perkins, Jr. and G. Leppert, *Int. J. Heat Mass Transfer* **7,** 143 (1964).
31. R. Bressler, *Kaeltetechnik* **11,** 365 (1958).
32. Ž. Kostič, "Int. Seminar Heat Transfer," Herzeg-Novi, Yugoslavia, 1969.
33. E. Achenbach, *Waerme- Stoffuebertrag.* **2,** 47 (1969).
34. Ž. Kostič and S. Oka, "Int. Seminar Heat Transfer," Herzeg-Novi, Yugoslavia, 1968.
35. A. Žukauskas, "Konvektivnyi Perenos v Teploonmennikakh" (Convective Transfer in Heat Exchangers), p. 472. Nauka, Moscow, 1982.
36. V. Makarevičius, "Teploobmen pri Fiziko-Khimicheskikh Izmenenyakh" (Heat Transfer in the Presence of Physical-Chemical Factors), p. 227. Mokslas, Vilnius, 1978.
37. A. A. Žukauskas, *Teploenergetika* **4,** 38 (1954).
38. V. Katinas, J. Žiugžda, and A. Žukauskas, *Heat Transfer—Sov. Res.* **3,** 10 (1971).
39. N. V. Kuznetsov and S. I. Turilin, *Izv. Vses Teplotekh. Inst.* **11,** 23 (1952).
40. J. K. Stasiulevičius and P. S. Samoška, *Inzh.-Fiz. J.* **7,** No. 11, 10 (1964).
41. M. A. Mikhe'yev, "Osnovy Teploperedachi" (Fundamentals of Heat Transfer). Gosenergoizdat, Moscow, 1956.
42. A. A. Šlančiauskas, R. V. Ulinskas, and A. A. Žukauskas, *Liet. TSR Mokslu Akad. Darb. Ser. B* **4**(59), 163 (1969).
43. G. N. Kruzhilin, *Zh. Tekh. Fiz.* **6,** 858 (1936).
44. E. R. G. Eckert, *VDI-Forschungsh.* 416 (1942).

45. K. Hiemenz, "Die Grenzschicht an einem in den gleich-formigen Flüssigkeitsstrom eingetauschten geraden Kreiszylinder." Drs., Göttingen, 1911.
46. H. J. Merk, *J. Fluid Mech.* **5**, 460 (1959).
47. N. Frössling, *Lunds. Univ. Arsskr. Avd. 2 [N.F.]* **36**, 4 (1940).
48. K. M. Krall and E. R. Eckert, Heat Transfer, *Proc. Int. Heat Transfer,* 1970, Vol. 3, FC 7.5 (1970).
49. E. R. Eckert and E. Soengen, *Trans. ASME* **74**, 343 (1952).
50. W. H. Giedt, *J. Aeronaut. Sci.* **18**, 725 (1951).
51. G. M. Zapp, M. S. Thesis, Oregon State College, Corvallis (1950).
52. J. Kestin, *Adv. Heat Transfer* **3**, 1 (1966).
53. E. P. Dyban and E. V. Epick, *Heat Transfer, Proc. Int. Heat Transfer, 1970,* Vol. 2, FC 5.7 (1970).
54. E. Achenbach, *Int. J. Heat Mass Transfer* **10**, 1387 (1975).
55. V. T. Morgan, *Adv. Heat Transfer* **11**, 199 (1975).
56. B. S. Van der Hegge Zijnen, *Appl. Sci. Res., Sect. A* **6**, 129 (1956).
57. D. S. Collis and M. J. Williams, *J. Fluid Mech.* **6**, 359 (1959).
58. P. O. Davies and M. J. Fisher. *Proc. R. Soc. London, Ser. A* **280**, 486 (1964).
59. L. J. S. Bradbury, and I.P. Castro, *J. Fluid Mech.* **51**, 487 (1972).
60. E. L. Piret, W. James, and E. Stacy, *Ind. Eng. Chem.* **39**, 1088 (1947).
61. R. M. Fand, *Int. J. Heat Mass Transfer* **8**, 995 (1965).
62. E. Schmidt and K. Wenner, *Forsch. Geb. Ingenieurwes.* **12**, 65 (1941).
63. L. N. Ilyin, *Teplo, Aerodin., Sb. TsKTI* **18**, No. 4. 3 (1950).
64. V. Gnielinski, *Forsch. Ingenieurwes.* **41**, 145 (1975).
65. G. A. Mikhaylov, *Sov. Kotloturbostroyeniye* **12**, 434 (1939).
66. F. P. Kazakevich, *Teploenergetika* **8**, 22 (1954).
67. F. M. Mayinger and O. Schad, *Waerme-und Stoffuebertrag.* **1**, 43 (1968).
68. R. A. Bortoli, R. E. Grimmble, and J. E. Zerbe, *Nucl. Sci. Eng.* **1**, 239 (1956).
69. C. C. Winding and A. J. Cheney, *Ind. Eng. Chem.* **40**, 1087 (1948).
70. P. S. Poškas, V. J. Survila, and A. A. Žukauskas, *Int. Chem. Eng.* **18**, No. 2, 337 (1978).
71. G. A. Omohundro, O. P. Bergelin, and A. P. Colburn, *Trans. ASME* **71**, 27 (1949).
72. O. P. Bergelin, G. A. Brown, H. L. Hull, and F. W. Sullivan, *Trans. ASME* **72**, 881 (1950).
73. O. P. Bergelin, G. A. Brown, and S. C. Doberstein, *Trans. ASME* **74**, 953 (1952).
74. V. P. Isachenko, "Teploperedacha i Teplovoe Modelirovanye" (Heat Transfer and Thermal Simulation), p. 213. Akad. Nauk SSSR, 1959.
75. A. A. Žukauskas, R. V. Ulinskas, and Č. J. Sipavičius, *Heat Transfer—Sov. Res.* **10**, No. 6, 90 (1978).
76. J. K. Stasiulevičius and P. S. Samoška, *Liet. TSR Mokslu Akad. Darb., Ser. B* **4**(35), 77, 83 (1963); **4**(55), 133 (1968).
77. A. A. Žukauskas, R. V. Ulinskas, and K. F. Marcinauskas. *Int. Chem. Eng.* **17**, No. 4, 744 (1977).
78. K. Hammecke, E. Heinecke, and F. Scholz, *Int J. Heat Mass Transfer* **10**, 427 (1967).
79. A. A. Šlančiauskas, A. A. Pedišius, and A. A. Žukauskas, *Liet. TSR Mokslu Akad. Darb., Ser. B* **3**(58), 193 (1969).
80. M. B. Lyapin, *Teploenergetika* **9**, 49 (1956).
81. O.E. Dwyer and T. V. Sheeman, *Ind. Eng. Chem.* **48**, No. 10, 1836 (1956).
82. A. F. Fritzsche, "Gestaltung und Berechnung von Ölkühlern." Leeman, Zürich, 1953.
83. E. M. Sparrow and A. A. Uanezmoreno, *Int. J. Heat Mass Transfer* **26**, 1791 (1983).
84. D. L. Peters, *Waerme- Stoffuebertrag.* **3**, 220 (1970).
85. R. L. Webb, E. R. G. Eckert, and R. W. Goldstein, *Int. J. Heat Mass Transfer* **14**, 25 (1971).

Thermal Conductivity of Structured Liquids†

A. DUTTA* AND R. A. MASHELKAR

Chemical Engineering Division, National Chemical Laboratory, Pune 411008, India

I. Introduction

Complex liquids, which are structured or non-Newtonian such as polymer solutions, filled and unfilled polymer melts, suspensions, emulsions, slurries etc., are frequently encountered in the chemical process industry. Owing to their structured nature these liquids behave quite differently from ordinary fluids such as air, water, etc. and their physicochemical and flow properties are of considerable interest in the analysis and design of industrial processes. Heat-transfer operations, in particular, play a critical role in processing and handling of these liquids and a knowledge of the thermal conduction phenomena in such liquids is essential for any analysis of the associated heat-transport phenomena.

In light of the above, it is not surprising that since the turn of the century, significant efforts have been devoted to the understanding of thermal conduction phenomena in complex liquids in general and estimation of their thermal conductivity in particular. It is natural to expect that a state-of-the-art review on thermal conductivity of such complex liquids would be readily available. However, a survey of the literature shows otherwise. Although several reviews in the past dealt with thermal conductivity of polymers [1–10], the focus has been primarily on the solid-state phenomena. The same observation, in general, holds true in the case of the recent treatise on heat conduction by Parrott and Stuckes [11] and Berman [12].

† NCL Communication No. 3907
* Present address: Shalimar Paints, Ltd., P.O. Danesh Shaikh Lane, Howrah 711109, India.

161

For non-Newtonian fluids, some elements of flow and heat transport phenomena were discussed by Skelland [13] in his book almost two decades ago. However, practically no information was provided regarding the thermal conductivity of these liquids. Even the review by Astarita and Mashelkar [14] on heat and mass transfer phenomena in non-Newtonian fluids only cursorily mention the likelihood of deviations from Fourier's law owing to the structured nature of these fluids. Similarly, even though heat conduction in heterogeneous media has been comprehensively reviewed by Gorring and Churchill [71], Meredith and Tobias [15] and by Baxley [16], these appeared about two decades ago, and considerable developments in this field have taken place since then. Therefore, it was deemed fit to undertake an effort to review the field to serve a dual purpose: first to critically review and summarize the existing body of information that allows definitive conclusions, and second to elucidate areas that are not clearly understood but are significant enough to warrant future research.

As a preamble to the more specific discussion on the thermal conductivity of structured liquids, we shall first outline the basic principles underlying the heat-conduction phenomenon along with a general classification of complex liquids. This will be followed by a brief section on experimental techniques that are commonly being used for determining thermal conductivity of liquids. A detailed discussion on thermal conductivity of specific classes of complex liquids will be then presented. For convenience, structured liquids will be dealt with in two broad categories, polymeric and nonpolymeric. To start with, nonpolymeric systems like suspensions (both solid–liquid and liquid–liquid), slurries, etc., will be considered following which the discussion will shift to polymeric (both single- as well as multi-component) systems. In all cases, pertinent theoretical developments and experimental observations will be highlighted and, wherever possible, the outcome of the above theoretical analyses will be compared with relevant experimental findings.

II. Background Information

A. BASIC PRINCIPLES OF HEAT CONDUCTION

The basic field equation for temperature which governs heat transfer in an incompressible liquid medium can be obtained from an overall energy balance for the system. Neglecting radiative heat transport, this balance equation can be written as

$$\rho \, DU/Dt = -\nabla \cdot \mathbf{q} + \text{Tr}(\boldsymbol{\tau} \cdot \nabla\mathbf{v}) \tag{1}$$

In Eq. (1), which is a direct formulation of the first law of thermodynamics, the last term represents the net rate of work done by internal stresses in a flowing medium only. For non-Newtonian liquids, the contribution of this term is likely to be quite different from that due to ordinary Newtonian liquids, not only because the stress tensor (τ) is a nonlinear function of the velocity gradient (∇v), but also because of the partially dissipative nature of the stress power if the liquid is viscoelastic, as most polymeric liquids are [14]. Fortunately, however, from a heat-transfer standpoint the accumulation of elastic energy occurs primarily through a decrease in conformational entropy, thereby implying that the stress power of such liquids can be treated as entirely dissipative. The validity of this concept has indeed been verified experimentally [17].

In order to express Eq. (1) in terms of temperature (the required dependent variable), a constitutive relationship is necessary that relates U to q, which in turn needs to be related to the temperature. A rigorous development of such a constitutive relation requires a thermodynamic description of the material [18]. Nevertheless, for the special case where the elastic energy is accumulated only through changes in the conformational entropy, U will be governed solely by its specific heat and temperature as in the case of Newtonian liquids:

$$U = U_0 + C_v(T - T_0) \tag{2}$$

Here the subscript 0 denotes some reference state. Regarding the relationship between the heat flux and the temperature, the classical theory of linear conduction as represented by Fourier's law is commonly assumed:

$$\mathbf{q} = -\mathbf{k}\,\nabla T \tag{3}$$

As is evident from Eq. (3), Fourier's law states that the heat-flux vector is directly proportional to the temperature gradient, the proportionality constant being a second-order thermal conductivity tensor ($\mathbf{k}$) for an anisotropic medium. This relationship works perfectly well for ordinary solids and liquids, but certain doubts regarding its validity arise for conduction in complex media like solids with memory [18–20] and liquids with structure. For structured liquids, it is quite plausible that nonlinear forms of the stress–strain equation will introduce complexities leading to departure from Fourier's law.

Some theoretical justification [21] as well as experimental evidence [22–24] of this effect have already been reported. Besides, as suggested by Astarita and Mashelkar [14] under certain conditions the specific forces on the microstructural elements within these liquids can be significant enough to cause departure from Fourier's law. In spite of the above reservations, however, no alternative to Fourier's law is presently available for analyzing heat conduction problems in structured liquids.

Equations (1) to (3), when combined, give the following working equation for the temperature field

$$\rho C_v\, DT/Dt = \nabla \cdot (\mathbf{k}\,\nabla T) + \mathrm{Tr}(\boldsymbol{\tau} \cdot \nabla\mathbf{v}) \tag{4}$$

where $\mathbf{k}$ is the symmetric second-order thermal conductivity tensor given as

$$\mathbf{k} = \begin{bmatrix} k_{11} & k_{12} & k_{13} \\ k_{21} & k_{22} & k_{23} \\ k_{31} & k_{32} & k_{33} \end{bmatrix} \tag{5}$$

If the thermal conductivity is independent of temperature, that is, $\mathbf{k} \neq \mathbf{k}(T)$, then in Cartesian coordinates Eq. (4) can be written as

$$\rho C_v \frac{DT}{Dt} = \sum_{i=1}^{3} \sum_{j=1}^{3} k_{ij} \frac{\partial^2 T}{\partial x_i\, \partial x_j}$$
$$+ \sum_{i=1}^{3} \sum_{j=1}^{3} \tau_{ij} \left(\frac{\partial u_i}{\partial x_j} + \frac{\partial u_j}{\partial x_i} \right) \tag{6}$$

and Chang [25] discusses procedures for solving general equation for stagnant media by the use of appropriate Green's function. As we shall see later, thermal conductivity of the structured liquids can be either isotropic or anisotropic depending on the particular nature and state of the medium. For an isotropic medium, Eq. (6) simplifies considerably and can be expressed as

$$\frac{DT}{Dt} = \alpha \sum_{i=1}^{3} \frac{\partial^2 T}{\partial x_i^2} + \sum_{i=1}^{3} \sum_{j=1}^{3} \tau_{ij} \left(\frac{\partial u_i}{\partial x_j} + \frac{\partial u_j}{\partial x_i} \right) \tag{7}$$

where $\alpha = k/\rho C_v$ is the thermal diffusivity. It should be noted that in the foregoing C_v is the specific heat at constant volume per unit mass whereas in practice the conducting medium is at a constant pressure and hence C_p is more representative. Fortunately, however, for most polymeric liquids the difference between C_p and C_v is rather small [7, 26] and therefore the use of C_p as a substitute for C_v in the above theoretical relationships is quite adequate for most practical purposes.

B. CLASSIFICATION OF COMPLEX LIQUIDS

As mentioned earlier, many fluids that are commonly encountered in industrial practice are structured as well as non-Newtonian in nature.

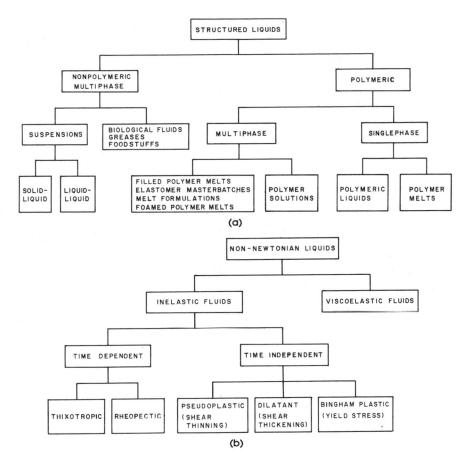

FIG. 1. (a) General classification of structured liquids. (b) Rheological classification of non-Newtonian liquids.

Typical examples would be polymeric melts, liquids, and solutions, elastomeric masterbatches, filled polymer systems, foams, suspensions, slurries, pastes, etc. As shown in Fig 1a, these fluids can be broadly classified in two distinct categories: polymeric and nonpolymeric. In the former category, the liquid can be either single component (e.g., polymeric liquids, melts, etc.) or it can be multicomponent (e.g., polymeric solutions and foams, filled polymer melts, etc.). By single component we mean that the liquid comprises a single constituent, whereas multiphase liquids are taken to consist of two or more components which may or may not be liquids. Unlike polymeric liquids, practically all nonpolymeric liquids are multicomponent in nature and most of them can be classified as suspensions. For suspensions, although the continuous phase is a liquid, the dispersed phase can either be a solid or another liquid and these accordingly will be referred to as solid–liquid or liquid–liquid suspensions. Other than conventional suspensions or dispersions, other nonpolymeric multicomponent liquids (e.g., biological fluids, greases, foodstuffs, etc.) also possess a certain degree of structure and hence their thermophysical properties are also structure dependent.

Owing to their structured nature, almost all of the above liquids exhibit non-Newtonian flow behavior. The precise nature of this nonlinear behavior, however, depends largely on the particular type of the liquid, and Fig. 1b summarizes the various rheological complexities that are possible. Typically, suspensions, slurries, etc. can either be pseudoplastic, dilatant, or Bingham plastic in nature, whereas polymer solutions and melts usually exhibit pseudoplasticity and filled polymer melts are commonly of the yield–pseudoplastic type. In certain cases, the flow behavior is further complicated by the viscosity being time dependent as in the case of thixotropic and rheopectic liquids. Furthermore, unlike their nonpolymeric counterparts, polymeric liquids are also viscoelastic in nature, exhibiting a combination of properties attributable to both viscous liquids and elastic solids. Consequently, phenomena like finite normal stress differences and stress relaxation are peculiar to these liquids.

To date, considerable effort has been devoted to understanding the rheological behavior of structured liquids and such efforts are still continuing. However, discussion of this area is outside the scope of the present review, since we will primarily focus attention on thermal conductivity of liquids possessing a certain degree of structure.

III. Experimental Techniques

Although the concept of thermal conductivity as a key thermophysical property in heat conduction processes is used for quite some time, accurate experimental determination of this property has proved to be somewhat difficult. Ideally, values of thermal conductivity (k) measured by different techniques should be the same within experimental error. Figure 2, taken from Hands [7], shows the experimental values of k reported for polystyrene using different experimental techniques. Considerable scatter in the data obtained by different research groups is evident and it is difficult to arrive at a consensus regarding the true variation of k with temperature for polystyrene. The same observation is true for other liquids also and such discrepancies are primarily due to the errors associated with different experimental methods. Since experimental techniques seem to significantly influence the k value obtained, in the following we shall briefly

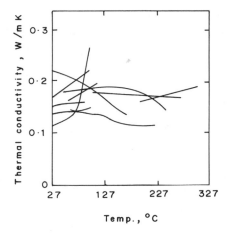

FIG. 2. Reported values of thermal conductivity of polystyrene melts (from Hands [7]).

consider some of the common techniques that are employed for thermal conductivity determination. This discussion, by no means comprehensive, will emphasize some of the key techniques used successfully for liquids and particularly for those liquids that are structured in nature. For a detailed discussion on experimental determination of liquid thermal conductivity, the reader is referred to an earlier review by Tait and Hills [47].

Unlike in solids, heat transfer within a liquid medium caused by the application of a temperature gradient can be both by means of conduction and convection. For the cases considered here, we consider heat transport by radiation to be negligible. In liquids, heat transfer by convection may be comparable to or greater than that by pure conduction and hence in any determination of thermal conductivity of liquids it is imperative to minimize, if not eliminate, the convection effects. Typically, the temperature gradient is applied across a vertical or horizontal fluid layer and, as we shall see shortly, the magnitude of the convection effects depend to a large extent on the precise configuration of the experimental arrangement [48]. Expectedly, for horizontal layers heated from top downward the convection effects are not present, whereas when the same layer is heated from below, the cellular motion due to natural convection is induced when the Rayleigh number (Ra) [the product of Grashof number (Gr $= g\rho^2\beta$ $\Delta T(\Delta x)^3/\mu^2$) and Prandtl number (Pr $= C_p\mu/k$)] exceeds a critical value [49–52]. Similar observations hold for heat transfer across liquid layers between two horizontal coaxial cylinders. In this case, convection effects are less than 2% if Ra $<$ 1000 [52, 53] and since the magnitude of convection effects is less in a vertical layer, the above criterion may also be considered as a conservative estimate for such geometries [54].

The basic philosophy in most experimental techniques devised for measuring thermal conductivity has been to subject a liquid layer to a temperature gradient by applying a heat flux. If the magnitude of the heat flux q is known and the temperature difference (ΔT) across the layer is then measured for steady conditions, Fourier's law [see Eq. (3)] allows the determination of thermal conductivity from a knowledge of ΔT along with the relevant geometrical parameters. For example, in the case of unidirectional heat conduction in a plane geometry (see Fig. 3) thermal conductivity can be obtained as

$$k = -q\,\Delta x/\Delta T \qquad (8a)$$

where Δx is the thickness of the fluid layer. In the case of coaxial cylindrical geometry, the expression analogous to the above equation is

$$k = -q\,\Delta r/\Delta T \qquad (8b)$$

where $\Delta r = r_0 - r_i$ and the heat flux q is based on the mean surface area $\bar{A}$ $(= 2\pi L\,r_{\mathrm{ln}}), r_{\mathrm{ln}}$ being the log-mean radius of the annular liquid layer. Note that Eq. (8) is applicable only when experiments are carried out under steady conditions. Measurements under transient conditions allow significant reduction in time of experimentation. For unsteady conditions, if x is the direction of heat conduction, the temperature field is described by the solution of the one-dimensional Laplace equation,

$$\frac{\partial T}{\partial t} = \frac{1}{x^a}\frac{\partial}{\partial x}\left(x^a\alpha\,\frac{\partial T}{\partial x}\right) \qquad (9)$$

subjected to appropriate boundary conditions, where $a = 0$ (plane geometry) or $a = 1$ (annular geometry). In general, for a known heat flux, if the temperature is measured at a given location x for different times, the solution of Eq. (9) allows the determination of thermal conductivity. For

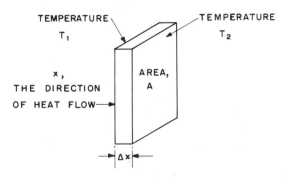

FIG. 3. Schematic representation of thermal conductivity measurement.

example, in the case of an infinitely long line source of heat in an infinite expanse of liquid (a situation that closely resembles the commonly used hot-wire method), the temperature at a radial distance x from the source is given as [7, 55]

$$T = \frac{Q}{4\pi k} \int_b^\infty \frac{\exp(-s)}{s} \, ds \qquad (10a)$$

$$\approx \frac{Q}{4\pi k} \left[\ln\left(\frac{4\alpha t}{x^2}\right) - 0.5772 \right] \qquad \text{if} \quad b \ll 1 \qquad (10b)$$

where $b = x^2/4\alpha t$ and Q is the rate of heat input per unit length. Assuming both α and k to be constant over the temperature interval T_1 to T_2, Eq. (10b) leads to the following expression for the thermal conductivity

$$k = \frac{Q}{4\pi(T_2 - T_1)} \ln\left(\frac{t_2}{t_1}\right) \qquad (11)$$

Thus, at a given location within the liquid medium, measurement of temperatures T_1 and T_2 corresponding to times t_1 and t_2, along with the knowledge of the rate of heat input, allows for the measurement of the thermal conductivity of the medium.

The above methods, both steady as well as transient, have been used to measure thermal conductivity of diverse types of structured liquids, and some of these are summarized in Table I [27–46]. It can be seen that among steady methods both hot-plate and concentric-cylinder techniques have been commonly used, whereas among transient methods both linear heating and hot-wire conductivity probes have been often used. We shall now briefly discuss the salient features of these commonly used techniques.

A. STEADY METHODS

1. Hot-Plate Technique

In the hot-plate technique, a thin layer of liquid is confined between two horizontal plates made of a highly conducting material (usually copper), which are maintained at constant temperatures. Typically the layer thicknesses are small (about 1 to 5 mm) and the convection effects are suppressed by heating the top plate electrically and by cooling the bottom plate with the aid of a circulating liquid from a constant temperature bath. The plate temperatures and heat input are measured and k is calculated from Eq. (8a). Although the entire assembly is insulated, appreciable heat loss takes place through the edges of the two plates, thereby affecting the accuracy. This drawback can be eliminated to a certain extent by the use of

TABLE I

MEASUREMENT OF THERMAL CONDUCTIVITY IN STRUCTURED LIQUIDS

Nature of the experimental technique	Experimental set-up	Conducting medium	Reference	Remarks
Steady	Concentric cylinder	Aqueous Methocel solution Aqueous gelatin solution Polybutene silicones	27	Concentration ≤10%
Quasi-steady	Enclosed apparatus	Polymer melts	28	Temperature ≤250°C
Steady	Guarded hot-plate	Polymer solutions	29	Concentration ≤1%
Steady	Unguarded hot-plate	Biological fluids	30	
Steady	Concentric cylinders	Greases	31	
Steady	Hot-plate	Silicones	32	
Steady	Split-bar	Foodstuffs	33	
Transient	Hot-wire	Polymer solutions	34	Concentration ≤1%
Steady	Hot-plate	Slurries	35	Solid concentration 20 wt %
Steady	Concentric spheres	Solid–liquid suspensions	36	
Steady	Inverted temperature baths	Solid–liquid suspensions	37	Maximum solid content 13 wt%
Steady	Hot-plate	Solid–liquid suspension	38	
Transient	Thermal needle	Pastes	39	Maximum liquid volume fraction 12%
Transient	Axial flow cell	Polymeric liquid	24	Strain rates 0 to 300/sec
Transient	Couette flow	Polymeric liquid	40	Strain rates <400/sec
Steady	Colora thermoconductometer	Reactive polymers	41	
Steady	Concentric cylinders	Polymer melts	42	Temperature ≤290°C
Steady	Concentric cylinders	Polymer melts	43	Temperature ≤300°C
Transient	Couette flow	Polymer melts	44	Strain rate ≤400/sec

TABLE I   *(Continued)*

Nature of the experimental technique	Experimental set-up	Conducting medium	Reference	Remarks
Pseudosteady-state	Hot-plate	Polymer melts	45	Temperature ≤300°C
Transient	Conductivity probe	Filled polymeric liquids	46	Solids concentration up to 20 wt %
Transient	Hot-wire	Greases	56	Anisotropic effect
Steady	Concentric spheres	Solid–liquid suspension	16	
Steady	Couette device	Solid–liquid suspensions	77	Sheared media $x_v \leq 0.25$
Steady	Guarded hot-plate	Polymer melt	109	Temperature ≤300°C
Steady	Couette device	Polymer solutions	119	Concentration up to 20%
Steady	Parallel disc	Solid–liquid suspensions	73,74	—

standard samples to correct for the side losses or by the use of guards to minimize the side losses. Figure 4 shows a schematic diagram of a guarded hot-plate cell. As above, the liquid sample is confined between an electrically heated plate A and another plate B cooled by circulating a constant temperature coolant. The liquid layer thickness is controlled by the size of the insulating spacers F. Parts D and E are the side and upper guard plates, which are also electrically heated to minimize end losses from the main heater plate A. Thermocouples, denoted by x, are usually located as shown in order to monitor the temperature equilibration and also to determine

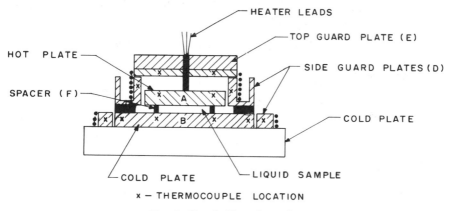

FIG. 4. Guarded hot-plate cell.

the final temperature difference across the sample. The heat input is generally obtained from the electrical power supplied to A.

The hot-plate method, with minor variations, has been used by different research groups to determine the thermal conductivities of most structured liquids excepting highly viscous polymer melts that pose sample loading problems. In particular, this method has been used for polymer solutions [29], liquid polymers [32], biological liquids [30], and solid–liquid suspensions [35, 37, 38]. Charm and Merrill [33] used the split bar technique for thermal conductivity measurements of food materials. This technique, quite similar to the unguarded hot-plate method, employs the end faces of two cylindrical rods in place of plates to confine the liquid sample layer.

2. Coaxial Cylinder Method

In coaxial cylinder arrangement, since heat transmission is effected in the radial direction, side losses from the nonisothermal sample boundaries are reduced. As shown schematically in Fig. 5, the apparatus consists of a cylindrical heat source and a concentric cylindrical shell that serves as a heat sink. The liquid sample is confined in the annular region between the heat source and the sink. The heat source is usually maintained at a constant temperature by electrical heating, whereas the sink is maintained at a constant lower temperature by means of a circulating liquid. Known

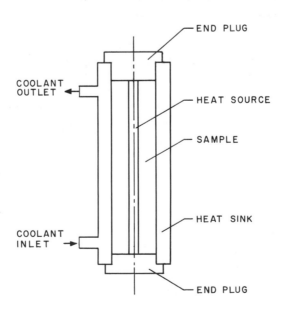

Fig. 5. Coaxial cylindrical cell.

power supply to the source determines the heat input to the sample, and measurement of the source and sink temperatures (usually by thermocouples) allows the determination of the thermal conductivity of the liquid sample from Eq. (8b). In order to minimize side losses, end-guard plates are frequently used. Increased length-to-diameter ratio reduces the magnitude of end losses as compared to the amount of heat transferred radially. This technique has been employed to determine the thermal conductivity of polymeric solutions and liquids [27], greases [31], and polymer melts [42, 43]. As in the case of the hot-plate method, highly viscous polymer melts pose considerable difficulty in loading the sample. To take care of this problem, Fuller and Fricke [43] used a barrell–plunger assembly to inject the sample into the thermal conductivity cell.

Concentric spheres, which also permit radial heat transfer and practically eliminate all side and end effects were used by Hamilton and Crosser [36] to measure thermal conductivity of solid–liquid suspensions. This method, though similar to the coaxial cylinder technique in principle, is not as popular primarily because of its need for mechanical support and for precise alignment of the inner sphere to ensure concentricity.

3. *Enclosed Apparatus*

The steady methods described above are mostly suited for solids and low-viscosity liquids. Hands and Horsfall [28] have developed a thermal conductivity apparatus suitable for both solids as well as molten polymers. Using this technique, they have reported conductivity data for plastics and elastomers for a temperature range from the ambient to 250°C. In this apparatus, the problem of heat loss from the exposed edges is eliminated by enclosing the heat source and the sample inside a heat sink. The device (see Fig. 6) consists of a disk-shaped heat source placed concentrically in a short cylindrical heat sink and supported by an insulating annulus thereby providing a disk-shaped sample space on either side of the source. The heat sink is formed by two end plates, which are bolted on each end plate provided with an expansion tube.

In this configuration, heat transfer occurs in more than one direction, and consequently the steady-state temperature distribution is given by the solution of the appropriate Laplace equation. Assuming constant thermal conductivity within a short temperature interval ΔT, dimensional analysis suggests that

$$k_s/k_A = f(W/(\Delta T \, \delta k_A) \tag{12}$$

The governing Laplace equation is solved numerically to generate the function given by Eq. (12), which represents an instrument curve. Experimentally, a known power is supplied to the heat source and the tempera-

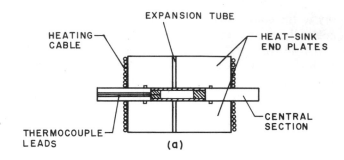

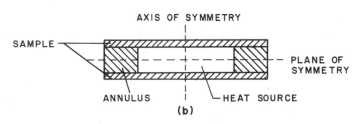

FIG. 6. (a) Enclosed apparatus. (b) Details of sample holder assembly (from Hands and Horsfall [28]).

ture difference is measured. For the particular value of $W/\Delta T \, \delta k_A$, thermal conductivity (k_s) is then obtained from the instrument curve. Prior to that, however, knowledge of k_A is necessary and is obtained by conducting preliminary measurements on samples of annular material. Hands and Horsfall suggest that, other than for molten polymers, this technique is also suitable for viscous liquids, but this has not yet been experimentally verified. They caution that this method is suitable for isotropic media only.

B. TRANSIENT METHODS

As mentioned earlier, transient methods allow faster determination of thermal conductivity. In these experiments, thermal conductivity is obtained from the rate of establishment of the thermal gradient in a liquid medium and not from the final thermal gradient as in steady methods. In particular, under transient conditions, the time-dependent temperature field is governed by the solution of Eq. (9) subject to appropriate boundary conditions. In general, these methods allow the determination of the thermal diffusivity α, which in turn yields the value of k provided the appropriate density and specific heat is known. Only under certain conditions, these techniques permit direct determination of k.

This basic principle holds for all the transient techniques, and differences arise only due to the change in the geometry or the nature of the boundary conditions for Eq. (9).

1. *Hot-Wire/Thermal Conductivity Probe Technique*

In the hot-wire technique, an electrically heated wire is usually stretched vertically in a relatively large bulk of the test liquid, and the temperature of the sample at a fixed location from the wire is measured at different times. The thermal conductivity is then obtained from either Eq. (10a) or (10b). If b is small, i.e., if time t is large, care must be taken to ensure that Ra < 1000 in order to avoid the convection effects, particularly for low-viscosity samples.

The thermal conductivity probe or "thermal needle," shown schematically in Fig. 7, is also based on this principle. The probe is inserted in the sample to be tested. Using Eq. (10b), the thermal conductivity is determined from the rate of heat addition per unit length of the heater coil and the increase in temperature as measured by the probe thermocouple within a given time period. The heat input is obtained from the electrical power input. A likely source of error in this rapid and convenient technique may be due to the varied depths to which the probe is inserted within the sample.

Both the above methods were used for thermal conductivity determina-

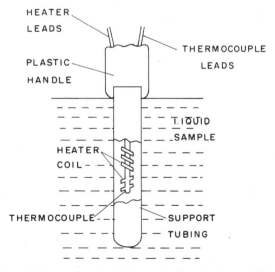

FIG. 7. Thermal conductivity probe (from Anderson [4]).

tion of structured liquids. In particular, Brunson [34] used the hot-wire technique for polymer solutions, whereas Shulman *et al.* [46] and Jackson and Black [39] have employed the thermal conductivity probe for metal-filled polymeric liquids and liquid-bound granular media, respectively.

2. *Linear Heating Technique*

The linear heating technique was used by Shoulberg [45] for determining thermal conductivity of molten polymers. He used two thin disks of polymer with a thermocouple sandwiched between them, and this assembly was placed in a cavity formed in two massive aluminium blocks that were securely bolted together. These blocks were heated electrically and the power was adjusted to give a linear temperature rise over an interval of 30°C.

For an infinite slab whose surface temperature is rising linearly with time, the time-dependent temperature (T_0) at the center of the slab is given by [46, 56]

$$T_0 = ht - \frac{h\delta^2}{2\alpha} + \frac{16\delta^2}{\alpha\pi^3} \sum_{n=0}^{\infty} \frac{(-1)^n}{(2n+1)} \exp\left\{-\frac{\alpha(2n+1)^2\pi^2 t}{4\delta^2}\right\} \quad (13)$$

where h is the rate of surface temperature increase. Clearly, for known h and δ, a single measurement of T_0 at a given t allows the determination of α from Eq. (13). Mathematically, however, this may prove cumbersome and thus an alternative procedure is sought. Interestingly, as t becomes large the summation term in Eq. (13) becomes negligible and the temperature difference between the surface and the center becomes constant. In that case, Eq. (13) can be simplified to

$$\alpha = h\delta^2/2(ht - T_0) = h\delta^2/2(\Delta T) \quad (14)$$

By experimentally monitoring the temperature difference between the surface and the center line, Shoulberg utilized eq. (14) to determine α for different polymer melts. The data obtained were comparable to those obtained by different experimental techniques. However, this method does not seem to have been used for other liquids, probably because in a liquid layer, locating the thermocouple precisely at the center line poses formidable difficulties.

C. TECHNIQUES FOR SHEARED LIQUIDS

The preceding experimental techniques dealt with thermal conductivity measurement of stagnant structured liquids. Since the structure is significantly altered when the fluid is subjected to a deformation field, considerable interest is centered on the behavior of these liquids under flowing conditions. Several studies on thermal conductivities of structured liquids under shear have been reported. Shulman *et al.* [56] employed a hot-wire method to investigate anisotropy of the thermal conductivity of several greases oriented by forcing the test liquid through a system of plane slots.

Cocci and Picot [24] used an annular axial flow test cell and a hot-wire probe to study the strain-rate effect on thermal conductivity of polymeric liquids. The test cell, immersed in a constant temperature bath, is a cylindrical tube provided with inlet and outlet at the bottom and the top to permit axial flow of the test liquid. A step input of heat is applied to the fluid by passing a current through a wire located at the center of the cell. The temperature of the wire is recorded as a function of time. For this flow arrangement, the governing equation for the fluid temperature is

$$\rho C_p \left[\frac{\partial T}{\partial t} + v_z \frac{\partial T}{\partial z} \right] = k \left[\frac{1}{r} \frac{\partial}{\partial r} \left(r \frac{\partial T}{\partial r} \right) \right] + \mu \left[\frac{\partial v_z}{\partial r} \right]^2 \tag{15}$$

If the temperature field $T(t, r, z)$ is split into its transient contribution $T_t(t, r)$ and its steady-state contribution $T_s(r, z)$, such that

$$T(t, r, z) = T_t(t, r) + T_s(r, z) \tag{16}$$

then Eq. (15) gives rise to the two following equations

$$\rho C_p \frac{\partial T_t}{\partial t} = k \left[\frac{1}{r} \frac{\partial}{\partial r} \left(r \frac{\partial T_t}{\partial r} \right) \right] \tag{17}$$

and

$$\rho C_p v_z \frac{\partial T_s}{\partial t} = k \left[\frac{1}{r} \frac{\partial}{\partial r} \left(r \frac{\partial T_s}{\partial r} \right) \right] + \mu \left[\frac{\partial v_z}{\partial r} \right]^2 \tag{18}$$

Equation (17) involves no convection term and thus allows a means to measure the thermal conductivity of the flowing liquid. For an initial condition of step change in heat flux at the wire surface, Picot and Frederickson [57] have demonstrated that after a small time transient, a plot of temperature rise of the wire versus ln t is linear with the slope being inversely related to the thermal conductivity of the test liquid. As a result, by measuring the wire surface temperatures T_1 and T_2 at two different times t_1 and t_2 and knowing the rate of heat input per unit length (Q), k can be determined as

$$k = Q \ln(t_2/t_1)/4\pi(T_2 - T_1) \tag{19}$$

In order to minimize any convection contribution due to the end effects, Cocci and Picot used only the central portion of the wire as the test section. Moreover, instead of thermocouples, the temperature rise was monitored by measuring corresponding electrical resistance changes in the heater wire. The accuracy of this extremely rapid technique (duration of the order of few seconds) is claimed to be within $\pm 1.5\%$.

The above method, however, suffers from several drawbacks [40]. These are (a) fragility of the hot wire in the midst of the flowing liquid, (b) variation of shear rate along the path of heat conduction, (c) the requirement of large quantities of test sample, and (d) difficulties in maintaining precise temperature control due to the large size of the apparatus. Referring to (b), since a Couette apparatus provides a constant shear rate within the annular gap, Picot and co-workers have used this arrangement also to measure the thermal conductivity of polymeric liquids [40] and polymer melts [44]. In this case the test fluid is located in a narrow annular gap formed by a stationary inner cylinder and a rotating jacketed outer cylinder. Following equilibration due to viscous heating, the inner cylinder is electrically heated at a constant rate and the temperature of the liquid at the inner surface is recorded as a function of time by monitoring the resistance of the heater wire wound round the inner cylinder. If q is the heat flux at the inner surface, Picot *et al.* [44] show that

$$\Delta T \propto t^{1/2}/[(k\rho C_p)^{1/2} + B] \tag{20}$$

where ΔT is the rise in the inner surface temperature and B is an instrument constant (related to the thermal conductivity and diffusivity of the inner cylinder) to be measured by using standard samples. Equation (20) suggests that a plot of ΔT versus $t^{1/2}$ should be linear and its slope should allow determination of k provided ρ, C_p, and B are known. This technique also is extremely fast (duration about 3 sec) and the accuracy is estimated to be of the order of 5% [44].

IV. Nonpolymeric Liquids: Suspensions

The microstructure of a suspension, be it of solid–liquid or liquid–liquid type, depends on a number of factors. One of these factors is the presence of a deformation field, which inevitably alters the microstructure of the flowing suspension. As a result of this microstructural change, it is expected that most physical properties of a suspension will depend, among many other factors, on whether the medium is stagnant or flowing. In the following we shall discuss the (effective) thermal conductivities of stagnant and flowing suspensions separately.

A. Stagnant Media

In heterogeneous liquids like suspensions, Fourier's law of heat conduction strictly holds for the individual phases and the geometries associated with it. Thus, a rigorous analysis of the heat conduction phenomenon in suspension would not only require the knowledge of the shape, size, location, and thermal conductivity of each individual particle within the system but would also involve solution of the governing Laplace equation for each phase. In doing so Fourier's law needs to be assumed and continuity of flux and temperature at the phase boundaries needs to be satisfied. This represents a formidable problem and most of the efforts to date incorporate considerable simplifications.

1. *Conventional Approach*

The usual theoretical approach has been to assume that the effective thermal conductivity of a real two-phase system is the same for a system of simple geometry having the same volume fraction of the two phases. The conductivity of this hypothetical arrangement is then found from the analytical solution of the Laplace equation with appropriate boundary conditions. The ratio (K_d) of the dispersed phase conductivity to that of the continuous phase and the volume fraction (x_v) of the dispersed phase are the two most important parameters controlling the effective thermal conductivity of a suspension (k_e). Aside from these, as we shall shortly see, k_e is also influenced by factors such as particle shape, size distribution, orientation, etc.

Since the turn of the century, the above approach has been used to develop theoretical expressions for k_e. This field of activity has been comprehensively reviewed by Meredith and Tobias [15], Gorring and Churchill [71], and Baxley [16] and therefore in this section we shall briefly highlight the significant contributions. The models developed to date for predicting k_e may be classified into two categories, those which assume linear heat flow and those based on nonlinear heat flow. Former models, summarized in Table II [58–61, 75, 76], assume that heat flows in straight lines perpendicular to parallel isothermal planes. Such an assumption is reasonable only if the individual phase conductivities k_d and k_c do not differ significantly. However, when k_c and k_d are quite different, the heat flux lines will be toward the dispersed phase if $k_d \gg 1$ and will be toward the continuous phase if $k_d \ll 1$. This represents a more general situation and therefore, for a model to be less restrictive, the assumption of nonlinear heat flow is more appropriate. Indeed most of the work has focused on nonlinear heat flow models and Table III summarizes some of the pertinent developments.

Derivations of more realistic and accurate nonlinear heat flow models have been an active area of research for the past several decades. Most of these efforts represent refinements and extensions of the classical works of Maxwell [62] and Rayleigh [63] for dispersion of uniformly sized particles.

TABLE II

SOME THEORETICAL EXPRESSIONS FOR EFFECTIVE THERMAL CONDUCTIVITY OF SUSPENSIONS: LINEAR HEAT FLOW MODEL

Basic assumption	Expression	Equation number	Reference	Remarks
Alternate layers of dispersed and continuous phases parallel to heat flow direction	$K_e = (1 - x_v) + K_d x_v$	(21)	58	Same expression results from "weighted average" based on x_v
As above but layers perpendicular to heat-flow direction	$K_e = \dfrac{K_d}{x_v + K_d(1 - x_v)}$	(22)	59	—
Cubic array of cubes of dispersed phase	$K_e = \left[\dfrac{x_v^{2/3} + (1 - x_v^{2/3})/K_d}{x_v^{2/3} - x_v + (x_v + 1 - x_v^{2/3})/K_d} \right]$ or $K_e = \left[1 - x_v^{2/3} + \dfrac{x_v^{2/3}}{x_v^{1/3}/K_d + 1 - x_v^{1/3}} \right]$	(23) (24)	60	Different expressions result for different method of calculation
Spherical particles enclosed in a unit cube of continuous phase	$K_e = 1 - \dfrac{\pi}{4(1+2n)^2} + \dfrac{\pi}{4(1+2n)^2} \ln K_d - \dfrac{(0.5+n)K_a}{0.5+nK_a}$ where $K_a = K_d \dfrac{2K_d}{(K_d-1)^2} \ln K_d - \dfrac{2}{K_d-1}$, $x_v < 0.52$, and $n = 0.403\, x_v^{-2/3} - 0.5$	(25)	61	
Randomly oriented particles	$K_e = \left[1 + A_2 + 2(A_1 A_3)^{-1/2} \tan^{-1}\left[\left(\dfrac{A_1}{A_4}\right)^{1/2} \dfrac{A_2}{2} \right] \right]^{-1}$ $x_v \le 0.66 \quad$ for $\quad K_d < 1$	—	75	
	$K_e = \left[1 - A_2 + \ln \dfrac{A_4 + (-A_1)^{1/2} A_2/2}{A_4 - (-A_1)^{1/2} A_2/2} \right]^{-1}$ for $\quad K_d > 1$ where $A_1 = A_3(K_d - 1)$, $A_2 = (3x_v/2)^{1/2}$, $A_3 = -4(2/3x_v)^{1/2}$, and $A_4 = 1 + A_2(K_d - 1)$	—	76	

TABLE III

SOME THEORETICAL EXPRESSIONS FOR EFFECTIVE THERMAL CONDUCTIVITY OF SUSPENSIONS: NONLINEAR HEAT-FLOW MODEL

Expression	Equation number	Reference	Remarks
$K_e = \dfrac{K_d + 2 - 2x_v(1 - K_d)}{K_d + 2 + x_v(1 - K_d)}$	(26)	62	Theoretically valid for very dilute dispersions of uniformly sized spherical particles.
$K_e = \dfrac{1 + K_d - x_v(1 - K_d)}{1 + K_d + x_v(1 - K_d)}$	(27)	63	As above excepting the dispersed phase consist of infinitely long cylinders
$\dfrac{K_e - 1}{K_e + X} = x_v\left[\dfrac{K_d - 1}{K_d + 2}\right]$ where X is a shape factor depending on the ratio of ellipsoid axes and K_d	(28)	64	For dilute dispersion of randomly oriented uniform ellipsoidal particles
$K_e = 1 - 3x_v\left[\dfrac{(2 + K_d)}{(1 - K_d)} + x_v - \dfrac{0.525(1 - K_d)x_v^{10/3}}{(\frac{4}{3} + K_d)}\right]^{-1}$	(29)	62, 65	For concentrated dispersion ($x_v < 0.5236$) of uniform spheres
$K_e = 1 - 2x_v\left[\dfrac{1 + K_d}{1 - K_d} + x_v - 0.3058\dfrac{(1 - K_d)x_v^4}{1 + K_d} - 0.1334\dfrac{(1 - K_d)x_v^8}{1 + K_d}\right]^{-1}$	(30)	62, 65	As above ($x_v < 0.7854$) but for dispersion of uniform cylindrical particles
$K_e = 1 - 3x_v\left[\dfrac{2 + K_d}{1 - K_d} + x_v - \dfrac{1.315(1 - K_d)x_v^{10/3}}{[\frac{4}{3} + K_d + 0.409(1 - K_d)x_v^{7/3}]}\right]^{-1}$	(31)	66	Improved version of Eq. (29)
$K_e = \dfrac{K_d + (N - 1) - x_v(N - 1)(1 - K_d)}{K_d + (N - 1) + x_v(1 - K_d)}$ where $N = 3/\psi$, ψ being the sphericity	(32)	36	For concentrated dispersion of arbitrarily shaped uniform particles
$\dfrac{K_e - K_d}{K_e^{1/3}(1 - K_d)} = 1 - x_v$	(33)	67	For concentrated suspension of spheres of different sizes

(Continued)

TABLE III (Continued)

Expression	Equation number	Reference	Remarks
$$K_e = \left[\frac{2(K_d + 2) + 2(K_d - 1)x_v}{2(K_d + 2) - (K_d - 1)x_v}\right] \times \left[\frac{(2 - x_v)(K_d + 2) + 2(K_d - 1)x_v}{(2 - x_v)(K_d + 2) - (K_d - 1)x_v}\right]$$	(34)	68	For two-phase suspensions containing mixed-size spheres
$$K_e = \left[\frac{2 + (XK_d - 1)x_v}{2 + (X - 1)x_v}\right] \times \left[\frac{2(1 - x_v) + XK_dx_v}{2(1 - x_v) + Xx_v}\right]$$	(35)	68	As above but for dispersion of spheroids of mixed size
$$K_e = \frac{(1 - x_v)}{1 + x} = \frac{K_d - K_e}{K_d - 1}$$	(36)	35	As above

Assuming that the dispersed phase consisted of uniform spherical particles, sufficiently displaced from one another so that the field surrounding a particle does not interact with that associated with another particle, Maxwell obtained Eq. (26). Rigorously, the assumptions leading to this equation make it valid strictly for dilute dispersions. However, in general, for $0 \leq K_d \leq \infty$ this expression works quite well for volume fractions (at least) up to 0.1 (see Figs. 8 and 9). Besides, when K_d is of the order of unity, there will be little field interaction and Maxwell's equation is expected to be adequate for larger volume fractions x_v. Expressions, analogous to that derived by Maxwell, were also obtained by Rayleigh [63] and Fricke [64] for randomly arranged uniform cylindrical and spheroidal particles, respectively. Fricke demonstrates the significant role of shape effects in controlling k_e. These shape effects, as determined by both negative as well as positive deviations from predictions for spherical particles, depend on K_d, x_v, and a geometrical constant characterized by the ratios of spheroid axes. Since most regular particle shapes can be represented by ellipsoids with different axis ratios (a/b), Fricke's analysis permits evaluation of k_e for a large variety of dispersed phases. This analysis, however, involves a shape factor (X) which depends on k_d and the spheroid axis ratio a/b. The factor X is a geometrical constant and its evaluation requires a somewhat in-

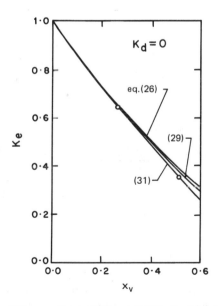

FIG. 8. Comparison of the equations of Maxwell (26), Rayleigh (29), and Meredith and Tobias (31) for spheres in a cubical lattice. Data are for nonconducting spherical particles in a cubic lattice (from Meredith and Tobias [15]).

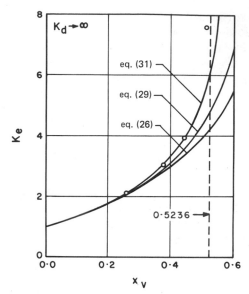

FIG. 9. A comparison of the equations of Maxwell (26), Rayleigh (26), and Meredith and Tobias (31) for spheres in a cubical lattice. Data are for infinitely conducting spherical particles in a cubic lattice (from Meredith and Tobias [15]).

volved mathematical procedure. Hamilton and Crosser [31], while investigating the effect of particle shape on k_e, proposed Eq. (32) for dispersion of nonspherical particles. In this equation, N is a shape factor (equal to $1 + X$) that has been found empirically to be

$$N = 3/\psi \tag{37}$$

where ψ is the sphericity given as

$$\psi = \frac{\text{surface area of a sphere of equal volume}}{\text{surface area of the particle}}$$

For $K_d < 100$, Hamilton and Crosser found the shape effect to be insignificant. In the case of higher K_d values, however, k_e decreased as ψ increased or the particles became spherical in shape.

For relatively concentrated suspensions, dispersed phase particles are located closely and the fields surrounding them begin to interact. Expressions for effective thermal conductivity, accounting for this interaction in ordered arrangements of parallel cylinders in a square array [Eq. (30) and uniform size spheres in a cubic lattice arrangement [Eq. (29)] were derived by Rayleigh [63] from a second-order treatment accounting for dipoles and octupoles only. The final expression, which had only small error was subsequently corrected by Runge [65]. The steric constraints limit the physical significance of these equations to maximum volume fractions of 0.5236 and 0.7854, respectively, concentrations corresponding to close packing of the particles. An improved version of Eq. (29), as represented by Eq. (31) in Table III, was obtained by Meredith and Tobias [66] by taking into account even higher-order multipoles. It can be seen from Fig. 9 that for an ordered arrangement of spheres in a cubical lattice, Eq. (31) proposed by Meredith and Tobias performs better (particularly as $x_v \rightarrow$ 0.5236) than Rayleigh's Eq. (29), which in turn, is expectedly better than Maxwell's equation for random arrangement. However, as $x_v \rightarrow 0.5236$ for infinitely conducting spherical inclusions ($k_d \rightarrow \infty$), k_e should also be infinitely large. Equation (31) proposed by Meredith and Tobias does not truly reflect this behavior. As we shall see in the following, this inadequacy can be removed by taking into account higher-order proximity effects.

In recent years, Rayleigh's method for predicting the thermal conductivity of regular lattices has been extended primarily by McKenzie and co-workers [130–132] and O'Brien [133] to account for higher-order interaction effects. For simple cubic lattice of identical spheres, McKenzie and McPhedron [130, 131] extended Rayleigh's method to take into account multipoles of arbitrarily high order and have clearly demonstrated the benefit of doing so. Figure 10 shows a comparison between the experimen-

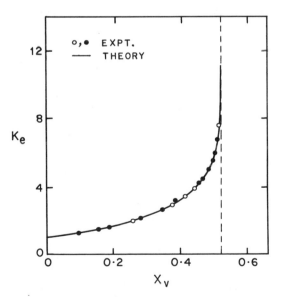

FIG. 10. Comparison of experimental measurements with exact theoretical prediction for simple cubic array of highly conducting spheres (from McKenzie and McPhedron [130]).

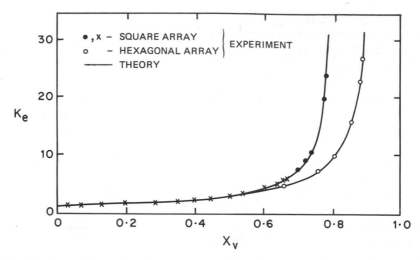

FIG. 11. Comparison of experimental conductivity with theoretical predictions for square and hexagonal array of highly conducting cylinders (from Perrins [137]).

tal and theoretical values of conductivity predicted by taking into account all possible orders of interaction, and clearly the agreement between them is excellent. Also note that now the theory predicts that $K_e \rightarrow \infty$ as $x_v \rightarrow$ 0.5236. Similar extensions of Rayleigh theory for other cubic lattices were also developed by McKenzie et al. [132]. The same results were also obtained by Sangani and Acrivos [135] employing the generalized function approach devised by Zuzovski and Brenner [136]. They, however, suggested that evaluation of the effective conductivity of cubic arrays allows estimation of K_e for random arrays of identical spheres. Regarding cylindrical inclusions, Perrins et al. [137, 138] presented results for square and hexagonal arrays and here too, as illustrated in Fig. 11, predictions from the extended theories agree remarkably well with the experimental data.

In the case of suspensions containing dispersed particles of different sizes, Bruggeman [67] hypothesized that in concentrated mixtures the field adjacent to the particles will be more realistically represented if the effective conductivity of the mixture as a whole is used rather than that of the continuous phase. This concept, coupled with an integral procedure for accounting for the contributions due to each individual volume fraction, leads to Eqs. (33) and (36) for K_e of a dispersion of differently sized spherical and spheroidal particles, respectively. Meredith and Tobias [15] argued that the surrounding medium may be treated as a continuum only if the successively added infinitesimal volume fractions of the dispersed phase consist of particles, which are much larger than the ones added previously. As a result, Eq. (33) is a good approximation only for particles

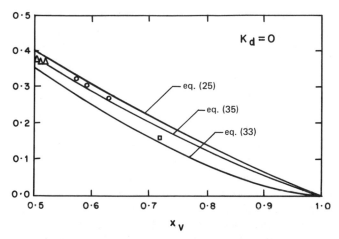

FIG. 12. A comparison of the equations of Maxwell (26), Bruggeman (33), and Meredith and Tobias (35) with experimental data for high volume fractions. Data for randomly dispersed spherical particles of uniform size are represented by (△) and (○), whereas (□) represents data for particles of mixed size (from Meredith and Tobias [15]).

with broad size ranges. Indeed, this has been confirmed experimentally by Pearce [69] and De La Rue and Tobias [70]. Figure 12 illustrates that the agreement between Bruggeman's Eq. (33) and experimental data is good only when the dispersed particles are of mixed sizes. Note that Eq. (33) holds for $x_v = 1$ since complete packing is possible due to a size distribution of the particles. For particles of mixed sizes, good agreement between experimental data and Eq. (33) was reported by De La Rue and Tobias [70] for solid dispersions and by Meredith and Tobias [68] for emulsions (see

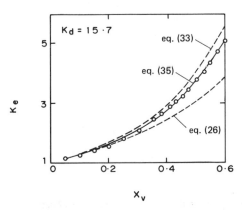

FIG. 13. Comparison of Maxwell (26), Bruggeman (33), and Meredith and Tobias (35) equations with experimental data (○) on emulsion (from Meredith and Tobias [15]).

Fig. 13). As shown in Fig. 12, Eq. (33) underpredicts K_e for a dispersion of uniform particles. This implies, for this particular case, that a particle "sees in its vicinity a medium of conductivity k_e." This is, perhaps, too idealized. In order to rectify this drawback, Meredith and Tobias [15], instead of an integral procedure, suggested a stepwise addition method in order to arrive at Eq. (35) for a dispersion of spheroidal particles of mixed sizes. For spherical particles, Eq. (35) reduces to Eq. (34). It can be seen from Figs. 12 and 13 that, both for uniform as well as mixed size particles Eq. (34) predicts K_e in good agreement with the experimental data. The predictive performance of the general equation (35), based on Fricke's work, was also assessed by Meredith and Tobias [15]. In particular, they compared the predicted conductivities with that of blood in which the dispersed phase consists of red blood cells resembling oblate spheroids. They found that the theoretical predictions for a spheroid (with axis ratio of about 3) compared very well with the experimental data reported by Stewart [72].

An approach, somewhat different from the above, was followed by Baxley [16]. He used a numerical simulation procedure to solve the general Laplace equation in order to determine the thermal conductivity of multi-dimensional heterogeneous media. The model predictions were found to be in agreement with an accuracy of about $\pm 20\%$ when compared to the experimental data. Kumada [38] conducted a comparison of different expressions for predicting thermal conductivity, and these are depicted in Figs. 14 and 15. It is seen that among the various models considered, Bruggeman Eq. (33) appears to work quite well for spherical particles within the range considered ($0 \le x_v \le 0.6$ and $0 \le K_d \le 1116$). For non-

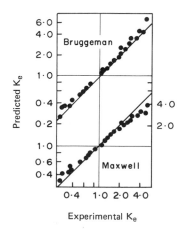

FIG. 14. Comparison between theoretical predictions and experimental data for spherical particles (from Kumada [38]).

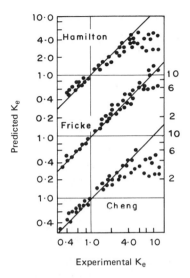

FIG. 15. Comparison between theoretical predictions and experimental data for nonspherical particles (from Kumada [38]).

spherical particles, Fricke's Eq. (28) was found to yield the predictions that were closest to the experimental data. Unfortunately, in both cases, Eqs. (34) and (35) by Meredith and Tobias were not considered for comparison.

Apart from the models discussed above, several other equations [139–142] have also been proposed for predicting conductivity of two-phase systems. Nielsen [143] argued that nearly all of these theories neglect several important factors, like the effects due to the particle shape, particle packing, and possible anisotropy. Instead, he suggested the following expression based on the theory of elastic moduli of composite materials:

$$K_e = \frac{1 + (C_e - 1)x_v\,[(K_d - 1)/(K_d + C_e - 1)]}{1 - [(K_d - 1)x_v/(K_d + C_e - 1)][1 + x_v(1 - x_m)/x_m^2]} \tag{38}$$

where C_e is the Einstein coefficient accounting for the particle shape and x_m is the maximum packing fraction of the dispersed phase. Nielsen showed that Eq. (38) predicts the thermal conductivity of multiphase (solid) systems quite accurately. In the case of liquid systems, however, the predictive ability of this equation has not yet been assessed.

The conventional approach outlined above has been quite successful in predicting thermal conductivity of heterogeneous media for moderately concentrated dispersions and regularly sized dispersed-phase particles. In practice, however, suspensions that are relatively concentrated and that contain particles of arbitrary shapes are also encountered. For such suspensions, several empirical expressions have been proposed. These were summarized by Meredith and Tobias [15]. A theoretical basis for many of these correlations, however, is not totally clear.

2. *Generalized Analysis*

It is evident from the foregoing discussion that the problem of determining effective thermal conductivity of suspensions has been considered by a large number of investigators for almost a century. Most of these studies appear to have been restricted, however, to special geometries with the result that no sufficiently general method has evolved from these for obtaining the effective thermal conductivity given the thermal properties of the constituents and the geometry of the dispersed phase. The bulk of these studies were inspired by Maxwell's work [62] undertaken more than a century ago.

Only during the last decade or so, efforts have been made to develop a generalized analytical framework for determining thermal conductivity in heterogeneous media. In particular, bulk of the early developments pertain to mixtures that comprise just two phases that are separately homogeneous. One of these phases is assumed to be in the form of discrete inclusions or particles and the other is a connected embedding or suspending matrix that may either be fluid or solid. Commonly, the relative position and orientation of the inclusion or particles vary randomly within the bulk of the mixture and hence the statistical description of the bulk material needs to be introduced. Early developments regarding transport properties in two-phase heterogeneous media have been discussed by Batchelor [78], and in the following we shall provide a brief outline of the pertinent developments (to date) in this field with particular emphasis on the thermal conductivity determination.

The generalized approach is concerned with the determination of thermal conductivity of a composite medium by analyzing the heat conduction process in a multiconnected domain H, consisting of a matrix of conductivity k_c and an inclusion of conductivity k_d. For simplicity, k_c and k_d are usually taken to be position and temperature independent. Several efforts at tackling this general problem or its specialized versions have been reported in the literature [78, 81–85]. For example, using the mathematical technique devised by Batchelor [78, 80], Jeffrey [81] analyzed the problem of effective thermal conductivity (k_e) of a stationary dilute suspension of spheres. In effect, Jeffrey extended Maxwell's Eq. (A) of Table III from $O(x_v)$ to $O(x_v^2)$ by incorporating the interaction of each inclusion with every other inclusion, taken separately. That is, the effect of any one particle on any other is calculated ignoring the presence of a third particle. McCoy and Beran [82], on the other hand, reported an analysis for a similar system of random homogeneous suspension of solid spheres by taking into account only the nearest-neighbor interaction. In doing so, they were able to circumvent the difficulty of conditionally convergent integrals (resulting from the averaging process) that necessitated the use of Batchelor's technique by Jeffrey. These analyses, however, were specific to spherical inclusions. A more general and elaborate theory has been developed by Batchelor [78] and Acrivos and co-workers [83–85]. Since presently this represents the most general approach, we shall consider this in some detail.

It may be added that recently Batchelor and O'Brien [144] have provided a generalized analysis of the problem of heat conduction in a granular medium, where the randomly arranged particles are in or are nearly in contact. Such a problem was subsequently considered by McPhedron and McKenzie [131] and Sangani and Acrivos [135] for spheres and by Perrins et al. [138] for cylinders. These analyses are most appropriate for highly filled solids or solidlike liquids (e.g., pastes, slurries, gels), media that are outside the scope of the present review and hence will not be considered further. Similarly, various analytical approaches like the generalized function approach [131–138], the generalized moment analysis [145], the multiscale technique [146, 147], and the volume averaging approach [148, 149] have been developed to predict the effective transport properties of two-phase systems and a detailed discussion of these can be a review topic by itself.

a. Basic equations for generalized analytical framework

As demonstrated by Batchelor [78, 80], the concept of a macroscopically homogeneous suspension implies the existence of a subdomain V of H large enough to contain many inclusions and yet small enough for the variation of the local statistical properties of the dispersion over V to be negligible. In that case, the local volume averages of heat flux and temperature gradient can be defined as

$$\langle q_i \rangle = \frac{1}{V} \int_V q_i \, dV \tag{39a}$$

and

$$\langle T_i' \rangle = \frac{1}{V} \int_V T_i' \, dV \tag{39b}$$

respectively, where $T_i' = \partial T / \partial x_i$ and x_i represent Cartesian coordinates. Following Batchelor [79], Eq. (22) can be combined to give

$$\langle q_i \rangle + k_c \langle T_i' \rangle = \frac{1}{V} \int_V (q_i + kT_i') \, dV \qquad (40)$$

Since the local heat fluxes for the matrix and the inclusion regions within V are

$$q_i = -k_c T_i' \quad \text{in} \quad V - \sum_{m=1}^{N} V_{pm} \qquad (41a)$$

and

$$q_i = -k_d T_i' \quad \text{in} \quad V_{pm} \qquad (41b)$$

respectively, where N is the total number of inclusions within V and V_{pm} is the volume of mth inclusion. Equation (40) can be written as

$$\langle q_i \rangle + k_c \langle T_i' \rangle = \frac{1}{V} (k_c - k_d) \sum_{m=1}^{N} \int_{V_{pm}} T_i' \, dV \qquad (42)$$

which, by the use of the divergence theorem, becomes

$$\langle q_i \rangle + k_c \langle T_i' \rangle = \frac{1}{V} (k_c - k_d) \sum_{m=1}^{N} \int_{S_{pm}} T n_i \, dS \qquad (43)$$

where S_{pm} is the surface of the mth inclusion and n_i is the unit normal to S_{pm}. Equation (43) is completely general and is valid even in case of unsteady heat conduction. Macroscopically, if the heat conduction for the bulk suspension is written in terms of Fourier's law, then

$$\langle q_i \rangle + k_{ij} \langle T_i' \rangle = 0 \qquad (44)$$

where k_{ij} is a second-order tensor representing the effective thermal conductivity of the suspension. Thus from Eqs. (43) and (44)

$$k_{ij} \langle T_i' \rangle = -\langle q_i \rangle = k_c \langle T_i' \rangle - \frac{1}{V} (k_c - k_d) \sum_{m=1}^{N} \int_{S_{pm}} T n_i \, dS \qquad (45a)$$

Equation (45a) is the general equation for determining the thermal conductivity of a suspension. It can be shown that for very dilute dispersions no interaction exists between individual inclusions and Eq. (45a) can be simplified considerably with k_{ij} becoming a symmetric tensor. In particular, for an infinitely dilute dispersion for identical inclusion shape, size, and orientation, Eq. (45a) simplifies to

$$k_{ij} \langle T_i' \rangle = k_c \langle T_i' \rangle - c_p (k_c - k_d) \int_{S_p} T n_i \, dS \qquad (45b)$$

where c_p is the number of inclusions per unit volume of the dispersion and S_p is the surface of an inclusion.

The basic Eqs. (45a) or (45b) as developed by Rocha and Acrivos [83] can now be used to determine $(k_e)_{ij}$ for inclusions of different types.

b. Expressions of $(k_e)_{ij}$ for different inclusion shapes [84]

i. First-order expressions for different inclusion shapes. As suggested by Fricke [64], most regularly shaped inclusions can be adequately represented by ellipsoids with appropriate adjustment of the axes sizes. For a dilute dispersion of such inclusions all aligned in the same direction and in such a way that their semi-axes a, b, c are parallel to the Cartesian coordinates x_1, x_2, and x_3, respectively, the steady-state temperature due to a single ellipsoidal inclusion in an infinite medium along with Eq. (45) gives the normalized effective thermal conductivity as

$$(K_e)_{ij} = \delta_{ij} + 2 \left\{ \frac{(K_d - 1)}{(K_d + 1)} x_v \left[\delta_{ij} + \sum_{m=1}^{\infty} \left\{ \frac{K_d - 1}{K_d + 1} \right\}^m \right. \right.$$
$$\left. \left. \times \{(1 - 2A_0)^m \delta_{i1}\delta_{j1} + (1 - 2B_0)^m \delta_{i2}\delta_{j2} + (1 - 2C_0)^m \delta_{i3}\delta_{j3}\} \right] \right\} \quad (46)$$

where δ_{ij} is the Kronecker delta and

$$A_0 = \frac{1}{2} abc \int_0^{\infty} \frac{ds}{(a^2 + s)\Delta(s)}$$

$$B_0 = \frac{1}{2} abc \int_0^{\infty} \frac{ds}{(b^2 + s)\Delta(s)}$$

$$C_0 = \frac{1}{2} abc \int_0^{\infty} \frac{ds}{(c^2 + s)\Delta(s)}$$

with $\Delta(s) = [(a^2 + s)(b^2 + s)(c^2 + s)]^{1/2}$. Equation (46) converges for all K_d since A_0, B_0, and C_0 always lie between zero and unity. Besides, it is evident from Eq. (46) that, though $(K_e)_{ij}$ is anistropic in general, it remains isotropic in the first order in $(K_d - 1)$. In the limit $x_v \to 0$, Eq. (46) reduces to Eq. (28) derived by Fricke.

For spherical particles $A_0 = B_0 = C_0$ and therefore Eq. (46) reduces to

$$(K_e)_{ij} = \delta_{ij} \left[1 + \frac{3(K_d - 1)x_v}{K_d + 2} \right] \quad (47)$$

Expectedly, $(K_e)_{ij}$ is isotropic for spherical inclusion and Eq. (47) reduces to Maxwell Eq. (26) as $x_v \to 0$.

In case of slender spheroid with $a \gg b$ or c, Eq. (46) becomes

$$(K_e)_{11} = 1 + (K_d - 1)x_v \left[1 + (K_d - 1) \left\{ \frac{bc}{a^2} \left(1 - \log \frac{4a}{b+c} \right) \right. \right.$$

$$\left. \left. + O \left(\frac{b^2c^2}{a^2} \log \frac{4a}{b+c} \right) \right\} \right] \tag{48a}$$

for $K_d(bc/a^2) \log[4a/(b+c)] \ll 1$ and

$$(K_e)_{11} = 1 + x_v \frac{a^2}{bc} \left(\log \frac{4a}{b+c} - 1 \right)^{-1}$$

$$\times \left[1 + O \left\{ \frac{a^2}{K_d b_c} \left(\log \frac{4a}{b+c} \right)^{-1} \right\} \right] \tag{48b}$$

when $K_d(bc/a^2) \log[4a/(b+c)] \gg 1$ and, for all α,

$$(K_e)_{22} = 1 + 2x_v \frac{K_d - 1}{K_d + 1} \left\{ 1 - \left(\frac{K_d - 1}{K_d + 1} \right) \left(\frac{b - c}{b + c} \right) \right\}^{-1}$$

$$\times \left[1 + \frac{K_d - 1}{K_d + 1} \left\{ \frac{bc}{a^2} \log \left(\frac{4a}{b+c} \right) + O \left(\frac{bc}{a^2} \right) \right\} \right] \tag{49}$$

$$(K_e)_{33} = 1 + 2x_v \frac{K_d - 1}{K_d + 1} \left\{ 1 - \left(\frac{K_d - 1}{K_d + 1} \right) \left(\frac{c - b}{c + b} \right) \right\}^{-1}$$

$$\times \left[1 + \frac{K_d - 1}{K_d + 1} \left\{ \frac{bc}{a^2} \log \frac{4a}{b+c} + O \left(\frac{bc}{a^2} \right) \right\} \right] \tag{50}$$

ii. *Inclusions of arbitrary shape.* Rocha and Acrivos [83] show that, for dilute dispersions containing inclusions of arbitrary shape, the basic equations can be adapted to reduce the problem of determining the steady-state temperature field to that of solving an integral equation of second kind. In particular, for $K_d \approx 1$, they have analytically derived the K_e expressions for axisymmetric as well as slender inclusions. In the case of axisymmetric inclusions, the problem simplifies further since the integral equation, which in the general case involves integration over the surface of the inclusion, reduces to an integral equation in a single variable. For slender inclusions (all aligned), irrespective of the particular inclusion geometry, the longitudinal conductivity is derived as

$$(K_e)_{11} = 1 + (K_d - 1)x_v[1 + (K_d - 1) O(\varepsilon^2 \log \varepsilon^{-1})]$$

$$\text{if} \quad K_d \varepsilon^2 \log \varepsilon^{-1} \ll 1 \tag{51a}$$

and

$$(K_e)_{11} = 1 + x_v O[\varepsilon^{-2}(\log \varepsilon^{-1})^{-1}]$$

$$\text{if} \quad K_d \varepsilon^2 \log \varepsilon^{-1} \gg 1 \tag{51b}$$

where $\varepsilon \ll 1$ is the slenderness ratio. For slender ellipsoidal inclusions, Eq. (51) reduces to those given by Eq. (48). In regard to the transverse conductivity for axisymmetric inclusions, irrespective of K_d, the results derived were

$$(K_e)_{22} = (K_e)_{33} = 1 + 2x_v \left[\frac{K_d - 1}{K_d + 1} \right] \left[1 + \left(\frac{K_d - 1}{K_d + 1} \right) O(\varepsilon^2 \log \varepsilon^{-1}) \right] \tag{52}$$

Here too, Eq. (52) reduces to eqs. (49) and (50) when $b = c$. Besides, as $\varepsilon \to 0$, Eqs. (51) and (52) appropriately reduce to Eq. (21) for longitudinal conductivity of two phases in parallel and to the expression for the transverse conductivity of a very dilute dispersion of parallel infinite cylinders. However, it must be realized that the above equations are valid only for very small volume fractions of the dispersed phase, i.e., $x_v \ll \varepsilon^2 \log \varepsilon$ and for slender inclusions with $k_d \approx k_c$.

In case of dilute ($x_v \ll 1$) dispersions of highly conducting slender inclusions all aligned in a common direction, the thermal conductivity of the dispersion is expected to follow the following form [87]:

$$(K_e)_{11} = 1 + A_1 x_v + A_2 x_v^2 + O(x_v^3)$$

$$(K_e)_{22} = 1 + B_1 x_v + B_2 x_v^2 + O(x_v^3) \tag{53}$$

$$(K_e)_{33} = 1 + C_1 x_v + C_2 x_v^2 + O(x_v^3)$$

Rocha and Acrivos [84] derived the expressions for A_1, B_1, and C_1 for uniformly sized, long slender cylinders (radius R_0 and length $2L$). These are

$$A_1 = \left(\frac{2}{3} \right) \frac{x_v}{\varepsilon^2} \left[\frac{1 + 0.64(\ln(2/\varepsilon))^{-1}}{\ln(2/\varepsilon) - 1} + 1.339 \left(\ln \left(\frac{2}{\varepsilon} \right) \right)^{-3} \right] \tag{54}$$

$$B_1 = C_1 = 2$$

where $\varepsilon = R_0/L \ll 1$. Equation (53) was verified experimentally by Rocha and Acrivos [86] and Figs. 16 and 17 show some typical results. Experiments were performed by measuring electrical conductivity of dilute suspensions of chopped stainless steel fibers in corn syrup leading to $K_d \approx 10^9$. Data presented in Figs. 16 and 17 were used to obtain values of A_1, A_2, B_1, and B_2 from a least-square fit, and these are summarized in Table IV along with the theoretical predictions. It is seen that theoretical predictions of A_1 are in reasonably good agreement with the experimental values. The mea-

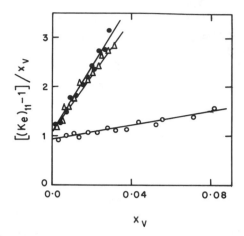

Fig. 16. Longitudinal conductivity against volume fractions of solids for different slenderness ratios; $\varepsilon = 0.0106$ (O), 0.0389 ($\triangle$), and 0.0897 ($\bullet$). $\Phi = x_v\, \varepsilon^2/\ln(2/\varepsilon)$ (from Rocha and Acrivos [86]).

sured values of B_1, however, are markedly different from the theoretical values. Rocha and Acrivos attribute this discrepancy to nonrandom particle distribution, irregular fiber shape and inadequate fiber alignment.

In a subsequent development, Chen and Acrivos [85] have extended the earlier analysis to derive expressions for A_2 [see Eq. (53)] by incorporating particle–particle interaction. Although cluster formation involving more than five particles was observed experimentally, they argued that for $x_v \ll 1$ the effect is primarily restricted to the interaction between two

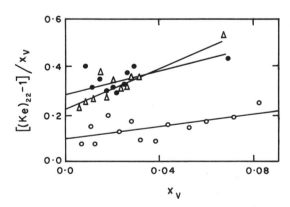

Fig. 17. Transverse conductivity against volume fraction of solids for different slenderness ratios. $\varepsilon = 0.0106$ (O), 0.0389 ($\triangle$) and 0.0897 ($\bullet$), $\Phi = x_v\varepsilon^2/\ln(2/\varepsilon)$ (from Rocha and Acrivos [86]).

TABLE IV

COMPARISON BETWEEN EXPERIMENTAL DATA AND THEORETICAL PREDICTIONS FOR
DILUTE DISPERSION OF SLENDER INCLUSIONS[a]

$K_d \times 10^2$	$A_1 \times 10^{-2}$		$B_1 \times 10^{-2}$		$A_2 \times 10^{-6b}$	
	Experiment	Eq. (54)	Experiment	Eq. (54)	Experiment	Eq. (55)
1.06	16.0	17.1	1.6	0.02	20.5	15.3
3.89	1.9	1.94	0.37	0.02	1.59	1.73
8.97	0.43	0.562	0.11	0.02	0.106	0.239

[a] From references 85 and 86.
[b] At t = 25 sec.

adjacent particles. Since structure build-up will be time dependent, at least initially, the effective thermal conductivity is also expected to be time dependent during this period. By analyzing the motion of pairs of two adjacent slender particles and their probability of being at a given location within the medium, they obtained the following

$$A_2 = \frac{1}{\varepsilon^4}\left[\frac{V(t)}{3\pi L^3}\left(\frac{4}{\ln(4/\varepsilon)-1} - \frac{1}{\ln(2/\varepsilon)-1}\right)\right.$$
$$\left. + \frac{0.4832}{(\ln 2/\varepsilon)^2(\ln(2/\varepsilon)-1)}\right] \tag{55}$$

where $V(t)$ is a time-dependent volume occupied by a particle pair and has to be determined numerically. As is evident from Table IV, values of A_2 predicted by Eq. (55) agree reasonably well with those measured experimentally [86]. Chen and Acrivos's analysis indicate the significance of the particle distribution in determining the effective thermal conductivity of suspensions when the interactions between particles are appreciable. Undoubtedly, for concentrated dispersions of arbitrarily shaped inclusions, effects of all kinds of aggregates like pairs, triplets, quadruplets, and even long chains need to be accounted for. Presently, however, no such analyses are available to extend the above generalized analytical framework to even moderately concentrated suspensions. Instead, only bounds on the K_e are available, which are valid for any given volume fraction of the dispersed phase.

3. Bounds on Thermal Conductivity of Suspensions

For moderately concentrated suspensions of arbitrarily shaped inclusions, very few reliable theoretical calculations are available and considerable work needs to be undertaken in this area. At the present state of development, only bounds on the effective thermal conductivity of such suspensions can be obtained. These bounds, valid for all x_v, are usually derived from variational principles based on the governing transport and constitutive equations, although a different mathematical formulation of this problem has been recently proposed by Bergmann [134]. In the case of thermal conductivity of isotropic suspensions Hashim and Shtrikman [88] obtained the bounds

$$K_1^* \leq K_e \leq K_2^* \quad \text{for} \quad K_d \geq 1$$

and (56)

$$K_2^* \leq K_e \leq K_1^* \quad \text{for} \quad K_d < 1$$

where

$$K_1^* = 1 + \frac{x_v(K_d - 1)}{1 + \frac{1}{3}(1 - x_v)(K_d - 1)}$$

$$K_2^* = K_d + \frac{3K_d(1 - K_d)(1 - x_v)}{3K_d + x_v(1 - K_d)}$$

Equation (56) involves no information about the geometry of the dispersed phase and thus as expected, Eq. (56) predicts that $K_e \to \infty$ as $K_d \to \infty$. More restrictive bounds incorporating shape information were considered by Miller [89]. In particular, he used a cell model to obtain the following bounds for $K_d \geq 1$:

$$K_{e_1} \geq K_e \geq K_{e_2} \tag{57}$$

where

$$K_{e_1} = 1 + x_v(K_d - 1) - \frac{\frac{1}{3}x_v(1 - x_v)(K_d - 1)^2}{1 + (K_d - 1)(x_v + 3G(1 - x_v))}$$

and

$$K_{e_2} = K_d \left[K_d - x_v(K_d - 1) - \frac{\frac{1}{3}4x_v(1 - x_v)(K_d - 1)^2}{1 + K_d + 3(K_d - 1)G(2x_v - 1)} \right]^{-1}$$

and G is a constant. Miller showed that $\frac{1}{9} \leq G \leq \frac{1}{3}$, G being equal to $\frac{1}{9}$, $\frac{1}{6}$, and $\frac{1}{3}$ for spherical, needlelike, and plate- or disklike inclusions, respectively. Because of cell model assumptions, bounds given by the above equation are applicable to a restricted class of inclusions and also are better suited for relatively low x_v. A further refinement in this approach was made by Beran [90] and Elsayed [91] by taking into account the particle positioning information. Using various order correlation functions (in order to represent the shape information) along with the Miller model, Elsayed demonstrated that in addition to the constant G, the bounds also depend on two additional shape factors and two positioning factors. Like G, the shape factors are bounded by constant values. No general limits on the positional factors, however, are possible and these values depend on the specific geometrical arrangement of the inclusions [82].

Recent bounds on thermal conductivity of multiphase systems are obtained by Milton and Phan-Thien. Milton [150] showed that if volume fraction of the components are known then Hashim–Shtrikman bounds on the thermal conductivity can be achieved not only for two-phase systems but also for systems containing m number of phases provided that

$$K_{\min} \leq K_2 \quad \text{and} \quad K_{\max} \geq K_{m-1} \tag{58}$$

where $K_1 \leq K_2 \leq \cdots K_m$ are the thermal conductivities of the m phases and $K_{\max}$ and $K_{\min}$ are the upper and lower bounds given by Eq. (56). This implies that if Eq. (58) is satisfied, Hashim — Shtrikman bounds (which are second order in $\delta K = K_m - K_1$) are the most restrictive bounds if only volume fractions $(x_{v_1}, x_{v_2}, \ldots, x_{v_m})$ of the phases are known. For higher-order bounds, Phan-Thien and Milton [151] developed a generalized treatment taking into account the effective microstructure of the composite assumed to possess periodic geometry. Expectedly, therefore, these bounds involve microstructural parameters which equal $m(m-1)^2/2$ or $m^2(m-1)^2/2$ in number for the third- and the fourth-order bounds, respectively. In the case of two-phase systems $(m = 2)$, the fourth-order bounds obtained are as follows [151, 152]:

$$\langle K \rangle - \frac{x_v(1 - x_v)(\delta K)^2}{\langle \tilde{K} \rangle + 2\Phi} \leq K_e \leq \langle K \rangle - \frac{x_v(1 - x_v)(\delta K)^2}{3\langle K \rangle}$$

$$+ \frac{(\delta K)^3 A_s}{3\langle K \rangle^2} \frac{(\delta K)^4 B_s}{3\langle K \rangle^3} \tag{59}$$

where

$$\delta K = 1 - K_d$$

$$\Phi = K_d(\langle K \rangle_\zeta + (1 - \zeta)(\delta K)B^*)/(K_d + (1 - \zeta)(\delta K)B^*)$$

$$\langle K \rangle = 1 - x_v + x_v K_d$$

$$\langle \tilde{K} \rangle = x_v + K_d(1 - x_v)$$

$$\langle K \rangle_\zeta = \zeta + K_d(1 - \zeta)$$

$$B^* = \frac{3(B_s - A_s^2/x_v(1 - x_v)}{2\zeta(1 - \zeta)x_v(1 - x_v)}$$

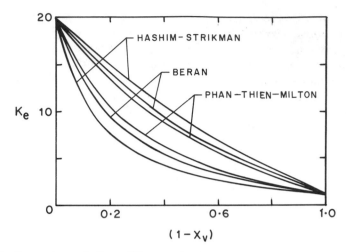

Fig. 18. Comparison of effects different order bounds on thermal conductivity of two-phase media. $K_d = 20$, $\zeta = 0.5$, and $B^* = 0.5$ (from Phan-Thien and Milton [151]).

and

$$\zeta = 1 - x_v + \frac{A_s}{x_v(1 - x_v)} - x_v/3$$

where A_s and B_s are microstructural parameters. Phan-Thien and Milton showed that the fourth-order bounds represented by Eq. (59) are always more restrictive than the third-order Beran bounds and second-order Hashim–Shtrikman bounds and this is more clearly demonstrated by the comparison of these bounds provided in Fig. 18 for a typical set of microstructural parameters and $K_d = 20$. For random media, similar bounds based on variational principles also have been obtained by Kohler and Papanicolaou [153].

B. Flowing Media

In the preceding section, we have discussed a generalized analytical framework for determining the effective thermal conductivity of stationary suspensions. A number of general results follow from these analyses. First, the bulk conductive heat flux is simply the ensemble average of the microscale conduction flux and thus the effective thermal conductivity differs from that of the continuous phase only if the conductivities of the dispersed and the continuous phases are different. Second, the bulk heat flux and the bulk temperature gradient are related by symmetric second-order effective conductivity tensor. Third, the effective thermal conductivity depends on the geometric microstructure of the suspension.

In light of the above characteristics of stationary suspensions, it is expected that since the microstructure can be considerably altered by imposition of a flow field, the effective conductivity of a flowing suspension will depend also on the type as well as the strength of the imposed flow field. Very little work has been undertaken in this area and a general overview regarding macroscopic transport properties in sheared suspensions has been provided by Leal [92]. In the following we shall briefly highlight some of the pertinent developments in the area of effective thermal conductivity of flowing suspensions.

1. *Generalized Analytical Framework*

Apart from the complexities introduced by the bulk flow of the medium, the analytical framework for determining K_e for flowing suspensions is developed along lines similar to that described in Section IV, A, 2 for stationary suspensions. The general relations between macroscopic heat flux and the microstructural state of the flowing suspension were obtained by Leal [23] and Nir and Acrivos [93]. The problem considered was that of a suspension of neutrally buoyant particles in presence of a general bulk shear flow and a bulk temperature field. It is assumed that particle dimensions are large compared to the molecular dimensions so that the dispersion can be treated as a continuum. The concept proposed by Leal was that at any arbitrary point within the suspension the local variables are random functions of time with their instantaneous values depending on the proximity and motion of the dispersed particles. In that case, instantaneous local values of velocity, temperature, and conductive heat flux may be expressed as a sum of the appropriate volume averaged quantity and an additional fluctuating component, i.e.,

$$u_i = \langle u_i \rangle + u_i'$$
$$T = \langle T \rangle + T' \tag{60a}$$
$$q_i = \langle q_i \rangle + q_i'$$

where i denotes direction in Cartesian coordinates and, by definition,

$$\langle u_i' \rangle = \langle T' \rangle = \langle q_i' \rangle = 0 \tag{60b}$$

If q_i^* is a bulk heat flux such that the overall heat-transfer problem within the suspension can be represented by the usual convection–diffusion equation

$$\rho\overline{C_p}\langle u_i \rangle \frac{\partial\langle T\rangle}{\partial x_i} + \frac{\partial q_i^*}{\partial x_i} = 0 \tag{61}$$

then it can be shown that

$$q_i^* = \langle q_i \rangle + \rho\langle C_p T' u_i' \rangle \tag{62}$$

where $C_p = (1 - x_v)C_{p_c} + x_v C_{p_d}$ is the volume average heat capacity.

Equation (62) implies that the bulk heat flux comprises a volume average of the local instantaneous conductive heat flux and an additional contribution due to the local instantaneous convective heat flux. By expressing the averages as appropriate volume integrals and following a procedure similar to that outlined in Section IV,A,2 for stationary suspensions, Eqs. (60)–(62) can be combined to give the following expression for the bulk "conductive" heat flux [23]:

$$q_i^* = -k_c \frac{\partial\overline{T}}{\partial x_i} + q_i^p \tag{63}$$

where

$$q_i^p = \frac{(k_c - k_d)}{V}\int_{\Sigma V_p}(\nabla T)_i\, dV + \frac{(C_{p_d} - C_{p_c})}{V}\int_{\Sigma V_p} u_i' T'\, dV$$

$$+ \frac{\rho C_{p_c}}{V}\int_V u_i' T'\, dV \tag{64}$$

V is the subdomain volume (within the medium) within which the suspension is statistically homogeneous, V_p the volume of the individual particles and $\partial\overline{T}/\partial x_i$ is the bulk temperature gradient. Equation (63) suggests that the bulk heat flux comprises the heat flux that would exist in the absence of particles and the combination q_i^p due solely to the presence of the particles. The particle contribution to the heat flux can further be seen to consist of that due to difference in thermal properties of the two phases and that caused by the imposition of a flow field. From Eq. (64), it is seen that for identical thermal properties ($k_c = k_d$ and $C_{p_c} = C_{p_d}$) only the first two terms vanish, thereby suggesting that, unlike the case of stagnant suspensions, the effective thermal conductivity of a flowing suspension will be different from that of a flowing homogeneous fluid even when the dispersed and continuous phase thermal properties are identical [92, 93]. As in the case of stationary suspensions, Eqs. (63) and (64) can be further simplified in the case of dilute dispersions and uniform-size particles. Under these conditions, the temperature distribution around any particle is identical to that around any other particle so that the summation over individual particle volume or surface area can be replaced by a single integral multiplied by the total particle volume or surface area per unit volume of the suspension. If V_p is the volume of a particle (assumed identical for all particles) then Eq. (64) simplifies to

$$q_i^p = \frac{(k_c - k_d)x_v}{V_p} \int_{V_p} (\nabla \overline{T})_i \, dV + \frac{\rho(C_{p_d} - C_{p_c})}{V_p} \int_{V_p} \overline{u_i' T'} \, dV$$

$$+ \frac{\rho C_p x_v}{V_p} \int_V \overline{u_i' T'} \, dV \tag{65}$$

where V is the fluid volume that is assumed to extend to infinity for a simple particle microscale problem [93].

2. Some Specific Cases

The above generalized framework relates the effective bulk conductive heat flux of a sheared suspension to the microscale velocity and temperature fields associated with each individual particle. Knowledge of this effective heat flux along with that of the corresponding temperature gradient permits evaluation of the effective thermal conductivity of the flowing suspension. Such evaluations have been reported by Leal and co-workers [23, 94, 95] and Nir and Acrivos [93] for sheared dilute ($x_v \ll 1$) suspensions under certain limiting conditions. In all cases, for mathematical simplicity, the suspensions were assumed to be subjected to a linear shear field and a linear temperature gradient.

a. Suspension of spherical drops/particles

Leal [23] derived the effective thermal conductivity of a dilute dispersion of spherical drops in the limit of low particle Peclet number ($Pe = d^2\rho C_{p_c}\dot{\gamma}/k_c \ll 1$). The velocity fields were obtained from the classical creeping flow solution of Taylor for a spherical drop in shear flow, and the microscale temperature field was calculated using the method of matched asymptotic expansions. The expression obtained was as follows:

$$k_{eff} = k_c \left[1 + x_v \left\{ \frac{3(k_d - k_c)}{k_d + 2k_c} + \left[\frac{1.176(k_d - k_c)^2}{(k_d + 2k_c)^2} - \frac{2\mu_c + 5\mu_d}{\mu_c + \mu_d} \right. \right. \right.$$

$$\left. \left. \left. \times \left(0.12 \frac{2\mu_c + 5\mu_d}{\mu_c + \mu_d} - 0.028 \frac{k_d - k_c}{k_d + 2k_c} \right) \right] Pe^{3/2} + O(Pe^2) \right\} \right] \tag{66}$$

where μ_c and μ_d are the viscosities of the continuous and dispersed phases respectively. Note that for solid spherical particles ($\mu_d \rightarrow \infty$), Eq. (66)

reduces to

$$k_{eff} = k_c \left[1 + x_v \frac{3(k_d - k_c)}{k_d + 2k_c} + \text{Pe}^{3/2} x_v \right.$$
$$\left. \times \left\{ \frac{1.176(k_d - k_c)^2}{(k_d + 2k_c)} + 3.0 - 0.14 \frac{(k_d - k_c)}{(k_d + 2k_c)} \right\} + O(\text{Pe}^2) \right] \quad (67)$$

Several interesting features can be noticed from the above expressions. Expectedly, the effective thermal conductivity comprises both a conductive and a convective contribution. The conductive contribution, which is present even in the absence of bulk flow, can either be negative or positive depending on the relative magnitudes of k_c and k_d. The convective contribution, on the other hand, is always positive and hence the effective thermal conductivity of a flowing suspension is always enhanced by increasing the bulk shear rate. However, it is only at $O(\text{Pe}^{3/2})$ that the presence of the disturbance velocity field is manifested in the effective thermal conductivity, even though the temperature field gets altered at $O(\text{Pe})$. This is most likely due to the symmetry induced in the temperature field by the spherical shape of the dispersed phase.

Nir and Acrivos [93] analyzed the opposite limit of large Peclet number ($\text{Pe} \gg 1$) for suspension of rigid spherical particles. Due to the existence of closed-stream surfaces near the particle, the calculation of the microscale temperature distribution around each particle is extremely involved and was not solved rigorously. Instead, certain simplifying assumptions regarding the integrability of the solution were made, as well as the following estimate for the flow-induced contribution to the effective thermal conductivity:

$$k_{eff} = k_c \{ 1 + x_v [\Omega \, \text{Pe}^{1/11} + O(1)] \} \quad (68)$$

where Ω is a constant that may be determined from the temperature distribution associated with each particle. Similarly, for a dilute dispersion of infinitely long cylinders at high Peclet numbers, Nir and Acrivos obtained

$$k_{eff} = k_c \{ 1 + 0.02813 x_v [(\ln \text{Pe})^2 + O(\ln \text{Pe})] \} \quad (69)$$

Unlike the low-Peclet-number case, the above expressions suggest that for $\text{Pe} \gg 1$, the effective thermal conductivities are independent of the dispersed-phase thermal conductivity k_d. Besides, it is also apparent that, as Pe increases, the flow-induced contribution to k_{eff} becomes relatively less significant.

b. *Suspension of deformed drops/particles*

In the foregoing [see Eq. (66)] we have seen that for a sheared suspension of spherical drops, the flow-induced contribution to the effective thermal conductivity is of the order of $Pe^{3/2}$. McMillen and Leal [94] maintained that this result is due to spherical symmetry and hence investigated the role of drop deformation on the effective thermal conductivity in the limit $Pe \ll 1$. In particular, the limiting cases of dominant interfacial forces ($\varepsilon_c \sim d\dot{\gamma}\mu_c/\sigma \ll 1$) and dominant internal (drop) viscosity effects ($\varepsilon_c \sim 1/\lambda \ll 1$) were considered where ε_c is the deformation parameter, which is small, $\lambda = (\mu_d/\mu_c)$ the viscosity ratio, d the undeformed drop radius, and σ the interfacial tension. For the case of surface-tension-controlled deformation, McMillen and Leal derived the following expression for the effective thermal conductivity

$$K_{eff} = \frac{k_{eff}}{k_c} = 1 + x_v \left\{ \frac{3(K_d - 1)}{(K_d + 2)} + \left(\frac{1.176(K_d - 1)^2}{(K_d + 2)^2} \right. \right.$$
$$\left. + \frac{5\lambda + 2}{\lambda + 1}\left(0.12 \left[\frac{5\lambda + 2}{\lambda + 1} \right] - 0.028 \left[\frac{K_d - 1}{K_d + 2} \right] \right) \right) Pe^{3/2}$$
$$\left. + I(K_d, \lambda, w)\varepsilon_c \, Pe + O(\varepsilon_c^2) + O(\varepsilon_c Pe^{3/2}) + O(Pe^2) \right\} \qquad (70)$$

where $w = C_{p_d}/C_{p_c}$, $K_d = k_d/k_c$, and I represents a complex function of K_d, λ, and w that can be either positive or negative.

In the case of internal viscous forces controlling deformation, the corresponding expression is as follows

$$K_{eff} = 1 + x_v \left\{ \frac{3(K_d - 1)}{K_d + 2} + \left(3.0 - 0.14\frac{K_d - 1}{K_d + 1} + 1.176\frac{(K_d - 1)}{(K_d + 2)^2} \right) Pe^{3/2} \right.$$
$$- \left(3.6\frac{(K_d - 1)^2}{(K_d + 2)^2} \right) \varepsilon_c + \left(1.411\frac{K_d(K_d - 1)^2}{(K_d + 2)^3} \left[\left(\frac{w}{K_d} \right)^{3/2} - 1 \right] \right.$$
$$\left. - 0.168\frac{(K_d - 1)^2}{(K_d + 2)^2} \right) \varepsilon_c \, Pe^{3/2} + O(\varepsilon_c^2) + O(Pe^2) + O(\varepsilon_c \, Pe)^2) \right\} \qquad (71)$$

It is seen that for both the limiting cases, terms additional to those obtained by Leal [see Eq. (66)] for spherical drops are introduced due to slight deformation. These additional contributions, representing small corrections, are of the order of $\varepsilon_c \, Pe$ for the case of deformation dominated by interfacial tension and of the order of ε_c and $\varepsilon_c \, Pe^{3/2}$ for the case of deformation dominated by drop phase viscosity. The difference in these two small deformation limits is most likely to be due to the differences in the nature of the particle shape in the two cases [94]. Detailed analyses of the various limiting cases of Eqs. (70) and (71) were reported by McMillen and Leal [94] in order to demonstrate that the presence of even a small degree of shape change can cause fundamental change in the nature of the dominant flow contribution to the effective thermal conductivity. In light of this finding, therefore, caution needs to be exercised in attempting to correlate experimental data for any suspension in which particles are not exactly spherical with theoretical results for spherical particles.

3. *The Role of Brownian Motion*

One source of randomness at the microscale that has been neglected in the preceding analyses is that due to Brownian motion for sufficiently small particles. Both the translational and rotational motion of a Brownian particle will affect the temperature fields associated with the particles, and hence the bulk heat flux will also be altered. This implies that for sufficiently small particles, Brownian motion will have some influence on the effective thermal conductivity of suspensions. Very little information regarding the magnitude of this effect is presently available, although some theoretical work in this direction has been reported by Leal and co-workers [92, 95].

Leal [92] discusses a generalized approach toward accounting for Brownian contributions to the effective thermal conductivity of suspensions. This general formulation has been used by McMillen and Leal [95] to analyze the effective thermal conductivity of rigid prolate spheroidal particles in simple shear flow with rotational Brownian motion and small Peclet number. This limiting case of Pe ≪ 1 was solved by an involved perturbation scheme valid for $D_r/\dot\gamma \gg 1$ where D_r is the rotational Stokes–Einstein diffusion coefficient for the particles. The results suggest that the effective thermal conductivity comprises three contributions: those due to (i) conduction as expected, (ii) preferential alignment of the particles caused by shear flow, and (iii) the combined convective effects of bulk and mean Brownian velocity fields. The last contribution, of the order of $(\dot\gamma \, \text{Pe}/D_r)$, becomes significant as the particles become slender and as the conductivity ratio becomes large ($K_d \gg 1$).

It is obvious that incorporation of the Brownian effect significantly complicates the problem of theoretically predicting the effective thermal conductivity of flowing suspensions. Prior to undertaking such an effort, the question naturally arises as to whether these Brownian motion contributions are significant enough to merit careful attention for practical purposes. Leal [92] provides some guidelines for assessing the significance of Brownian motion effects. In particular, for flowing suspensions the direct Brownian heat flux is negligible compared to the convective effects of the bulk motion only if

$$\rho l_p^3 \gg K_B T / V_{bulk} \qquad (72)$$

where l_p is the characteristic particle dimension, K_B the Boltzmann constant, T the temperature, and V_{bulk} the bulk suspension velocity. The criterion represented by Eq. (72) is only valid for flowing suspensions. For a stationary or nearly stationary suspension, the magnitude of the Brownian motion effects is negligible compared to the pure conduction terms when

$$\rho l_p^3 \gg K_B T l_p^2 / (k_c / \rho C_{p_c})^2 \qquad (73)$$

For typical situations, Eq. (73) is satisfied and predictions for stationary suspensions (see Section IV,A) without a Brownian motion contribution should serve as adequate first approximations. This is evident from reasonably good correspondence between the theoretical predictions and experimental data as discussed in Section IV,A.

4. Comparison with Experimental Data

Theoretical analyses, though limited in number, suggest that the effective thermal conductivity of a suspension is augmented by the presence of a flow field. However, there are practically no data in the literature that permit quantitative comparison between the theory and the experimental data. Only recently, Chung and Leal [77] experimentally investigated the thermal conductivity behavior of suspensions of rigid spheres at low Peclet number with the primary objective of verifying Leal's result, Eq. (67). A Couette flow device was used and the test medium was a dispersion of polystyrene latex spheres in a lubricant oil. For this system K_d is of the order of unity and for $K_d = 1$ Eq. (67) simplifies to

$$\text{enhancement factor (EF)} = \frac{\Delta k}{k_c x_v} = \frac{k_{eff} - k_c}{k_c x_v} = 3.0 \, \text{Pe}^{1.5} \qquad (74)$$

Figure 19 shows EF as a function of Peclet number for two different volume fractions x_v. It is seen that EF predicted by Eq. (74) is of the same order of magnitude as that measured experimentally. Besides, as predicted theoretically, the EF–Pe relationship is independent of the sphere size. Surprisingly, Leal's result is expected to be valid for Pe < 0.01 and x_v < 0.01, but reasonable agreement is observed even for moderately concentrated solutions ($x_v \leq 0.25$) and relatively large Peclet numbers (Pe ~ $O(1)$). The poor agreement at low x_v is perhaps due to the inadequacy of the experimental technique to detect small changes in conductivity very

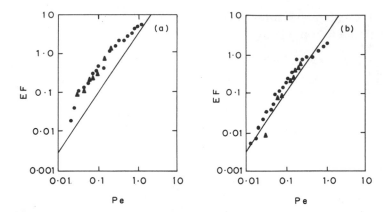

FIG. 19. Relative augmentation in effective thermal conductivity as a function of Peclet number. Sphere sizes: 17.3 ($\bullet$), 26.7 ($\blacktriangledown$), 61($\bullet$) m. Solid line represents Eq. (74). (a) $x_v = 0.02$, (b) $x_v = 0.20$ (from Chung and Leal [77]).

accurately. The results in Fig. 19 also indicate that for low Peclet number the data follow a power-law relationship of the form

$$EF = C \, Pe^e \qquad (75)$$

From Eq. (74), C and e are expected to be about 3 and 1.5, respectively, for $K_d \sim 1$. Measured values of C and e, however, ranged from 30 to 1 and 1.9 to 0.6, respectively. Thus, Chung and Leal [77] inferred that the theory is adequate only for an order of magnitude estimate of effective thermal conductivity of sheared suspensions, in spite of the fact that it obviously does not give a complete physical description of the transport behavior. Considerable experimental information will be needed both to aid theoretical developments as well as to validate them.

Although Chung and Leal's study is the only one to date that allows some comparison between theory and experiment, other indirect evidence of flow effects on effective thermal conductivity of sheared suspensions may be briefly mentioned. These studies pertain to electrical and mass transport in flowing suspensions, and since these transport processes are somewhat analogous to heat transport the general features regarding the effect of shear on the transport coefficient are expected to be similar. Keller [96] has studied the effect of fluid shear on electrical conductivity of flowing blood. The blood samples, with hematocrit of 20 to 40% could be considered as concentrated suspensions, and a significant increase in conductivity was observed with increasing flow. A similar increase in effective diffusion coefficient due to shear was also reported by Colton et al. [97] for diffusion of urea in blood (16% hematocrit). These observations, however, must be viewed with caution since the shear rate effects may be due to migration of suspended particles away from the high-shear wall regions resulting in apparent slip. Further elaboration regarding the possibility of anomalous shear-rate effect on transport coefficients has been provided by Mashelkar and Dutta [98] for structured liquids.

V. Other Nonpolymeric Liquids

Apart from suspensions, other nonpolymeric liquids like biological fluids, greases, foodstuffs, etc. also have a structured nature, and hence their thermal conductivity behavior is also somewhat different from that of ordinary liquids. In comparison to suspensions, thermal conductivity behavior of these liquids has not yet received adequate attention in terms of developing predictive capability, and in the following we shall briefly discuss the information (mostly experimental) available for such liquids.

A. BIOLOGICAL LIQUIDS

Problems pertaining to heat transfer in biological materials necessitate the knowledge of thermal conductivity of these materials. Several efforts have been made to measure the thermal conductivity of biological liquids,

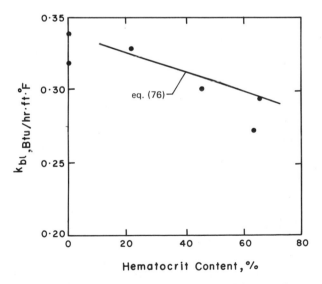

FIG. 20. Thermal conductivity of blood as a function of hematocrit content. Symbols represent experimental data (from Poppendiak [30]).

primarily blood. Poppendiak *et al.* [30] used a hot-plate technique to determine the thermal conductivity of various biological fluids and tissues. For human blood in the temperature range of 80 to 92°F, the thermal conductivity of blood was found to decrease by about 14% as the hematocrit content increased from 0 to 80% (see Fig. 20). Both series and parallel heat flow models were used to predict the data and, as shown in Figure 21, the parallel flow model adequately represented the data for most practical purposes.

A more rigorous approach, employing a microcontinuum model, was developed for predicting heat conductivity of blood by Riha [99, 100]. The final result for blood conductivity (k_{bl}) was as follows

$$k_{bl} = \frac{k_p + (k_m - k_p)(H/H_m)^2}{1 - \dfrac{(k_m - k_p)(H/H_m)^2(1 - H/H_m)}{k_p + (k_m - k_p)(H/H_m)}} \tag{76}$$

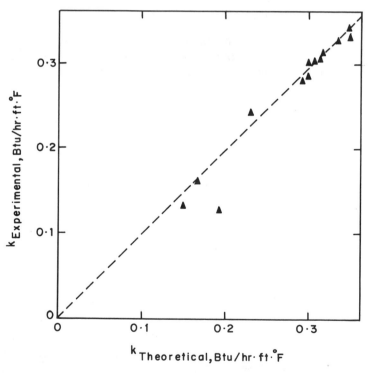

FIG. 21. Comparison between experimental data and parallel heat-flow model predictions for biological liquids and tissues (from Poppendiak *et al.* [30]).

where k_p is the thermal conductivity of plasma, k_m the thermal conductivity of blood corresponding to maximum hematocrit content (H_m) in percent, and H is the hematocrit content. Figure 20 shows a comparison between theoretical prediction of Eq. (76) with $k_p = 0.33$, $k_m = 0.28$ (Btu/ hr ft °F), $H_m = 85\%$. It also shows the experimental data reported by Poppendiak et al. It can be seen that Eq. (76) provides a good estimate of blood conductivity for different hematocrit content. These results, however, are valid only for stagnant blood samples. As mentioned earlier for flowing blood, Keller's [96] observations imply that the thermal conductivity is likely to be a strongly increasing function of the strength of the applied flow field. No direct theoretical treatment, however, is presently available that enables the prediction of such significant flow-induced augmentation of thermal conductivity in blood flow.

B. GREASES AND FOODSTUFFS

In the literature, studies on thermal conductivity of greases and food materials are scarce, and hence very little is known about thermal conduction phenomena in these structured liquids. Few experimental studies have been reported and these too were primarily motivated by the need to analyze various heat transfer processes experienced by these liquids in practical operations. In the absence of more detailed and thorough investigation, in the following we summarize these observations in order to highlight some similarities that thermal conductivity of these liquids exhibit with that of other structured liquids.

Hamilton and Crosser [31], using a concentric cylinder arrangement, measured the thermal conductivity of (stationary) greases made from different neutral oils and soaps. Figure 22 shows a typical set of results. It can be seen that the grease conductivity k_g is always greater than that of the base oil since the thermal conductivity of the solid soaplike material in the grease is considerably higher than that of the hydrocarbon oil. In order to predict grease conductivity, Hamilton and Crosser suggest that the grease structure be modeled as a composite medium with clumps of soap particles evenly distributed within the oil and that the following equation, analogous to that proposed by Fricke [64], be used

$$k_{sp} = k_{ol} \frac{2(k_s - k_{ol}) + V_s(k_g + 2k_{ol})}{k_{ol} - k_g + V_s(k_g + 2k_{ol})} \tag{77}$$

where k_{sp} and k_{ol} are the soap and oil thermal conductivity and V_s is the volume fraction of soap. As a check on the validity of Eq. (77), it has been used to determine k_s from the measured values of k_g and k_{ol}. The k_s values obtained compared quite favorably with the literature values, thereby sug-

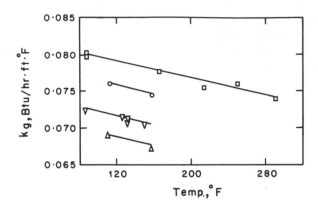

FIG. 22. Typical data for grease and oil conductivities. Symbols: 13.8% calcium soap grease (□), oil for 13.8% soap grease (○), 9.5% calcium soap grease (▽), oil for 9.5% soap grease (△) (from Hamilton and Crosser [31]).

gesting the appropriateness of Eq. (77) as a first approximation for estimating thermal conductivity of greases.

Greases, in their undeformed relaxed state, are isotropic in nature. When subjected to deformation, however, alteration in structure occurs due to orientation of the dispersed soap particles. As a result, the medium becomes anisotropic and so does its thermal conductivity. Shulman *et al.* [56] have experimentally observed this behavior and have shown that the thermal conductivities of sheared greases are about 30% higher and 10% lower than that of the undeformed greases when measured in directions parallel and transverse to the direction of the applied shear field, respectively. Experimental data were found to obey the following relationship between isotropic (k_0) and anisotropic ($k_\parallel$, $k_\perp$) thermal conductivities:

$$3/k_0 = 1/k_\parallel + 2/k_\perp \qquad (78)$$

Incidentally, Eq. (78) has also been used quite successfully in describing heat-conduction anisotropy of linear polymers oriented by stretching [101–103].

With a specific view to examining non-Newtonian food materials (like mayonnaise, apple sauce, banana puree, ammonium alginate solutions, etc.), Charm and Merrill [33] conducted experiments to determine heat transfer coefficients in tubular flow. A split bar technique was used to measure the thermal conductivity of these materials. Unfortunately, however, no details regarding thermal conductivity data were presented even though the heat-transfer coefficients predicted from the Sieder–Tate equation were within ±24% of the experimental values. To date, no effort at relating the structure of these food materials to their effective thermal conductivity has been made. Undoubtedly, such efforts, both theoretical as well as experimental, would be necessary in order to develop acceptable predictive capability as regards the thermal conductivity of these materials, thereby avoiding detailed experimentation whenever such information is necessary for routine heat-transfer calculations.

VI. Polymeric Liquids: Single-Component Systems

Liquid polymers and (unfilled) polymer melts are typical examples of single-component polymeric liquids, and we shall focus attention on thermal conductivity of these liquids in this section. The bulk of the information available in the literature pertains to polymer melts, and hence emphasis will be primarily on thermal conductivity of molten polymers. This, perhaps, is somewhat natural because of the relative practical importance of polymer melts and also because of the important role that melt thermal conductivity plays in common polymer-forming operations. Before proceeding any further, however, it is necessary to clarify the term polymer melt as it will be used here. This is particularly relevant for thermoplastic polymers, which can be either amorphous or semicrystalline in nature. For amorphous polymers (like polystyrene, polyvinylchloride, polymethylmethacrylate, etc.) there is no sharp melting point T_m; instead the polymer gradually softens as the temperature T exceeds the glass transition temperature T_g. On the other hand, semicrystalline polymers (like polyethylene, isotactic polypropylene, nylon, etc.), exhibit a relatively sharp melting point and the polymer is in its molten form for $T > T_m$. In the following, we shall use the term polymer melt for amorphous polymers when $T > T_g$ and for semicrystalline polymers when $T > T_m$.

Thermal conductivity of single-component polymeric liquids depends on several factors such as temperature, molecular weight and its distribution, chain branching, strain rate, etc. Several attempts [104, 105] at explaining these effects based on theoretical considerations have been made, most of which, however, pertain to solid polymers. These theories, developed about two decades ago, have been reviewed earlier by Knappe [6] and Hands [7] and, therefore, will not be discussed here in detail. Instead, we shall focus our attention on some recent developments in this area.

A. EFFECT OF TEMPERATURE

Several attempts to explain the temperature dependence of the thermal conductivity of polymers on a theoretical basis have been made in the literature. Überreiter and Nens [106] assumed that thermal energy is transferred along the length of the polymer chain and the transverse waves set up above T_g tend to dampen the process of heat transfer resulting in decreased conductivity above the glass transition temperature T_g. A slightly different approach was adopted by Eiermann [3]. He modeled the amorphous polymer as a network of resistances in which the primary valence forces offered a negligible resistance to transfer of thermal energy in comparison to that offered by the secondary valence forces.

A different approach was adopted by Sheldon and Lane [107] who considered thermal conduction in polymers to be a result of molecule-to-molecule transfer of energy by translational, rotational, or vibrational modes. These are basically diffusion processes and, thus, with an increase in temperature, the density of the polymers decreases whereas the segmental mobility increases. The observed temperature dependence of thermal conductivity would therefore depend on the relative magnitudes of these two factors. Similar views were expressed in subsequent studies on thermal conductivity of polymer melts [42, 43]. These theories are suggestive of the fact that segmental motion might play an important role in thermal conduction in polymers.

1. WLF Correlation

Primarily, two parameters control the role of segmental motion in transport properties of polymeric liquids. These are the available free volume and the ease of rotation around the carbon–carbon bond. This realization prompted Kulkarni and Mashelkar [108] to argue that, similar to viscosity and diffusivity, it should be possible to correlate the thermal conductivity data within a free volume framework. According to the free volume model, Mashelkar and Kulkarni [121] suggested the following equation, analogous to the Doolittle equation for viscosity:

$$k = A_k \exp(B_k/f) \qquad (79)$$

where f is the fractional free volume and A_k and B_k are constants. Assuming a linear dependence of free volume on temperature [124]

$$f(T) = f(T_{ref}) + \alpha_k(T - T_{ref}) \qquad (80)$$

and combining Eqs. (79) and (80), it was shown that the temperature dependence of thermal conductivity can be described by the equation

$$\ln \frac{k(T)}{k(T_{ref})} = \frac{-(B_k/f_r)(T - T_{ref})}{(f_r/\alpha_k) + (T - T_{ref})} \qquad (81)$$

where f_r is the fractional free volume at the reference temperature T_{ref}, and

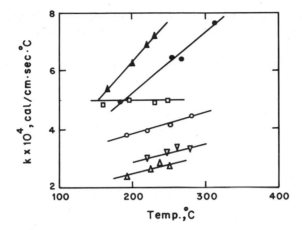

FIG. 23. Thermal conductivity of polymer melts. Symbols: Polystyrene (O), nylon 6 (∇), nylon 6,10 (△), linear polyethylene (▲), branched polyethylene (●), polypropylene (□) (from Fuller and Fricke [43]).

α_k is the expansion coefficient. Note that Eq. (81) is quite similar to the well known WLF equation. For polymer melts, Kulkarni and Mashelkar [108] have taken $T_{ref} = T_g$ in order to correlate the thermal conductivity data for a given melt at different temperatures. That is,

$$\ln \frac{k(T)}{k(T_g)} = \frac{-C_1(T - T_g)}{C_2 + (T - T_g)} \tag{82}$$

where C_1 and C_2 are constants. Based on the data presented in Fig. 23, Fig.

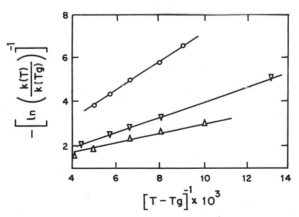

FIG. 24. WLF plot for thermal conductivity of polymer melts; legend is same as in Fig. 23 (from Kulkarni and Mashelkar [108]).

24 illustrates a typical WLF type plot for polymer melts. It is clear that Eq. (82) fits the data well and can be used to determine the melt conductivity k at any temperature T, provided the constants C_1 and C_2, the glass transition temperature T_g, and the conductivity $k(T_g)$ at T_g are known. Kulkarni and Mashelkar conjecture that, as in the case of the WLF correlation for other transport properties, it should be possible to correlate the constants C_1 and C_2 with the microcharacteristics of the system. Thus, $C_1 = B/2.303f_g$ and $C_2 = f_g/\alpha_f$ where f_g denotes the free volume at T_g, α_f is the coefficient of thermal expansion of free volume about T_g and B is a parameter in the Doolittle equation that is close to unity. Further validation of such ideas appears to be difficult since the available conductivity data on polymeric liquids are insufficient for developing any meaningful correlation.

2. Modified Bridgman Equation

Kraybill [109] has proposed a different procedure for predicting the temperature dependence of thermal conductivity of polymer melts. The approach is based on the Bridgman theory [104, 110] of energy transport in pure liquids. The Bridgman model assumes that molecular energy is transferred at a velocity equal to the local sonic velocity, and thus

$$k = 2.8 \times 10^{-5}(N_{av}/\overline{V})^{2/3}K_B v_s \quad \text{W/m K} \tag{83}$$

where N_{av} is the Avogadro number, $\overline{V}$ the volume of a single interaction unit, K_B the Boltzmann constant, and v_s the sonic velocity. The velocity of low-frequency sound is given as [110]

$$v_s = \left[\frac{C_p}{C_v}\left(\frac{\partial P}{\partial \rho}\right)_T\right]^{1/2} \quad \text{cm/sec} \tag{84}$$

where

$$C_v = C_p - T\,\alpha_c^2/\beta_c\rho \tag{85}$$

For polymers, the partial derivative $(\partial P/\partial \rho)_T$ can be obtained from isothermal compressibility data as compiled by Bernhardt [111] or from an equation of state as proposed by Spencer and Gilmore [112]

$$(v - b_i)(P + \pi_i) = RT/M = R'T \tag{86}$$

In Eqs. (84) and (85), α_c is the coefficient of volume expansion $= 1/v$ $(\partial v/\partial T)_P$, β_c the compressibility $= -(1/v)(\partial v/\partial P)_T$, v the specific volume, and π_i and b_i constants to be determined experimentally. Equations (83) and (86) have been used to calculate the thermal conductivity of polyethylene resins at different temperatures. It was found that Eq. (83) predicts the melt thermal conductivity with reasonable accuracy. However, whether this approach will also be valid for other polymers as well is presently not clear.

3. *Experimental Observations*

In general, data on thermal conductivity of polymeric liquids are relatively few and usually there is poor agreement between the data from different sources. This is especially true in relation to the temperature dependence of thermal conductivity. Fuller and Fricke [43] reported thermal conductivities of several amorphous and semicrystalline polymers (see Fig. 23) and similar data were also presented by Ramsey *et al.* [42] and Hands and Horsfall [28] as shown in Fig. 25 and 26, respectively. Considerable discrepancies between these data are apparent. It can be seen that the nature of the temperature dependence not only varies from polymer to polymer owing to the difference in segmental mobility of the chains but also differs for a given generic type of polymer. For example, in the case of molten polyethylene, thermal conductivity increases with temperature (Fig. 23), it increases and then decreases with temperature (Fig. 25), and it may decrease and then increase with temperature (Fig. 26). Such apparently erratic behavior is likely due to the differences in the molecular structure of the macromolecules.

It therefore appears that the temperature effect on the thermal conductivity of the polymer melts could perhaps only be qualitatively predicted. Ideally, conductivity should reduce with decreasing density (that is, in-

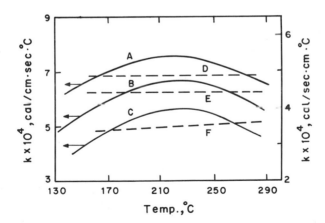

FIG. 25. Thermal conductivity of polyethylene (———) and propylene (---------) melts as a function of temperature. $M_w \times 10^4$: (A) 24.76, (B) 28, (C) 38.3, (D) 28.5, (E) 54, and (F) 47.5. M_w/M_n: (A) 23.57, (B) 23.17, (C) 27.68, (D) 8.2, (E) 7.76, and (F) 6.73.

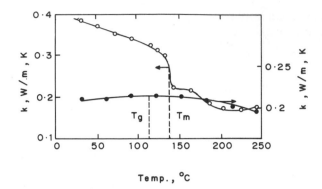

Temp., °C

FIG. 26. Thermal conductivity of semicrystalline polyethylene (O) and amorphous poly-methylmethacrylate (●) as functions of temperature (from Hands and Horsfall [28]).

creasing temperature). However, segmental mobility is also enhanced with increased temperature, which can result in increased conductivity. Generally these competing influences are nonlinear functions of temperature and thus the dependence on temperature may take various forms. As a result, *a priori* prediction of temperature dependence of thermal conductivity has still remained an elusive problem.

In the case of liquid polymers, data are available only for highly viscous liquid silicones of polydimethylsiloxane (PDMS) polymers. Bates [32] and Boggs and Sibbitt [27] have measured thermal conductivity of these liquids at different temperatures. The data suggest that conductivity is a linearly decreasing function of temperature. No quantitative theoretical interpretation of thermal conduction in liquid polymers, however, has yet been attempted.

B. EFFECT OF MOLECULAR CHARACTERISTICS

Molecular characteristics, such as type of monomer unit, molecular weight, molecular weight distribution, degree of chain branching, etc. exhibit profound influence on the thermal conductivity of polymeric liquids. By increasing the size of any substituent unit connected to any polymer backbone, disorder in the structure is promoted, and this is manifested as reduced conductivity. This is evident from the data presented in Fig. 23, which show that conductivity decreases in the following order: linear polyethylene > branched polyethylene > polypropylene > polystyrene > nylon 6 ≈ nylon 6, 10 [43]. This comparison, however, is only qualitative since the molecular weights are not identical and, as we shall discuss shortly, thermal conductivity varies rather significantly with molecular weight.

It is well known that energy transfer within a molecule is much more effective than that between molecules. Since increasing the molecular weight decreases the number of chain ends, thermal conductivity should increase with increasing molecular weight. However, because the coefficient of energy transfer along a chain is finite, the effect of molecular weight will be limited, and beyond a certain molecular weight (which varies from polymer to polymer) there will be no further increase in conductivity. Thus it is expected that polymer melt conductivity will increase with molecular weight at low molecular weights and eventually reach "saturation" or a limiting value [7]. Indeed such a behavior has been observed for thermal conductivity of different molecular-weight liquid polymers like PDMS [32].

Several efforts at predicting the molecular-weight dependence of thermal conductivity of polymers were undertaken in the past [3, 113, 114], among which the theory developed by Hansen and Ho [114] appears to be most successful. Based on a freely jointed–freely rotating model for the polymer molecules, they showed that the thermal conductivity k is proportional to one-half power of the molecular weight M_w at low molecular weights with the effect of M_w on k becoming negligible at high molecular weights. Figure 27 illustrates the qualitative comparison between the Hansen–Ho theory

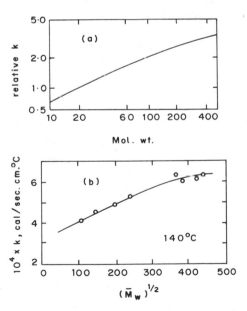

FIG. 27. (a) Calculated dependence of thermal conductivity on molecular weight as predicted by Hansen and Ho theory. (b) Thermal conductivity of molten polyethylene as a function of molecular weight. (From Hansen and Ho [114].)

and the experimental data. The data indicate a one-half power dependence of thermal conductivity on molecular weight, and this effect becomes negligible for $M_w > 90,000$. The experimental data, however, have been questioned [43] since different molecular weight samples were obtained by blending low-molecular-weight wax with polyethylene, but the accompanying variation in the molecular-weight distribution (MWD) has not been accounted for. Lohe [115] determined the effect of M_w on the thermal conductivity of polyethylene glycol for different degrees of polymerization. For this relatively low-molecular-weight material with a rather narrow MWD, conductivity increased with increasing molecular weight. Ramsey et al. [42] also experimentally investigated the molecular-weight dependence of thermal conductivity of polymer melts. Melt thermal conductivity is found to monotonically decrease with M_w for polyethylene and to exhibit a minimum in the $k-M_w$ relationship for polypropylene. This certainly is contrary to the expected behavior and no satisfactory explanation for this discrepancy is presently available.

Regarding the influence of MWD, it is reasonable to assume that, for samples of the same M_w, those with broader distribution will have the lower conductivity since the greater number of chain ends would increase the free volume. Thus, even if conductivity is expected to increase with M_w it is also anticipated that it will decrease with MWD, and hence k will be effectively governed by both M_w and M_n [28]. Indeed, Hands [116] showed that polymer conductivity is linearly related to the product $M_w M_n$. Similarly, melt conductivity of a branched polymer is expected to be lower than that of an unbranched polymer of the same molecular weights. This can be predicted from Hansen–Ho theory [117], and this behavior has been experimentally observed [43, 118] for polyethylene melts (see Fig. 23).

The above discussion shows that a number of gaps exist in our understanding of the dependence of the thermal conductivity of polymer melts on its molecular parameters. Analysis of the limited data available indicates that the thermal conductivity is likely to be a complex function of molecular parameters, and carefully controlled experimentation will be necessary in order to elucidate the effect of each individual parameter.

C. STRAIN-RATE EFFECT

Earlier discussions (in Sections IV and V) suggest that for structured liquids, imposition of a deformation field alters the structural state of the liquid and hence its thermal conductivity. As a result, thermal conductivity becomes strain-rate dependent as well as anisotropic. Owing to their structured nature, polymeric liquids are also expected to show this trend; the only studies that deal with this effect are by Picot and co-workers [24, 40, 44, 154].

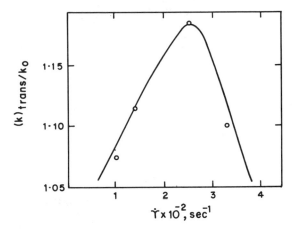

FIG. 28. Variation of transverse thermal conductivity with rate of shear for Dow 200 silicone liquid (from Chitrangad and Picot [40].)

Cocci and Picot [24] and Chitrangad and Picot [40] investigated the shear-rate effect on the thermal conductivity of liquid polymers, particularly PDMS. They observed that at low shear rates, the effective thermal conductivity ($k_\perp$, transverse to the shear direction) increases with increasing shear rate. Although no such effects were observed with low-molecular-weight liquids like toluene and polybutene ($M_w \sim 920$), the effect was more or less independent of molecular weight for the silicone liquids and became more predominant at lower temperatures [24]. Beyond a critical shear rate ($\sim 300/\mathrm{sec}$), however, the thermal conductivity of the silicone liquids decreased with further increase in shear rate, as shown in Figure 28. Similar observations were also reported by Picot et al. [44] for the effect of shear rate on k of molten polyethylene at 150°C. A tentative explanation of the observed behavior has been provided by Picot et al. [40, 44]. They maintained that, at low shear rates the liquid behaves as a nearly Newtonian liquid and orientation in the 45° (to the flow) direction is enhanced. This contributes to orientation in a direction transverse to the flow direction, and k increases. As strain rate increases, the molecular entanglements start to play a dominant role and the preferred orientation of the polymer chains is in a direction parallel to that of the flow, and naturally the transverse conductivity decreases.

More recent experimental observations from Picot's group, however, are quite different from their earlier results just discussed. In particular, Walace et al. [154] has reported results regarding shear dependence of the thermal conductivity of molten polyethylenes of two different molecular weights. As shown in Fig. 29, the thermal conductivity of the lower-molec-

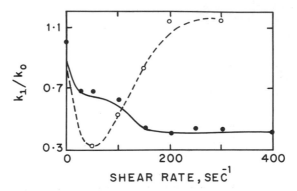

FIG. 29. Normalized thermal conductivity of molten polyethylene at 160°C as a function of shear rate; molecular weight: 29,423 (O) and 2,910 (●) (from Wallace *et al.* [154]).

ular-weight polyethylene decreased rapidly (up to 55%) with shear rate up to 150/sec and thereafter remained more or less the same. In contrast, for the higher molecular weight grade, conductivity decreased by a similar amount, reached a minimum at 50/sec and then increased with shear rate. Wallace *et al.* provided some tentative mechanistic interpretation for the observed behavior. They argued that the conductivity decrease at low shear rates is due to an orientation effect, whereas the rise in conductivity at higher shear rates for the higher-molecular-weight polyethylene is attributable to the formation of rotating units of entangled clusters. It should be mentioned that the recent findings of these authors for the same generic materials are at variance with their previous findings, and there is at present no explanation for this difference.

The work by Picot *et al.* nevertheless, demonstrates that for flowing polymer melts thermal conductivity is strain dependent, although it does not provide any idea regarding the degree of strain-induced anisotropy of the medium. For that, we believe, thermal conductivity in the flow direction needs to be determined also. As in the case of orientation effects in solid polymers [101–103], the latter conductivity ratio is expected to be greater than unity and an increasing function of shear rate. Since, in most polymer processing operations, the polymer melt is subjected to a deformation field, rigorous heat transfer analyses of these processes necessitate a detailed understanding of the melt thermal conductivity dependence on the deformation rate. Such an understanding is presently lacking and studies to resolve this deserve considerable attention in order to assess the relevance of the strain-induced anisotropy in thermal conductivity of polymer melts.

VII. Polymeric Liquids: Multicomponent Systems

In the foregoing, we have considered the thermal conductivity of polymeric liquids comprising a single component possessing a certain degree of order. Apart from these, polymeric liquids consisting of more than one component are also encountered frequently. Polymer solutions, polymer melt formulations, filled/foamed polymer melts, elastomeric masterbatches, etc. are typical examples of such systems. For convenience, we shall refer to these liquid systems as multicomponent polymeric systems. Polymers, by themselves, possess a degree of structure and when additional phases are introduced, complex alterations of this structure are expected that should be manifested in changes in transport coefficients. In the following, we shall discuss such alterations, but our attention will be restricted to the thermal conductivity of two major classes of heterogeneous polymeric liquid systems: polymer solutions and filled polymeric liquids, for which at least a certain amount of information is available.

A. POLYMER SOLUTIONS

Estimation of thermal conductivity of polymer solutions is important in processes like free coating, dry spinning, solvent casting, etc., and hence a fundamental understanding of this property is of considerable pragmatic importance. To date, very little work has been done in this direction, and only recently efforts at developing some theoretical understanding of the thermal conduction phenomena in polymer solutions have been initiated.

1. Experimental Observations

In the literature, experimental studies of thermal conductivity of polymer solutions are few and those that exist fail to provide a clear picture of the conductivity behavior. This is particularly true as regards the effect of polymer concentration on thermal conductivity. Data obtained by Boggs and Sibbitt [27] for aqueous gelatin and methocel solutions and by Bellet *et al.* [119] for aqueous carbopol and carboxymethyl cellulose solutions are shown in Figs. 30 and 31, respectively. It is seen that the thermal conductivity is an increasing function of temperature and a decreasing function of polymer concentration. However, it appears that the polymer concentration is far more significant than that of temperature in controlling conductivity of polymer solutions. Investigations by Lee *et al.* [29] for aqueous solutions of six different polymers and by Brunson [34] for polystyrene

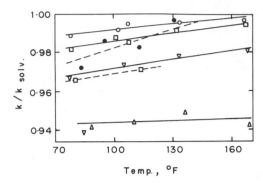

Fig. 30. Normalized thermal conductivity of aqueous polymer solutions as a function of temperature. Solid lines and open symbols are for gelatin solutions and broken line and closed symbols are for methocel solutions. Polymer concentration: 1 (O), 2 (□, ●), 5 (▽, ■), 10 (△, ▼) wt % (from Boggs and Sibbitt [27]).

solutions in toluene provide results that are somewhat different from the above. Although there is no disagreement regarding the temperature effect, they found that the thermal conductivity of the solutions were enhanced due to an increase in polymer concentration (see Fig. 32).

Presently, the reasons for discrepancies between different observations are not clear. However, it may be noted that the studies reported by Lee *et al.* and Brunson are for dilute solutions (polymer concentrations less than one weight percent), whereas those by Boggs and Sibbitt and Bellet *et al.* involve polymer concentrations ranging from 1 to 20 wt %. In view of the sudden change in the nature of the concentration dependence, it is quite likely that the solution conductivity is a net result of two or more competing effects. The observed change in the nature of the concentration dependence may be due to one or the other of the influences dominating depending on whether the solution is dilute or concentrated. This certainly is speculative and a careful study is necessary for a better understanding of the reasons behind these discrepancies.

Apart from polymer concentration and temperature, the polymer molecular weight, molecular-weight distribution, nature of the monomer units, degree of branching, etc. also affect thermal conductivity of polymer solutions, but information regarding these aspects is virtually nonexistent. Aside from this, most industrial operations involve polymer solutions under flow conditions, and since for most structured liquids thermal conductivity is strain-rate dependent, it is also necessary to understand the role of strain rate on the thermal conductivity. In doing so, however, care must be taken to avoid possible falsification by stress-induced migration effects as discussed by Mashelkar and Dutta [98, 120].

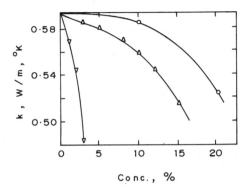

FIG. 31. Effect of concentration on thermal conductivity of polymer solutions. Symbols: CMC, 25°C (O), CMC, 20°C (△), Carbopol-960, 20°C (▽) (from Bellet *et al.* [119]).

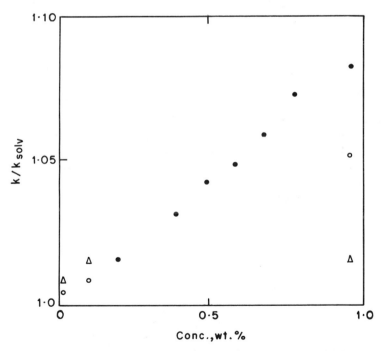

FIG. 32. Effect of concentration on thermal conductivity of polymer solutions. Symbols: PS-toluene at 30°C (●), Carbopol-960 in water (O), polyethylene oxide in water (△) (after Lee *et al.* [29] and Brunson [34]).

2. *Altered Free-Volume State Model*

Presently, no acceptable theoretical basis to interpret the limited thermal conductivity data in polymer solutions is available. Recently, however, Kulkarni and Mashelkar have developed a unified altered free-volume state (AFVS) model for analysis and correlation of a large variety of transport properties in polymeric media [121–123]. This unified framework has been used to explain thermal conduction in polymer solutions, plasticized polymers, semicrystalline polymers, and filled polymers [121]. In the present context, the central concept of the AFVS model is that the free-volume state of the medium is the key factor in determining its thermal conductivity. If a given parent medium has a particular state of free volume, this state gets altered by either physical or chemical modifications or by the change in the state variables such as temperature, pressure, etc. In the case of polymer solutions, such modification could also be done through changes in the polymer concentration. As a result the polymer solution will now be in an altered free-volume state as compared to that of the solvent. The key step then is the calculation of the alteration of the free-volume state of the solvent with respect to a carefully defined reference state.

For a given solvent, reduction in its free volume occurs due to addition of a polymer, which in turn increases the thermal conductivity of the solution. This happens because the monomers constituting the polymer are linked together by primary valence bonds, whereas the solvent molecules exert only secondary valence forces, and it is well known (see [3]) that the secondary valence forces offer a higher resistance to thermal conduction than the primary valence forces. Thus, the thermal conductivity of a polymer solution is expected to increase with polymer concentration. Quantitatively, Mashelkar and Kulkarni [121] show that if k and k_{solv} are the thermal conductivities of the solution and pure solvent, respectively, then the following relationship holds

$$\left[\ln\left(\frac{k}{k_{\text{solv}}}\right)\right]^{-1} = \frac{f_0^2}{B_k\delta_c}\frac{1}{\phi_p} + \frac{f_0}{B_k} \tag{87}$$

where f_0 is the fractional free volume of the pure solvent, δ_c the difference in fractional free volume of the polymer and the solvent, ϕ_p the polymer concentration, and B_k is a constant in the constitutive relation (79). Equation (87) suggests a linear relationship between $[\ln(k/k_{\text{solv}})]^{-1}$ and ϕ_p^{-1} and, as shown in Fig. 33, this prediction has been found to be true for experimental data reported by Brunson [34].

The AFVS model implies that the thermal conductivity of polymer solutions increases with polymer concentration. Available data indicate

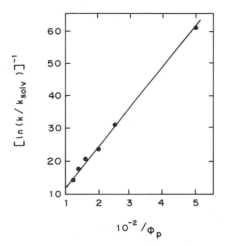

FIG. 33. Comparison of AFVS model with experimental data (from Mashelkar and Kulkarni [121]).

that this indeed is true only for dilute solutions (concentrations less than 1 wt %) and it appears that the model is valid only for dilute systems. Certainly, comparison with more experimental data will be required even to establish this fact. In case of relatively concentrated solutions, however, experimental data suggest that solution conductivity is a decreasing function of concentration. This is contrary to AFVS prediction and it is quite likely that the AFVS model is an oversimplification of the complex interaction process that may be present in this case. Nevertheless, the AFVS model provides a worthwhile starting point for analyzing thermal conductivity of polymeric liquids, and further work along these lines may be fruitful.

B. FILLED POLYMERIC SYSTEMS

In certain applications, polymer melts are dispersed with fillers in order to serve the dual purpose of cost dilution and property improvement. These fillers can be particulate (in the case of typical filled systems), other polymer melts (in the case of polymer blends), or gaseous inclusions (as in polymer foams). Expectedly, thermal conductivity of these multicomponent systems depends largely on the individual conductivities of the filler and the polymer matrix along with the filler loading.

In general, filled polymeric liquids are highly viscous suspensions, and only a few limited theoretical developments have been along lines similar to the suspension theories discussed in Section IV,A,1. The review article by Progelhof *et al.* [125] summarizes and compares a number of theoretical and empirical models for predicting thermal conductivities of mixtures of polymers and fillers. For low filler loading (ϕ_f) and low conductivity ratio (k_f/k_{pm}), a modified form of Maxwell's expression can be used to predict thermal conductivity of the filled system. The particular expression is

$$k_{pf} = k_{pm} \left[\frac{2k_{pm} + k_f + 2x_f(k_f - k_{pm})}{2k_{pm} + k_f - x_f(k_f - k_{pm})} \right] \qquad (88)$$

where

$$x_f = (\rho_{pf} - \rho_{pm})/(\rho_f - \rho_{pm}) \qquad (89)$$

Equation (88) has been used successfully to correlate the effect of MgO powder and glass spheres on the thermal conductivity of polyethylene. This equation is likely to be appropriate only for low ϕ_f ($< 10\%$) and for a k_f/k_{pm} ratio not very much greater than unity. In fact, Kusy and Corneliussen [126] show that, although Eq. (88) is adequate for nickel fillers for $\phi_f < 3\%$, it does not work equally well for more highly conducting copper fillers. All these data, however, are for solid polymers. For molten filled polymers, Kraybill [109] suggests the use of Eq. (88), with k_{pm} obtained from Eq. (83). Indeed, such a predictive approach was shown to work adequately well for polyethylene melts containing 10 wt % of TiO_2. Similarly, Sundstrom and Lee [127] studied the thermal conductivities of polystyrene and polyethylene containing several different particulate solids over a range of concentrations. In all cases, k_f/k_{pm} was greater than unity and the thermal conductivity of the filled system increased with the filler loading. The experimental data were compared with the results predicted by theoretical models for two-phase media. It was found that for $0.15 \leq \phi_f \leq 0.3$ the equations of Bruggeman [67] and Cheng and Vachon [76], as provided in Tables II and III, both gave better agreement with the measured values than that provided by the Maxwell equation. For $\phi_f \leq 0.10$, however, the Maxwell equation yielded the best agreement.

In a separate study, Shulman *et al.* [46] theoretically derived the effective thermal conductivity of a medium with anisodiametric particles. For n types of randomly oriented particles, the result obtained was

$$\frac{k_e - k_b}{k_e + 2k_b} = \sum_{j=1}^{n} \frac{4}{9} I_j \phi_j \qquad (90)$$

where k_e is the effective thermal conductivity of the filled system, k_b the binder conductivity, I_j the first invariant of the particle polarizability tensor, and ϕ_j the volume concentration of the jth inclusion. In Eq. (90),

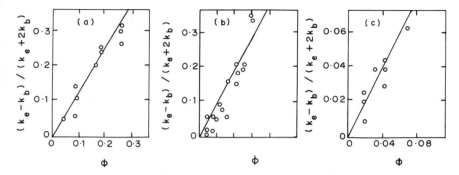

FIG. 34. Dependence of $(k_e - k_b)/(k_e + 2k_b)$ on the volume fraction of filler, ϕ. Filler: (a) magnesium, $\Lambda \approx 1.294$; (b) aluminum–magnesium, $\Lambda \approx 0.953$; and (c) γ-iron oxide, $\Lambda \approx 1.000$ (from Shulman et al. [46]).

$I_j > 0$ for $k_d > k_b$; $I_j < 0$ for $k_d < k_b$. When $k_d \gg k_b$ or $k_d \ll k_b$, $|I_j|$ is independent of k_d and is determined solely by the shape of inclusions. For a single type of inclusion ($n = 1$), Eq. (91) simplifies to a Maxwell-type equation:

$$\frac{k_e - k_b}{k_e + 2k_b} = \Lambda\phi \tag{91}$$

where ϕ is the particle volume fraction and $\Lambda = 4\pi I/9$ is a constant depending on particle shape and size distribution. Shulman et al. verified Eq. (7.5) for dispersion of several highly conducting (metallic) powders in polyisobutylene solutions. Their results, shown in Fig. 34, indicate that Eq. (91) is indeed satisfactory for different filler systems. The value of Λ, however, is dependent on the filler type. This is expected because each filler type has its own shape and size distribution and these two factors are predicted to affect Λ even if Λ is independent of k_d for $k_d \gg k_b$.

A somewhat different approach based on the AFVS model (see Section VII,A,2), has been proposed by Mashelkar and Kulkarni [121] for interpreting thermal conductivity in filled polymers. They maintained that fillers do not contribute to the free volume of the pure polymer matrix but act as more conductive inclusions. This physical obstruction effect reduces the free volume and the net result is an increase in thermal conductivity with increased filler content. Based on arguments similar to those leading to Eq. (87), Mashelkar and Kulkarni suggest the following expression for correlating thermal conductivity in filled polymer (k_f) with respect to that in the unfilled polymer (k_{pm})

$$\left[\ln\left(\frac{k_f}{k_{pm}}\right)\right]^{-1} = \frac{f_0^2}{B_k \delta_{cf}} \frac{1}{\phi_f} + \frac{f_0}{B_k} \tag{92}$$

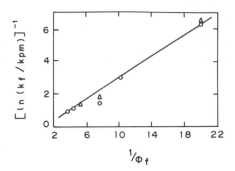

FIG. 35. Thermal conductivity of polystyrene as a function of CaO (O) and MgO (△) volume fraction (from Mashelkar and Kulkarni [121]).

where f_0 is the fractional free volume of the polymer matrix, δ_{cf} the difference between the free volume fraction of the unfilled polymer and the filler, and ϕ_f is the filler volume fraction. Figure 35 shows the verification of Eq. (92) for thermal conductivity of polystyrene–calcium oxide and polystyrene–magnesium oxide filled systems at about 32°C. Interestingly, the data for the two fillers coincide, thereby suggesting the possibility of an unique relationship between conductivity ratio and filler loading for a given polymer.

In the open literature, the experimental data for thermal conductivity of filled solid polymers are rather scant and those for filled polymer melts are practically nonexistent. As a result the semitheoretical developments discussed above pertain essentially to the solid systems. However, since most of these theoretical approaches deal with the effect of filler on thermal conductivity in relative terms, it is quite likely that with minor modifications they could also apply to the liquid systems. Aside from this, these developments are limited to rather low filler loadings ($\phi < 0.25$). Regarding highly filled polymeric systems, there is very little theoretical understanding and few experimental data are available. Highly filled polymeric liquids are very similar to granular media held together by a (polymeric) liquid binder. Recently, considerable progress in developing a theoretical framework for analyzing thermal conduction phenomena in such systems has been made by several research groups [39, 144, 148], and this may be suitably modified to develop a theoretical understanding of the thermal conductivity of highly filled systems.

VIII. Concluding Remarks

In the foregoing, an attempt has been made for the first time to review the current state of development in the general field of the thermal conductivity of structured liquids. Although such developments are significant and definitive conclusions in certain areas can be arrived at, there still exist a vast number of areas where additional work needs to be undertaken. In the following, we shall summarize some of these areas that appear to be both interesting and challenging from pragmatic as well as academic viewpoints.

As we have seen, the term "structured" liquid encompasses a wide variety of both single and multicomponent polymeric as well as nonpolymeric liquids. The conductivity depends not only on the liquid composition and the individual component conductivities but also on the details of the microstructural states of these liquids. Thus, the role of various factors (such as temperature, composition, deformation rate, etc.) in controlling the fluid microstructure has to be appreciated. It appears that the work done to date is not totally adequate for such an understanding. Figure 36 summarizes the current state of development for some key structured liquids. In most cases, the state of development is rather unsatisfactory.

The work summarized in Section IV,A indicates that thermal conductivity of stagnant suspensions, both solid–liquid and liquid–liquid, is relatively well understood. Many expressions, such as those listed in Tables II and III, have been derived to predict the thermal conductivity for different regularly shaped inclusions of various sizes and for a wide range of the conductivity ratio K_d. Sufficient experimental data are also available for these systems, and comparison with theoretical predictions suggests that the theory is quite adequate up to moderately concentrated dispersions. In the case of highly concentrated dispersions of irregularly shaped particles, these expressions are generally not adequate and only empirical equations are now available for the purpose of prediction. From theoretical consider-

SYSTEM	STAGNANT MEDIA		FLOWING MEDIA	
	DATA	THEORY	DATA	THEORY
SUSPENSIONS	S	G	U	U→S
POLYMER SOLUTIONS	U→S	U	U	U
POLYMER MELTS	S	U	U	U
G – GOOD , S – SATISFACTORY , U – UNSATISFACTORY				

FIG. 36. Current status of the understanding of thermal conductivity of structured liquids.

ations, it has been possible only to derive the bounds on the thermal conductivity of such systems.

The above expressions, developed over a period of more than a century, are essentially first- or second-order theories neglecting the general anisotropy of thermal conductivity of such systems. Only during the last decade have efforts at developing a generalized framework for thermal conductivity in heterogeneous media been initiated. Some progress has been made, but considerable work still needs to be done in order to make the theoretical treatment applicable to systems of practical interest. Presently, only dilute systems have been analyzed using this framework, whereas from a pragmatic viewpoint, concentrated dispersions are of equal interest. As an example, for dilute dispersions of highly conducting slender inclusions, accounting only for the interaction between two adjacent particles yields an improved prediction of A_2, the second-order coefficient of the thermal conductivity expression Eq. (53). Hence, it is expected that further work dealing with more concentrated dispersions must account for the particle interaction between pairs, triplets, quadruplets, etc. Again, for the same system, even though theory and experimental data agree quite well for the longitudinal conductivity, the situation is not as satisfactory in the case of conductivity. It is not clear whether such discrepancies are due to problems involved in experimentation or some inadequacy in the theoretical treatment. However, the generalized framework has been applied only to uniformly sized arbitrarily shaped particles, whereas most practical situations involve particles with a nonuniform size distribution.

In comparison to stationary suspensions, thermal conduction in flowing suspensions has received relatively little attention both theoretically as well as experimentally (see Section IV,B). Only recently has an analytical treatment relating the effective bulk conductive heat flux of a deforming suspension to the microscale velocity and temperature fields been available, which in principle allows evaluation of the thermal conductivity of the medium. Using this analysis, it has been possible to investigate few limiting cases of shear-rate effects on the thermal conductivity of very dilute dispersions subjected to simple shear flow. In particular, the limits of very high and very low Peclet numbers were analyzed along with a preliminary analysis of the role of the dispersed phase deformation and Brownian motion. Comparison between theory and experimental data for sheared dilute suspensions of rigid particles at very low Peclet numbers, suggests that the theory is only adequate for order of magnitude estimates of the thermal conductivity. This valuable first effort towards developing an understanding of the complex process of thermal conduction in deforming suspensions is to be developed further so that it can be applied to situations of practical interest, especially more concentrated dispersions and fields with higher shear rates (i.e., higher Peclet numbers). Aside from this, the effect of deformation modes other than shear on the suspension conductivity also needs to be investigated. Experimentally, with the sole exception of the work by Chung and Leal [77], practically no quantitative data regarding flow effects on thermal conductivity of suspensions is available. Other than suspensions, thermal conductivity of other nonpolymeric structured liquids has been discussed briefly in Section V. In the literature, these have been treated as isolated areas. This may perhaps be justified, since each individual class of liquids has a physical and structural peculiarity of its own. Since such systems are structurally quite similar to suspensions, it would be worthwhile to explore the possibility of analyzing the available data for these liquids within the framework of the suspension theories both in stagnant and deforming media.

Thermal conductivity of single-component polymeric liquids and unfilled polymer melts has been discussed in Section VI. Because of its pragmatic importance in polymer processing operations, most of the work reported in the literature pertains to polymer melts. The essential features of the thermal conduction process in liquid polymers, however, appear to be quite similar to those in polymer melts. For polymer melts, considerable effort has been devoted toward understanding and predicting temperature dependence of thermal conductivity. Bridgman's method for ordinary liquids has been modified by Kraybill [109] in order to estimate the temperature dependence of thermal conductivity of polyethylene, but whether such an approach works equally well for other polymers is still to be investigated. Similarly, WLF correlation proposed by Kulkarni and Mashelkar [108] is adequate only if the conductivity at T_g is known; the latter quantity is not readily available and development of a procedure for its determination would be desirable. The effect of molecular weight on thermal conductivity of polymers appears to be relatively well understood, but presently there exists practically no quantitative relationship between the other molecular characteristics (like molecular-weight distribution, degree of branching, etc.) and thermal conductivity. For flowing polymer melts, the deformation rate is an additional factor that controls effective thermal conductivity by introducing anisotropy. Presently, very little information, with the exception of the studies by Picot et al. [44, 154] are available.

In the case of polymer solutions, conflicting experimental evidence regarding the effect of polymer concentration on thermal conductivity was reported at concentrations above and below 1 wt %. In the dilute range, the AFVS model developed by Mashelkar and Kulkarni [121] appears to describe the limited thermal conductivity data, but further confirmation is still needed. For concentrated polymer solutions, the present AFVS model fails and no other theoretical approach is available.

Strain rates in flowing liquids alter the free volume (and hence the conductivity) and quantitative understanding of the relationship between free volume and strain rate is therefore necessary. Lamantia and Titomanlio [128] have introduced the concept of free-volume-dependent relaxation times in order to interpret rheological behavior of polymer melts and solutions, and efforts along similar lines may be helpful in analyzing the strain-rate effect on polymer liquid conductivity. While investigating the effect of strain rate, care must be taken to account for the phenomenon of stress-induced migration [98] that is likely to be present in flows of viscoelastic polymer solutions in nonuniform velocity fields. Mashelkar and Dutta [98] have shown the dangers of neglecting such phenomena [129].

Finally, the phenomenon of thermal conduction in structured liquids appears to offer a number of challenging and interesting opportunities for future research. It is hoped that the present review will stimulate work in some of these areas.

NOMENCLATURE

$\bar{A}$	mean surface area	G	constant in the definitions of K_{e_1} and K_{e_2} of Eq. (57)
A_1, A_2	constants in Eq. (53)		
A_k	constant in Eq. (80)	Gr	Grashoff number
A_s	microstructural parameter in Eq. (59)	g	acceleration due to gravity
		H	hematocrit content
a	radius of the rotational axis of a spheroid	H_m	maximum hematocrit content
		h	rate of temperature rise
B	instrument constant for Couette apparatus	I	first invariant of polarizability tensor
		K_1^*	bound defined by Eq. (56)
B_1, B_2	constants in Eq. (53)	K_2^*	bound defined by Eq. (56)
B_k	constant in Eq. (80)	K_B	Boltzmann constant
B_s	microstructural parameter in Eq. (59)	K_d	k_d/k_c
		K_e	k_e/k_c
b	radius	K_{max}	Hashim–Shtrikman upper bound
b_i	constant in Eq. (85)	K_{min}	Hashim–Shtrikman lower bound
C	constant in Eq. (75)	k	thermal conductivity
C_1, C_2	constants in Eq. (81)	$\mathbf{k}$	thermal conductivity tensor
C_e	Einstein coefficient in Eq. (38)	$k_\perp$	thermal conductivity in a direction perpendicular to orientation or shear
C_p	specific heat at constant pressure		
C_v	specific heat at constant volume		
c_p	number of inclusions per unit volume	$k_\parallel$	thermal conductivity in a direction parallel to orientation or shear
D_r	rotational diffusion coefficient	k_A	conductivity of the annulus
D/Dt	substantial derivative	k_b	thermal conductivity of binder
d	diameter of a spherical drop or particle	k_{bl}	thermal conductivity of blood
		k_g	thermal conductivity of grease
EF	enhancement factor	k_m	k_{bl} at $H = H_m$
e	constant in Eq. (75)	k_0	thermal conductivity for a stationary medium
f	fractional free volume		
f_0	fractional free volume of a parent matrix	k_p	thermal conductivity of plasma
		k_s	thermal conductivity of the sample

k_{solv}	thermal conductivity of the solvent	T	temperature
k_u	thermal conductivity of unplasti-	T_g	glass transition temperature
	cized polymer	T_m	melting temperature
L	cylinder length	t	time
l_p	characteristic particle dimension	U	internal energy
M_n	number avg. molecular weight	u	velocity
M_w	weight avg. molecular weight	V	fluid volume
N	$1 + X$	V_{bulk}	bulk velocity
N_{av}	Avogadro's number	V_p	volume of a particle
n	number of particle types	V_{pm}	volume of mth inclusion
P	pressure	V_s	volume fraction of soap
Pe	Peclet number	v	specific volume
Pr	Prandtl number	v_s	sonic velocity
q	rate of heat input per unit length	v_z	axial velocity
$\mathbf{q}$	heat-flux vector	W	power supply to heat source
R	gas constant	X	shape factor of dispersed particles
Ra	Raleigh number	x	coordinate direction
r	radial direction	x_f	constant defined by Eq. (90)
r_i	inner radius	x_m	maximum packing fraction
r_o	outer radius	x_v	volume fraction of dispersed phase
S_p	surface area of an inclusion	z	axial direction
S_{pm}	surface area of mth particle	∇	vector differential operator

Greek Symbols

α	thermal diffusivity	ψ	sphericity
α_c	coefficient of volume expansion	μ	liquid viscosity
α_k	expansion coefficient	ν	kinematic viscosity
β	coefficient of thermal expansion	ζ	parameter in the definition of Φ of
β_c	compressibility		Eq. (59)
$\dot{\gamma}$	shear rate	π	3.14159 . . .
Δr	$r_0 - r_i$	π_i	constant in Eq. (85)
Δx	layer thickness	ρ	density
δ	sample thickness	σ	interfacial tension
δ_c	fractional free volume difference	$\mathbf{\tau}$	stress tensor
δ_{ij}	Kronecker data	Φ	parameter in Eq. (59)
ε	slenderness ratio	ϕ	volume fraction
ε_e	dimensionless drop deformation	ϕ_p	polymer/plasticizer concentration
Λ	constant in Eq. (92)	Ω	constant in Eq. (68)
λ	μ_d/μ_c	ω	C_{pd}/C_{pc}

Subscripts

c	continuous phase	f	filler
d	dispersed phase	pf	filled polymer
e,eff	effective	pm	polymer matrix

Others

∇	vector differential operator

REFERENCES

1. K. Eiermann and K. N. Hellwedge, *J. Polym. Sci.* **57**, 99 (1962).
2. D. Hansen and C. C. Ho, *J. Polym. Sci., Part A* **3**, 659 (1965).
3. K. Eiermann, *Rubber Chem. Technol.* **39**, 841 (1966).
4. D. R. Anderson, *Chem. Rev.* **66**, 677 (1966).
5. W. Reese, *J. Macromol. Chem. Sci.*, **A3**, 1257 (1969).
6. W. Knappe, *Adv. Polym. Sci.* **7**, 477 (1971).
7. D. Hands, *Rubber Chem. Technol.* **50**, 480 (1977).
8. C. L. Choy, *Polymer* **18**, 984 (1977).
9. I. Prepechenko, "low-temperature Properties of Polymers." Pergamon, Oxford, 1980.
10. M. Pietralla, *Colloid Polym. Sci.* **259**, 111 (1981).
11. J. E. Parrott and A. D. Stuckles, "Thermal Conductivity in Solids." Pion Ltd., London, 1975.
12. R. Berman, "Thermal Conduction in Solids." Oxford Univ. Press (Clarendon), London and New York, 1976.
13. A. H. P. Skelland, 'Non-Newtonian Flow and Heat Transfer.' Wiley, New York, 1967.
14. G. Astarita and R. A. Mashelkar, *Chem. Eng. (London),* Feb., p. 100 (1977).
15. R. E. Meredith and C. W. Tobias, *Adv. Electrochem. Electrochem. Eng.* **2**, 15 (1962).
16. A. L. Baxley, Ph.D. Thesis, University of Arkansas, Little Rock (1967).
17. G. Astarita and G. Sarti, *J. Non-Newtonian Fluid Mech.* **1**, 39 (1976).
18. M. E. Gurtin and A. C. Pipkin, *Arch. Ration. Mech. Anal.* **31**, 113 (1968).
19. J. W. Nunziato, *Q. J. Appl. Math.* **29**, 187 (1971).
20. E. Lorenzini and M. Spiga, *Waerme- Stoffuebertrag.* **16**, 113 (1982).
21. R. R. Huilgol, *Acta Mech.* **7**, 252 (1969).
22. L. N. Novichanyok, *Prog. Heat Mass Transfer* **5**, 293 (1972).
23. L. G. Leal, *Chem. Eng. Commun.* **1**, 21 (1973).
24. A. A. Cocci and J. J. C. Picot, *Polym. Eng. Sci.* **13**, 337 (1973).
25. Y. P. Chang, *Int. J. Heat Mass Transfer* **20**, 1019 (1977).
26. C. L. Choy, *J. Polym. Sci., Polym. Phys. Ed.* **13**, 1263 (1975).
27. J. H. Boggs and W. L. Sibbitt, *Ind. Eng. Chem.* **47**, 289 (1955).
28. D. Hands and F. Horsfall, *J. Phys. E.* **8**, 687 (1975).
29. W. Y. Lee, Y. I. Cho, and J. P. Hartnett, *Lett. Heat Mass Transfer* **8**, 255 (1981).
30. H. F. Poppendiak, R. Randall, J. A. Breeden, J. E. Chambers, and J. R. Murphy, *Cryobiology* **3**, 318 (1966).
31. R. L. Hamilton and O. K. Crosser, *J. Chem. Eng. Data* **7**, 59 (1962).
32. O. K. Bates, *Ind. Eng. Chem.* **41**, 1966 (1949).
33. S. E. Charm and E. W. Merrill, *Pap. 18th Annu. Meet. Inst. Food Technol., 1958* (1958).
34. R. J. Brunson, *J. Chem. Eng. Data* **20**, 435 (1975).
35. F. A. Johnson, *U.K. At. Energy Res. Estab., Rep.* **AERE-R2578** (1958).
36. R. L. Hamilton and O. K. Crosser, *Ind. Eng. Chem. Fundam.* **1**, 187 (1962).
37. C. Orr and J. M. Dallavale, *Chem. Eng. Prog., Symp. Ser.* **9**, 50 (1954).
38. T. Kumada, *Bull. JSME* **18**, 1440 (1975).
39. K. W. Jackson and W. Z. Black, *Int. J. Heat Mass Transfer* **26**, 87 (1983).
40. B. Chitrangad and J. J. C. Picot, *Polym. Eng. Sci.* **21**, 782 (1981).
41. S. Y. Pusatcioglu, A. L. Fricke, and J. C. Hassler, *J. Appl. Polym. Sci.* **24**, 947 (1979).
42. J. C. Ramsey, A. L. Fricke, and J. A. Caskey, *J. Appl. Polym. Sci.* **17**, 1597 (1973).
43. T. R. Fuller and A. L. Fricke, *J. Appl. Polym. Sci.* **15**, 1729 (1971).
44. J. J. C. Picot, G. J. Goobie and G. S. Mawhinney, *Polym. Eng. Sci.* **22**, 154 (1982).
45. R. H. Shoulberg, *J. Appl. Polym. Sci.* **7**, 1597 (1963).

46. Z. P. Shulman, L. N. Novichyonok, E. P. Belskaya, B. M. Khursid, and V. V. Melnichenko, *Int. J. Heat Mass Transfer* **25**, 643 (1982).
47. R. W. F. Tait and B. A. Hills, *Ind. Eng. Chem.* **56**, 29 (1964).
48. Lord Rayleigh, *Philos. Mag.* [6] **32**, 529 (1916).
49. H. Jeffrey, *Proc. R. Soc. London, Ser. A* **118**, 195 (1928).
50. A. R. Low, *Proc. R. Soc. London, Ser. A* **125**, 180 (1929).
51. A. H. Saunders, M. Fishenden, and H. D. Mansion, *Engineering, (London) 139*, 483 (1935).
52. W. Beckmann, *VDI-Forschungsh.*, **2**, 165, 213 (1931).
53. M. Jacob, *Trans. ASME* **68**, 189 (1946).
54. E. F. M. Van der Held and F. G. Van Drunen, *Physica (Amsterdam)* **15**, 865 (1949).
55. H. S. Carslaw and J. C. Jaeger, "Conduction of Heat in Solids." Oxford Univ. Press., London and New York, 1959.
56. Z. P. Shulman, L. N. Novichenok, G. S. Klumel, and G. V. Gnilitsky, *Prog. Heat Mass Transfer* **4**, 159 (1970).
57. J. J. C. Picot and A. G. Frederickson, *Ind. Eng. Chem. Fundam.* **7**, 84 (1968).
58. C. L. Griffis, N. C. Nahas, and J. R. Couper, *Res. Rep. — Univ. Arkansas, Eng. Exp. Stn.* **5** (1964).
59. M. Jakob, "Heat Transfer," Vol. 1, p. 88. Wiley, New York, 1950.
60. H. W. Russell, *J. Am. Cera. Soc.* **18**, 1 (1935).
61. T. B. Jefferson, O. W. Witzwell, and N. L. Sibitt, *Ind. Eng. Chem.* **50**, 1589 (1958).
62. J. C. Maxwell, "Electricity and Magnetism," 1st ed., p. 365, Oxford Univ. Press (Clarendon), London and New York, 1873.
63. Lord Rayleigh, *Philos. Mag.* [5] **34**, 481 (1892).
64. H. Fricke, *Phys. Rev.* **24**, 575 (1924).
65. I. Runge, *Zh. Tekh. Fis.* **6**, 61 (1925).
66. R. E. Meredith and C. W. Tobias, *J. Appl. Phys.* **31**, 1270 (1960).
67. D. A. G. Bruggeman, *Ann. Phys. (Leipzig)* [5] **24**, 636 (1935).
68. R. E. Meredith and C. W. Tobias, *J. Electrochem. Soc.* **108**, 286 (1961).
69. C. A. R. Pearce, *Br. J. Appl. Phys.* **6**, 113 (1955).
70. R. M. De La Rue and C. W. Tobias, *J. Electrochem. Soc.* **106**, 827 (1959).
71. R. L. Gorring and S. W. Churchill, *Chem. Eng. Prog.* **57**, 53 (1961).
72. G. N. Stewart, *J. Physiol. (London)*, **24**, 356 (1899).
73. C. H. Lees, *Philos. Trans R. Soc. London, Ser. A* **191**, 339 (1898).
74. A. McAlister and C. Orr, *J. Chem. Eng. Data* **9**, 71 (1964).
75. G. T. Tsao, *Ind. Eng. Chem.* **53**, 395 (1961).
76. S. C. Cheng and R. I. Vachon, *Int. J. Heat Mass Transfer* **12**, 249 (1969).
77. Y. C. Chung and L. G. Leal, *Int. J. Multiphase Flow* **8**, 605 (1982).
78. G. K. Batchelor, *Annu. Rev. Fluid Mech.* **6**, 227 (1974).
79. G. K. Batchelor, *J. Fluid Mech.* **41**, 545 (1970).
80. G. K. Batchelor, *J. Fluid Mech.* **52**, 245 (1972).
81. D. Jeffrey, *Proc. R. Soc. London, Ser. A* **335**, 355 (1973).
82. J. J. McCoy and M. J. Beran, *Int. J. Eng. Sci.* **14**, 7 (1976).
83. A. Rocha and A. Acrivos, *Q. J. Mech. Appl. Math.* **26**, 217 (1973).
84. A. Rocha and A. Acrivos, *Q. J. Mech. Appl. Math.* **26**, 441 (1973).
85. H. S. Chen and A. Acrivos, *Proc. R. Soc. London, Ser. A* **349**, 261 (1976).
86. A. Rocha and A. Acrivos, *Proc. R. Soc. London, Ser. A* **337**, 123 (1974).
87. G. K. Batchelor and J. T. Green, *J. Fluid Mech.* **56**, 401 (1972).
88. Z. Hashim and S. Shtrikman, *J. Appl. Phys.* **33**, 3125 (1962).
89. M. Miller, *J. Math. Phys.* **10**, 1988 (1969).

90. M. Beran, "Statistical Continuum Theories." Wiley, New York, 1968.
91. M. Elsayed, *J. Math. Phys.* **15,** 176 (1974).
92. L. G. Leal, *J. Colloid. Interface Sci.* **58,** 296 (1977).
93. A. Nir and A. Acrivos, *J. Fluid Mech.* **78,** 33 (1976).
94. T. J. McMillen and L. G. Leal, *Int. J. Multiphase Flow* **2,** 105 (1975); errata: **2,** 363 (1975).
95. T. J. McMillen and L. C. Leal, *J. Colloid Interface Sci.* **69,** 45 (1979).
96. K. H. Keller, *Fed. Proc. Fed. Am. Soc. Exp. Biol.,* **30,** 1591 (1971).
97. C. K. Colton, K. A. Smith, E. W. Merrill, and S. Friedman, *AIChE J.* **17,** 800 (1971).
98. R. A. Mashelkar and A. Dutta, *Chem. Eng. Sci.* **37,** 969 (1982).
99. P. Riha, *Int. J. Eng. Sci.* **14,** 529 (1976).
100. P. Riha, *Biorheology* **13,** 185 (1976).
101. J. Hennig and W. Knappe, *J. Polym. Sci.* **6,** 167 (1963).
102. K. K. Hellwege, J. Hennig, and W. Knappe, *Kolloid Z.* Polym. **188,** 121 (1963).
103. D. Greig, *Dev. Oriented Polym.* **I,** 79 (1982).
104. P. Bridgmann, *Proc. Am. Acad. Arts Sci.* **59,** 141 (1923).
105. W. Reese, *J. Macromol. Sci. Chem.* **A3,** 1257 (1969).
106. K. Uberreiter and S. Nens, *Kolloid-Z.* **123,** 92 (1951).
107. R. P. Sheldon and K. Lane, *Polymer* **6,** 77 (1965).
108. M. G. Kulkarni and R. A. Mashelkar, *Polymer* **22,** 867 (1981).
109. R. R. Kraybill, *Polym. Eng. Sci.* **21,** 124 (1981).
110. R. B. Bird, W. E. Stewart, and E. N. Lightfoot "Transport Phenomena," p. 260. Wiley, New York, 1960.
111. E. C. Bernhardt, ed., "Processing of Thermoplastics Materials," p. 547. Van Nostrand-Reinhold, Princeton, New Jersey, 1959.
112. R. S. Spencer and G. D. Gilmore, *J. Appl. Phys.* **21,** 523 (1950).
113. P. Lohe, *Kolloid-Z.* **210,** 111 (1965).
114. D. Hansen and C. C. Ho, *J. Polym. Sci., A* **3,** 659 (1965).
115. P. Lohe, *Kolloid Z. Z. Polym.* **203,** 115 (1965).
116. D. Hands, Ph.D. Thesis, University of Bradford (1976).
117. D. Hansen, R. C. Kantayya, and C. C. Ho, *Polym. Eng. Sci.* **6,** 260 (1966).
118. J. N. Tomlinson, D. E. Kline, and J. A. Saner, *SPE Trans.* **5,** 44 (1965).
119. D. Bellet, M. Sengelin, and C. Thirriot, *Int. J. Heat Mass. Transfer,* **18,** 1177 (1975).
120. A. Dutta and R. A. Mashelkar, *J. Appl. Polym. Sci.* **27,** 2739 (1982).
121. R. A. Mashelkar and M. G. Kulkarni, *Pure Appl. Chem.* **55,** 737 (1983).
122. M. G. Kulkarni and R. A. Mashelkar, *Chem. Eng. Sci.* **38,** 925 (1983).
123. M. G. Kulkarni and R. A. Mashelkar, *Chem. Eng. Sci.* **38,** 941 (1983).
124. J. D. Ferry, "Viscoelastic Properties of Polymers," 3rd ed. Wiley, New York, 1981.
125. R. C. Progelhof, J. L. Throne, and R. R. Ruetsch, *Polym. Eng. Sci.* **15,** 107 (1975).
126. R. P. Kusy and R. D. Corneliussen, *Polym. Eng. Sci.* **15,** 107 (1975).
127. D. W. Sundstrom and Y. D. Lee, *J. Appl. Polym. Sci.* **16,** 3159 (1972).
128. F. P. Lamantia and G. Titomanlio, *Rheol. Acta* **18,** 469 (1979).
129. A. Dutta and R. A. Mashelkar, *Chem. Eng. Commun.* **33,** 181 (1985).
130. D. R. McKenzie and R. C. McPhedron, *Nature (London)* **265,** 128 (1977).
131. R. C. McPhedron and D. R. McKenzie, *Proc. R. Soc. London, Ser. A* **359,** 45 (1978).
132. D. R. McKenzie, R. C. McPhedron, and G. H. Derrick, *Proc. R. Soc. London, Ser. A* **362,** 211 (1978).
133. R. W. O'Brien, *J. Fluid Mech.* **91,** 17 (1979).
134. D. Bergmann, *Phys. Rep. C* **43,** 377 (1978).
135. A. S. Sangani and A. Acrivos, *Proc. R. Soc. London, Ser. A* **386,** 263 (1983).

136. M. Zuzovski and H. Brenner, *J. Appl. Math. Phys.* **28,** 979 (1977).
137. W. T. Perrins, R. C. McPhedron, and D. R. Mckenzie, *Thin Solid Films* **57,** 321 (1979).
138. W. T. Perrins, D. R. Mckenzie and R. C. McPhedron, *Proc. R. Soc. London, Ser. A* **369,** 207 (1979).
139. E. H. Kerner, *Proc. Phys. Soc., London, Sect B* **69,** 802, 808 (1956).
140. G. S. Springer and S. W. Tsai, *J. Compos. Mater.* **1,** 166 (1967).
141. D. Sundstrom and S. J. Chen, *J. Compos. Mater.* **4,** 113 (1970).
142. G. E. Zinemeister and K. S. Purohit, *J. Compos. Mater.* **4,** 278 (1970).
143. L. R. Nielsen, *Ind. Eng. Chem. Fundam.* **13,** 17 (1974).
144. G. K. Batchelor and R. W. O'Brien, *Proc. R. Soc. London, Ser. A* **355,** 313 (1977).
145. H. Brenner, *Philos. Trans. R. Soc. London, Ser. A* **297,** 81 (1980).
146. H. C. Chang, *Chem. Eng. Commun.* **15,** 83 (1982).
147. H. C. Chang, *AIChE J.* **29,** 846 (1983).
148. I. Nozad, R. G. Carbonell, and S. Whitaker, *Chem. Eng. Sci.* **40,** 843 (1985).
149. I. Nozad, R. G. Carbonell, and S. Whitaker, *Chem. Eng. Sci.* **40,** 857 (1985).
150. G. W. Milton, *J. Appl. Phys.* **52,** 5286, 5294 (1981).
151. N. Phan-Thien and G. W. Milton, *Proc. R. Soc. London, Ser. A* **380,** 333 (1982).
152. G. W. Milton and N. Phan-Thien, *Proc. R. Soc. London, Ser. A* **380,** 305 (1982).
153. W. Kohler and G. C. Papanicolaou, *Lect. Notes Phys.* **154,** 111 (1982).
154. D. J. Wallace, C. Moreland, and J. J. C. Picot, *Polym. Eng. Sci.* **25,** 70 (1985).

Transition Boiling Heat Transfer

E. K. KALININ, I. I. BERLIN, AND V. V. KOSTIOUK

Moscow Orjonikidze Aircraft Institute, Volokolamskoe Chaussee, Moscow A-80, USSR

I. Introduction

Transition boiling is a field of boiling heat transfer that has not received much attention. This is explained by the fact that practical interest in transition boiling has appeared only during recent decades. In conducting experiments, researchers have faced great methodological difficulties, and it has proved impossible to develop a satisfactory *a priori* theory of this complex process. Up to now there are relatively few works containing interesting information on various aspects of transition boiling.

According to the existing classification of boiling regimes [1, 2], the transition region is intermediate between the nucleate and film boiling regions, and its boundaries are the extrema of the boiling curve $q(\Delta T)$ from the nucleate boiling crisis $(\Delta T_{cr1}, q_{cr1})$ and the film boiling crisis $(\Delta T_{cr2}, q_{cr2})$. The main specificity of the boiling curve over the temperature potential range $(\Delta T_{cr1}, \Delta T_{cr2})$ is its negative slope $(\partial q/\partial \Delta T < 0)$. For steady-state boiling with a constant heat flux $(q = \text{const})$, this section of the boiling curve is not realized. Only for the boiling crises, when the wall temperature is sharply increased (first crisis) or decreased (second crisis), the unsteady-state process covers the temperature head range $(\Delta T_{cr1}, \Delta T_{cr2})$. In this case, the heat-transfer intensity in the transition boiling section determines the rate of varying wall temperature.

The transition boiling region [3, 4] may be determined in another way, assuming that in this region neither nucleate nor film boiling is stable (with regard for hydrodynamic or thermodynamic considerations). In this case, the transition boiling boundaries are the points 1′ and 2′ (as shown in

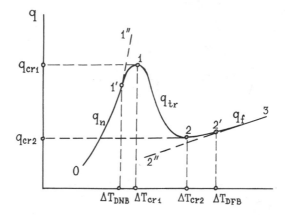

FIG. 1. Transition boiling region boundaries. 1, 2: Boundaries of the section of the boiling curve with a negative slope ($dq/d_L \Delta T < 0$); 1′: nucleate boiling stability boundary; 2′: film boiling stability boundary.

Fig. 1), at which the nucleate and film boiling becomes unstable. Point 1′, the departure from nucleate boiling (DNB), is characterized by the appearance of unstable local dry spots on the heating surface at increasing temperature head ΔT (or by their disappearance at decreasing ΔT). Point 2′, the departure from film boiling (FB), is characterized by the appearance of unstable local cold spots on the heating surface at decreasing ΔT (or by their disappearance at increasing ΔT).

The asymptotic behavior of the transition boiling curve near the transition region boundaries is a shortcoming of the choice of points 1′ and 2′ as these boundaries, as this choice raises practical difficulties associated with the exact determination of points 1′ and 2′ using the experimental data (while the points of extrema 1 and 2 are more easily fixed). However, the exact position of points 1′ and 2′ on the boiling curve is insignificant, just as in the boundary layer theory the exact determination of the boundary thickness is unimportant. Of more importance is the fact that the transition from nucleate to film boiling with increasing temperature head ΔT (or the opposite transition with decreasing ΔT) is smooth and gradual, and the first and second boiling crises are interior points of the transition region. The crisis nature of heat transfer at these points (a sharp increase or decrease in the wall temperature) appears only at a prescribed heat flux. In this case, as the temperature head changes, the proportion between the contributions of film and nucleate boiling to total heat transfer varies gradually.

II. Statement of the Problem

A. TRANSITION BOILING MECHANISM

The transition boiling mechanism is the basis for understanding the laws of heat transfer and its dependence on numerous governing parameters, developing theoretical models for the transition boiling process, and obtaining correlations.

The experimental results on the transition boiling mechanism and the estimates of heat-transfer rates show that, with transition boiling, at each time instant some part of the heating surface is wetted by the liquid and the remainder is covered by a vapor film. In this case, each point of the heating surface is alternately in contact with the liquid and vapor phases of the boiling medium. The mean duration of the heating surface contact with the liquid depends on the temperature difference, other parameters of the process, boiling substance properties, and wall material. Since the rate of heat transfer to the liquid is higher than that to the vapor, the processes at the wall–liquid contacts are dominant in the case of transition boiling.

From the point of view of the heat transfer mechanism, it is advisable to distinguish three zones in the transition boiling region on the wetted part of the heating surface:

1. A low-temperature head zone near the nucleate boiling crisis region where the duration of the liquid–wall contact is rather large and nucleate boiling occurs at the contact place;

2. a high-temperature head zone near the film boiling crisis region where nucleate boiling cannot develop because of the small contact time, and the heat transfer from the wall to the liquid dominates due to unsteady heat conduction;

3. a mean-temperature head zone where the contributions of nucleate boiling and unsteady heat conduction are comparable.

With increasing temperature head, the duration of the periodic liquid–wall contacts is decreased, and the heat transfer rate ranges from high values typical of nucleate boiling to low ones typical of film boiling.

The disturbance of the hydrodynamic stability of the vapor film and the conservation of the thermodynamic stability of the liquid at the contact place are necessary conditions for the liquid–wall contact. Consider how these conditions are satisfied when film boiling is replaced by a transient process in the case of slowly decreasing wall temperature.

It is evident that stable equilibrium of the vapor–liquid interface in the gravitational field is possible only when the less dense phase is above the more dense one (film boiling under the lower side of a horizontal heating surface). In all other cases (film boiling on vertical, inclined, cylindrical, spherical surfaces and above a horizontal heating surface), the interphase boundary is unstable as the more dense phase is above or to the side of the less dense one. The instability initiates a transverse motion of the interphase boundary (vibrational or aperiodic). However, at high-temperature heads, when the vapor film thickness is high, the liquid does not approach the wall. With decreasing temperature head, the value of the vapor film thickness is decreased and is consistent with the vibration amplitude of the interphase boundary, and liquid–wall contact becomes possible from a hydrodynamic viewpoint. Whether this contact occurs depends on the thermodynamic conditions or their combination with the hydrodynamic ones. When the wave comb approaches the wall, within $T_v > T_s$, the intense vaporization on the wave comb and the resulting reactive forces can throw the liquid from the wall, and the contact will not occur.

If contact between liquid and wall having different initial temperatures takes place, then some intermediate temperature T_b appears at the contact boundary. If this temperature exceeds a limiting metastable liquid heating temperature T_{lim}, then the explosive boiling of the thinnest layer occurs, and the liquid is thrown from the heating surface. If $T_b < T_{lim}$, then the thermodynamic condition for liquid wall wetting is satisfied, and there occurs more or less long liquid–wall contact accompanied by increasing heat removal due to unsteady heat conduction in the liquid and in the case of still longer contact, due to nucleate boiling.

Thus, of two necessary liquid–wall contact conditions, the condition of the smaller temperature head is the determining one. In particular, with boiling below a horizontal heating surface, when the interphase boundary is most stable, the film boiling is delayed at low temperature heads and a *hydrodynamic* film boiling crisis should be expected. With boiling above a horizontal heating surface and on vertical, cylindrical, and spherical surfaces it is more probable that the possible liquid–wall contact depends on the conditions of *thermodynamic* liquid stability.

III. Theoretical Studies of Transition Boiling

The existing mathematical models for transition boiling heat transfer may be divided into two groups depending on the type of process:

1. Heat removal due to unsteady heat conduction with periodic wall–liquid contacts and

2. Liquid film drying under a rising vapor bubble with the intense heat

removal area decreasing due to appearing and expanding dry sections of the heating surface.

Bankoff and Mehra [5] were the first to propose the transient boiling heat-transfer model allowing for heat conduction of a liquid in contact with the heating surface. The following assumptions were made:

1. The vapor film is hydrodynamically unstable, and at each heating surface point the discontinuous liquid–wall contact occurs with a frequency f for a time t_L and

2. when the wall contacts the vapor for the time t_v the heat transfer is negligibly small, and the initial wall temperature is recovered at the end of this period due to heat conduction.

The liquid temperature field was determined by solving the one-dimensional heat conduction equation for two semi-infinite contacting bodies at $t = 0$:

$$T_L(y, t) = T_{L0} + (T_b - T_{L0}) \operatorname{erfc}(y/2 \sqrt{a_L t});$$

where y is the distance from the heating surface and T_b is the liquid–wall contact boundary temperature

$$T_b = T_{L0} + \frac{T_{w0} - T_{L0}}{1 + \sqrt{(\rho c \lambda)_L / (\rho c \lambda)_w}} \tag{1}$$

and T_{w0} and T_{L0} the initial wall and liquid temperature, respectively.

The mean heat-flux density for a contact time t_L is determined by

$$\bar{q}_T = 2\lambda_L (T_b - T_{L0}) / \sqrt{\pi a_L t_L} \tag{2}$$

The mean heat-flux density for the vibration period, $t_L + t_v = 1/f$ is given by

$$\bar{q} = \bar{q}_T \frac{t_L}{t_L + t_v} = \frac{2\lambda_L (T_b - T_{L0})}{\sqrt{\pi a_L}} \sqrt{t_L} f \tag{3}$$

Prescribing the frequency of the liquid–wall contact by Zuber's theory [6] for the first and second boiling crises and determining it at the remaining transition region points through linear–logarithmic coordinate interpolation, Bankoff and Mehra [5] treated Berenson's data [7] for n-pentane with regard to t_L. The treated results are shown in Fig. 2. However, neither Bankoff and Mehra [5], nor other investigators referring to this work paid attention to the substantial discrepancy in the results obtained. The dashed line in Fig. 2 stands for the vibration period $t_L + t_v = 1/f$ as a function of the temperature head [5]. It is seen from this figure that in the region close to the first boiling crisis, the calculated values of t_L exceed those of $(t_L + t_v)$.

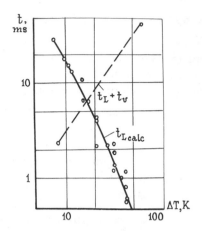

FIG. 2. Comparison of the predicted time of the liquid–wall contact with the contact period, both versus ΔT (from Bankoff and Mehra [5]).

In our opinion, this discrepancy results from the incorrect choice of the liquid–wall contact frequency as a function of the temperature head rather than the model being wrong [5]. In addition, the considered heat transfer mechanism is dominant not in the whole transition boiling region but only in the zone adjacent to the film boiling region. The discrepancy in the results of Bankoff and Mehra [5] arises in the other transition boiling zone near ΔT_{crl} where another heat-transfer mechanism prevails.

Aoki and Welty [8] determined t_L, f, and T_b through heating surface temperature pulsations and obtained, in high ΔT_W zones, a satisfactory agreement between the experimental values of the heat-flux density and those calculated by the formula resulting from (1) and (3):

$$\bar{q} = \frac{2\lambda_W(T_{W0} - T_b)}{\sqrt{\pi a_W}} \sqrt{t_W} f \tag{4}$$

The authors [8] considered that their data supported the model proposed in Bankoff and Mehra [5] but probably did not notice that the measured values of the wall temperature decrease $[T_{W0} - T_b)$, being ~40% of the temperature head $(T_{W0} - T_{L0})$, exceeded by as much as 3 to 4 times the value calculated by formula (1) for n-pentane and copper. Thus, the calculation by Eq. (3) would underestimate values of the heat-flux density. However, this underestimation might be avoided if in determining t_L the authors [8] allowed not only for a period of decreasing wall temperature but also for a longer interval between the end of the decreasing wall temperature and the start of increasing T_W, since the liquid–wall contact continues for this time.

Katto *et al.* [9] developed a transition boiling heat transfer model based on the liquid film drying process occurring above the vapor bubble assuming that

1. Nucleate boiling takes place on the liquid-wetted sections of the heating surface; in this case, the heat flux density on these sections may be determined by extrapolating the developed nucleate boiling curve to the region $\Delta T > \Delta T_{\text{DNB}}$;

2. at the vapor bubble departure, the heating surface is wetted with the liquid, and a liquid microlayer is formed with an initial thickness δ_{L0}, which may be found by extrapolating the relation obtained at $q_n < q_{\text{crl}}$ in Gaertner and Westwater [10, 11]

$$\delta_{\text{L0}} \propto q_n^{-3/2} \tag{5}$$

to the region, $q_n' > q_{\text{crl}}$ (i.e., $\Delta T > \Delta T_{\text{crl}}$);

3. the liquid microlayer evaporates during t_{L}, which depends on the initial thickness and heat flux as

$$t_{\text{L}} \propto \delta_{\text{L0}}/q_n' \tag{6}$$

In this case the heating surface remains dry during t_{V}, and the dry wall-to-vapor heat transfer is negligibly small;

4. at the nucleate boiling crisis, the liquid microlayer evaporation time is equal to that for vapor cloud departure (i.e., $t_{\text{L}} = t_0$, $t_{\text{V}} = 0$) and with increasing ΔT, the bubble departure period $t_0 = t_{\text{L}} + t_{\text{V}}$ does not vary.

Under these assumptions, the time – mean heat-flux density is determined as

$$q = q_n' \frac{t_{\text{L}}}{t_{\text{L}} + t_{\text{V}}}$$

Hence, taking into account relations (5) and (6) and noting that $(t_{\text{L}} + t_{\text{V}})$ does not vary, it follows that

$$q/q_{\text{crl}} - (q_n'/q_{\text{crl}})^{-3/2} \tag{7}$$

As shown in Fig. 3, relation is in bad agreement with the experimental data [9]. Hence, the model described may be considered only as a qualitative one.

For several years the authors of this review have been developing a mathematical model for transition boiling and DNB based on a theoretical analysis of the processes causing liquid – superheated wall contacts or their disturbance.

The theoretical analysis relies upon the following ideas relating to the transition boiling mechanism. At high values of the wall temperature T_{W}, a

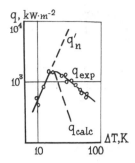

FIG. 3. Comparison of the predicted values of the heat-flux density in the transition boiling region with the experiment according to the data (from Katto *et al.* [9]).

vapor film separates the liquid and the surface (steady-state film boiling). As T_W decreases, the vapor wall becomes thinner and the interface fluctuations may produce liquid–heating surface contact. The liquid is heated at the place of contact, and, upon achieving a certain superheat of the liquid layer adjacent to the wall, stable vapor nucleates are formed. Then vapor bubbles grow, coalescing to form a continuous film that separates the liquid and heating surface. The vapor film becomes hydrodynamically unstable, which causes alternate contact between the liquid and the wall and repetition of the cycle. As the wall temperature decreases, the time of liquid-heating surface contact increases, thus enhancing the transition boiling heat transfer from the low values typical of film boiling to the high ones characterizing nucleate boiling. With a further drop of the heating-surface temperature, the number of vapor nucleates decreases to such an extent that the growing vapor bubbles achieve the separation diameter before they coalesce, and transition boiling is changed to steady-state nucleate boiling. In the transition boiling region the mean-time heat-transfer rate for each section of the heating surface depends on the mean contact time between the section and the liquid or vapor phases of the boiling medium as well as on the heat-transfer characteristics at each stage of the cycle.

The transition boiling theory based on the above mechanism includes mathematical models of three processes responsible for liquid-heating surface contacts or their disturbance. These are

1. development of the hydrodynamic instability of the vapor–liquid interface,

2. liquid heating around the liquid-heating surface contact, and

3. generation and growth of vapor bubbles till they coalesce to form a continuous film.

The analysis of the hydrodynamic instability of the interface during boiling has been made by Kutateladze [1], Chang [12], Zuber [6], Berenson [13], and Lienhard et al. [14] as applied to film boiling and boiling crises. The law of the interface fluctuations in the form

$$\tilde{\delta}_V = A_\delta e^{-i(\omega t - kx)} \qquad (8)$$

yields the dependence of the complex circular fluctuation frequency ω, the wave number k and the process parameters from the set of equations for liquid and vapor hydrodynamics.

The possibility of the liquid–heating surface contact depends on the dominant wave, for which the coefficient of amplitude increase β (the imaginary part of the complex angular frequency ω) attains the highest value:

$$\beta_d = \underset{k}{\mathrm{Sup}}\, \mathrm{Im}(\omega) \qquad (9)$$

The dependence of β_d on the governing process parameters has been obtained by the authors for some particular cases (film pool boiling of saturated or highly subcooled liquid on a flat horizontal surface).

The analysis of these dependences has shown that for saturated liquid the heat-transfer effect on vapor film fluctuations is negligible, whereas an increase in liquid subcooling induces higher stability of the interface (the latter effect becoming weaker as the wall thermal conductivity decreases). Surface tension and viscosity are also stabilizing factors, whereas vapor slipping along the liquid destabilizes the process. The force of gravity enhances the interface fluctuations with boiling on upward-facing heating surfaces and attenuates them on surfaces facing downwards.

For circular and rectangular horizontal heating surfaces, possible modes of standing waves have been analyzed and the film boiling heat-transfer similarity boundaries have been determined in relation to the horizontal heater dimensions as $D \geq 3.45 l_{cr}$ for a disk and $L \geq (3.46 - 3.67) l_{cr}$ for a rectangular plate. These data are in good agreement with the results reported by Grigorev [2].

The development of the dominant wave at the interface may result either in liquid–wall contact (in case of a thin vapor film) or splitting of the wave comb into drops (in case of a thick film). Assuming that the largest amplitude corresponds to the ultimate Stokes wave gives the condition for wave comb–wall contact

$$l_d = 2\pi/k_d \geq 14\bar{\delta}_V \qquad (10)$$

where $\bar{\delta}_V$ is the mean vapor-film thickness.

When determining the duration of the unstable vapor film t_V it has been

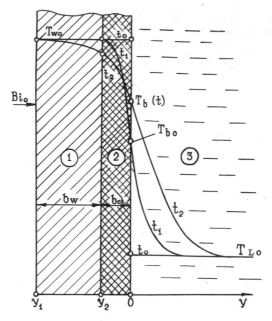

FIG. 4. Temperature field at the contact of the liquid with the two-layer heating surface.

assumed that t_v is in inverse proportion to the coefficient of the dominant wave amplitude increase β_d and depends on the temperature head, sub-cooling, and other conditions specifying the mean vapor-film thickness.

An analysis of liquid heating around the liquid–wall contact has been made by the current authors by calculating the temperature field that appears on contact of a semi-infinite liquid layer at an initial temperature T_{L0} and a two layer plate (wall material thickness b_w and coating thickness b_{ct} at the initial temperature T_{w0} (Fig. 4). The problem was reducible to solve

$$\frac{\partial^2 T_j}{\partial y^2} = \frac{1}{a_j} \frac{\partial T_j}{\partial t}, \quad j = 1, 2, 3 \tag{11}$$

where $j = 1$ represents the wall material, $j = 2$ the coating, and $j = 3$ the liquid with the initial conditions

$$t = 0, \qquad T_1 = T_2 = T_0, \qquad T_3 = T_{L0} \tag{12}$$

and boundary conditions in the bulk liquid ($y \to \infty$)

$$T_3 = T_{L0} \tag{13}$$

at the coating-liquid boundary ($y = 0$)

$$T_2 = T_3, \qquad -\lambda_2 \partial T_2/\partial y = -\lambda_3 \partial T_3/\partial y \tag{14}$$

at the wall-coating boundary ($y = -b_{ct}$)

$$T_1 = T_2, \qquad -\lambda_1 \, \partial T_1/\partial y = -\lambda_2 \, \partial T_2/\partial y \qquad (15)$$

and at the wall rear side ($y = -b_{ct} - b_{W}$)

$$\partial T_1/\partial y = 0 \qquad \text{or} \qquad T_1 = T_{W0} \qquad (16)$$

which corresponds to transition boiling under cooling ($\text{Bi}_0 = 0$) or steady conditions at high heat-transfer rate from the additional heat carrier ($\text{Bi}_0 \to \infty$).

Solution of system (11)–(16) gives expressions for the liquid–heating surface contact temperature

$$\frac{T_b - T_{L0}}{T_{W0} - T_{L0}} = \frac{1}{1 + \varepsilon_L/\varepsilon_{ct}}$$

$$\times \left[1 + \frac{2}{\varepsilon_{ct}/\varepsilon_L + 1} \sum_{m=0}^{\infty} \sum_{n=1}^{\infty} A_{mn} \, \text{erfc}\left(\frac{nb_{ct}}{\sqrt{a_{ct}t}} + \frac{mb_W}{\sqrt{a_W t}} \right) \right] \qquad (17)$$

and the mean heat-flux transferred from the wall to the liquid for the contact time

$$\bar{q}_b = \frac{2\varepsilon_L(T_{W0} - T_{L0})}{(1 + \varepsilon_L/\varepsilon_{ct}) \sqrt{\pi t_L}} \left[1 + \frac{2\sqrt{\pi}}{\varepsilon_{ct}/\varepsilon_L + 1} \right.$$

$$\left. \times \sum_{m=0}^{\infty} \sum_{n=1}^{\infty} B_{mn} \, \text{ierfc}\left(\frac{nb_{ct}}{\sqrt{a_{ct}t_L}} + \frac{mb_W}{\sqrt{a_W t_L}} \right) \right] \qquad (18)$$

where $\varepsilon = \sqrt{\rho c \lambda}$ is the thermal activity of the medium erfc(x) the complementary Gauss error function, and ierfc(x) its first integral.

The analysis of these relationships shows that at the initial moment the temperature

$$T_{b0} = T_{L0} + \frac{T_{W0} - T_{L0}}{1 + \varepsilon_L/\varepsilon_{ct}} \qquad (19)$$

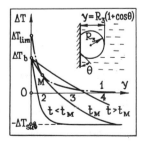

FIG. 5. Relation between necessary (1) and real (2, 3, 4) liquid superheatings.

is attained at the liquid–coating contact and then the thermal disturbance front reaches the coating border at the moment

$$t_1 = b_{ct}^2/4a_{ct} \tag{20}$$

and the rear-wall side at the moment

$$t_2 = \frac{b_w^2}{4a_w}\left(1 + \frac{b_{ct}}{b_w}\sqrt{\frac{a_w}{a_{ct}}}\right)^2 \tag{21}$$

For simpler calculations of heat transfer to the liquid when in contact with the heating surface, the effective wall thermal activity ε_{ef} is introduced and determined by dimensionless nomograms (Fig. 6).

The liquid–wall contact time in the first stage (liquid heating due to heat conduction) is found from the condition of the superheating necessary for the formation of stable vapor nuclei

$$t_T = \frac{\sigma^2 T_s^2(1 + \cos\theta)^2}{\rho_v^2 r^2 \, \Delta T_b^2 a_L} \cdot F\left(\frac{\Delta T_{sub}}{\Delta T_b}\right) \tag{22}$$

Analysis of the vapor bubble growth and evanescence relies on the known empirical relations for the surface density of the nucleation sites and bubble growth rate. The condition of bubble coalescence (covering the entire heating surface) yielded the duration of the second stage of the liquid–wall contact (from nucleation to the formation of a continuous vapor film):

$$t_{col} = \frac{\sigma^2 T_s^2}{\rho_v^2 r^2 \, \Delta T_b^2 a_L \, Ja^2} F\left(\frac{\rho_L C_{pL} \, \Delta T_{sub}}{\rho_v r}\right) \tag{23}$$

By comparing the duration of the bubble growth to the departure radius t_{dep} and bubble coalescence t_{col} for saturated liquid pool boiling on a horizontal heating surface, the authors arrived at an expression for the

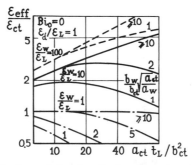

FIG. 6. Nomogram for determining the effective thermal activity of a two-layer wall.

temperature head for the border of stable nucleate boiling

$$\Delta T_{\text{DNB}} = C(\rho_\text{V} r q)^{1/7}(\sigma T_\text{S})^{3/7}\varepsilon_\text{L}^{-4/7} \tag{24}$$

The mathematical transition boiling heat transfer model includes three basic physical mechanisms acting at each heating surface section:

1. Heating of a thin liquid layer around the liquid–wall contact due to heat conduction leading to the formation of stable vapor nuclei,

2. a mechanism similar to bubble boiling during the growth of vapor bubbles till they coalesce to form a continuous film, and

3. a mechanism similar to film boiling during vapor insulation of the heating surface till the hydrodynamic instability of the interface causes the liquid–wall contact.

The mean heat flux in transition boiling is represented as a sum of three components, taking into account the fraction of time for each process

$$q_{\text{tr}} = \frac{q_\text{T} t_\text{T} + q_\text{n} t_{\text{col}} + q_f t_\text{V}}{t_\text{T} + t_{\text{col}} + t_\text{V}} \tag{25}$$

For boiling on a thick high-heat conducting heating surface, this yields the relationship

$$q_{\text{tr}} = \frac{\alpha_f K_\text{V} \Delta T_\text{W}^2 + 2\varepsilon_\text{L}\sqrt{K_\text{T}/\pi} + d_\text{n} K_{\text{col}} \Delta T_\text{W}^{-3}}{K_\text{V} \Delta T_\text{W} + K_\text{T} \Delta T_\text{W}^{-2} + K_{\text{col}} \Delta T_\text{W}^{-4}} \tag{26}$$

where

$$K_\text{T} = t_\text{T} \Delta T_\text{W}^2 = K_1 \frac{\sigma^2(1+\cos\theta)^2 T_\text{S}^2}{a_\text{L}\rho_{\text{VS}}^2 r^2}\left(1 + K_\text{S}\frac{\Delta T_{\text{sub}}}{\Delta T_b}\right) \tag{27}$$

$$K_{\text{col}} = t_{\text{col}} \Delta T_\text{W}^4 = K_2 \frac{\sigma^2 T_\text{s}^2}{\varepsilon_\text{L}^2}\left(1 + K_\text{S}'\frac{\rho_\text{L} C_{\text{pL}} \Delta T_{\text{sub}}}{\rho_\text{V} r}\right) \tag{28}$$

$$K_\text{V} = \frac{t_\text{V}}{\Delta T_\text{W}} = K_3 \frac{C_{\text{pV}}}{r\beta_\text{A}}\left(1 + K_\text{S}''\frac{\rho_\text{L} C_{\text{pL}} \Delta T_{\text{sub}}}{\rho_\text{V} r}\right)^{-1} \tag{29}$$

For boiling on thin and low-heat conducting walls, the dependence $q_{\text{tr}} = F(\Delta T_\text{W})$ is more complicated since the heating surface temperature changes with time following formula (17). In this case, as the ratio $\varepsilon_\text{L}/\varepsilon_{\text{ef}}$ increases, the transition boiling curve shifts to higher temperature heads.

The minimum function condition (26) provides an approximate formula to estimate the film boiling crisis parameters:

$$\Delta T_{\text{cr2}} \approx \sqrt{\frac{2\varepsilon_\text{L} C_{\text{pL}}}{\alpha_f K_\text{V} C_{\text{pV}}}}\sqrt{\frac{K_\text{T}}{\pi}} \tag{30}$$

$$q_{\text{cr2}} \approx 2\alpha_f \Delta T_{\text{cr2}} \tag{31}$$

An analysis of the theoretical relationships has shown that they correctly represent the effect of the thermophysical properties of the heating surface, the liquid properties and its subcooling to the saturation temperature, orientation of the heating surface in the field of body forces and the velocity of the forced flow on transition boiling heat transfer and film boiling crisis.

The cited theoretical model includes some empirical constants (three for saturated liquid and six for subcooled liquid) related to the random nature of three factors such as the distribution of the nucleation sites over the heating surface, the moment of nucleation and initial disturbance of the vapor film surface.

IV. Experimental Studies of Transition Boiling

A. METHODS OF EXPERIMENTAL STUDIES OF TRANSITION BOILING

Experimental study of transition boiling heat transfer involves specific difficulties. First of all since $q = F(\Delta T)$ the negative sign of the derivative $\partial q / \partial T_{\mathrm{w}}$ requires special measures to provide the process stability. In addition, the comparatively high absolute value of $\partial q / \partial T_{\mathrm{w}}$ hampers isothermality of the heating surface. Secondly, the unsteady-state and space nonuniformity of transition boiling require special attention.

The widespread simple method of electric heating in which current is allowed to pass through the test section does not provide stable steady transition boiling states in the temperature head range $\Delta T_{\mathrm{cr1}} < \Delta T < \Delta T_{\mathrm{cr2}}$. It is seen from Fig. 7 that at constant heat release the random wall temperature drop at the point A causes prevalence of heat removal over heat release, which results in a further decrease temperature that ceases only at steady state A'. In the same way, for the random wall temperature increase at the point A, heat release prevails over heat removal and the temperature continues to increase up to the steady state A''.

Petukhov and Kovalev [15] have pointed out that some equilibrium heat flux takes place corresponding to a stable coexistence of nucleate and film boiling on the heating surface due to heat conduction over the test section. In this case the wall temperature changes continuously along the surface from values typical of nucleate boiling to those corresponding to film boiling. Thus, the temperature heads at certain sections of the heating surface correspond to transition boiling. However, the interdependence of the process mechanisms in adjacent boiling regions and comparable heat removal and heat transfer rates along the heating surface does not allow $q(\Delta T)$ to be determined with proper accuracy.

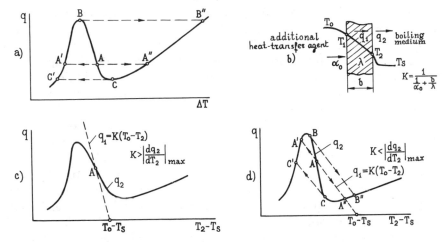

FIG. 7. Transition boiling stability conditions: (a) instability at $q = $ const; (b) heating by the additional heat-transfer agent, (c) stability at a high value of the coefficient K; (d) instability at a low value of K.

Heating of the test section with an additional heat carrier held at a prescribed temperature [16, 17], in particular, with condensing vapor [7, 8, 18–23] is the preferred method for achieving stable transition boiling. Kovalev [20] and Stephan [24] have shown that in such cases it is possible to provide spontaneous process stability with rather low thermal resistance within the additional heat carrier–boiling surface section. The condition for stable transition boiling on surface 2 of a flat wall with additional heat carrier on side 1 has the form [20]

$$dq_2/dT_2 > dq_1/dT_1 \qquad (32)$$

The dependence of q_2 on T_2 corresponds to the transient section of the boiling curve, and the relationship between q_1 and T_2 is governed by the equation

$$q_1 = \frac{T_0 - T_2}{1/\alpha_0 + b_\mathrm{w}/\lambda_\mathrm{w}} = k(T_0 - T_2), \qquad (33)$$

where T_0, α_0 are the temperature and heat transfer coefficient of the additional heat carrier, b_w and λ_w the wall thickness and thermal conductivity, and k the additional heat carrier to boiling surface heat-transfer coefficient.

Equation (33) implies $dq_1/dT_2 = -k$ and for transition boiling $dq_2/dT_2 < 0$; thus stability condition (32) assumes the form

$$|dq_2/dT_2| < k = (1/\alpha_0 + b_\mathrm{w}/\lambda_\mathrm{w})^{-1} \qquad (34)$$

If the test wall is neither uniform nor flat, then condition (34) must include the effective values of α_0 and b_w/λ_w.

In Fig. 7, stable transition boiling at state A is shown. With a random wall temperature increase (decrease), the heat transfer to the boiling liquid will be higher (lower) than the heat removal rate from the additional heat carrier, thus promoting the return of the wall temperature to its values at the point of equilibrium A.

In Fig. 7, the state of equilibrium A is unstable since in the case of a random wall temperature deviation from the equilibrium, the difference between the transferred and the removed heat quantities leads to greater deviation, which will cease upon achieving stable states A' or A''. This condition occurs when transition boiling is studied using an additional heat carrier with insufficiently high heat-transfer coefficient k (Fig. 7) and one fails to obtain stable transition boiling for the section BC (Fig. 7). This occurs in references [7, 20]. The studies under a step increase or decrease of T_0 show hysteresis, i.e., an erroneous overestimation of q_{cr2} and ΔT_{cr2} with heating or underestimation of q_{cr1} and ΔT_{cr1} when the wall temperature is decreasing.

The study of transition boiling with heating by an additional heat carrier is convenient for high-boiling liquids and cannot be applied to cryogenic fluids [2]. This is associated with the fact that the temperature range of the additional heat carrier is limited by the triple point–critical point interval (below the triple point the condensing vapor is a solid and above T_{cr} the heat-transfer coefficients α_0 drastically decrease). Therefore the choice of the additional heat carrier for any high-boiling liquid is not difficult, whereas it is hampered for cryogenic fluids in some temperature ranges.

Recently a combined heating method that involves direct electric heating and the use of an additional heat carrier has been suggested by Poletavkin et al. [25] and has been widely applied [9, 24, 26–29]. The main energy source is heat generation in the wall. Heat transfer to an additional working body provides a stable process, the stability condition being the same (34) as for heating with an additional heat carrier. Most interesting is the case where the temperature of the additional heat carrier T_0 is kept equal to the mean temperature of the wall surface:

$$T_0 = \overline{T}_1 = \overline{T}_2 + \frac{q_h b_w^2}{2\lambda_w} \tag{35}$$

where q_h is the volumetric heat release density.

Here the mean heat flux to the additional heat carrier is zero, thus ensuring a simple measurement of the heat flux on the boiling surface:

$$q_2 = q_h b_w$$

since q_h is easily determined by the current intensity and voltage.

In some works [4, 30–33] transition boiling was provided by electric heating with manual or automatic control of the heat generation with feedback from the temperature sensor at the heating surface. The accuracy of the wall temperature measurements and the time constants of the control systems are not indicated in the studies. Sakurai and Shiotsu [31, 32] have used a high-speed analog computer in an automatic system. However, the characteristics of the system evidently did not allow stable transition boiling as the experimental data are obtained for the heating surface temperature increasing or decreasing at the rate of 2 K/sec.

Denisov [34, 35] has studied transition boiling with slow electronic heating and cooling of a large-size heat source (the time of nucleate-to-film boiling transition was 120 sec, while the reverse transition took 2000 sec).

The unsteady-state calorimeter method is widely used to study transition boiling as well as other boiling modes, particularly for cryogenic fluids. In this case boiling occurs on the surface of a body the heat content of which decreases due to heat removal to the boiling medium. In this case the entire boiling curve including the transition region is realized with monotonic decrease of the temperature drop [26–28, 36–42]. The unsteady calorimeter method is known for simple experimental equipment and procedure but imposes large requirements on data processing.

In the above-mentioned studies the heat flux was measured by

1. the rate of the condensate accumulation when studying saturated liquid boiling [8, 9, 18, 20, 21, 29],

2. measuring the heat content, mass flow, and inlet and outlet temperature differences [16, 17] or by the condensation rate of heating vapor [19, 23] in case of heating from an additional heat carrier,

3. electric power [7, 22, 24, 31, 34, 43],

4. the temperature gradient in the test section when measuring temperature at different points deep into the body [4, 7, 26, 34, 44], and

5. using a heat balance [36, 37, 39, 42, 45, 46], from the solution of the inverse heat-conduction problem [28, 40, 47] or direct heat conduction problem with a boundary condition of the first kind [37] when using the unsteady calorimeter method.

The wall temperature was measured by

1. thermocouples near the heating surface (in most of the studies),

2. temperature field extrapolation when measuring temperature at different points deep in the test section [4, 26, 44];

3. measuring the electric resistance of the test section [16, 23, 31];

4. measuring the heating vapor temperature taking into account the wall temperature drop [19],

5. measuring the adiabatic surface temperature taking into account the

wall temperature drop determined from the inverse heat conduction problem solution in the case of the unsteady calorimeter method and small thickness of the test wall [28, 40], and

   6. using an optical pyrometer [33].

The following methods have been used to study the physical mechanism of the process:

   1. visual observations, photography and filming, and optical methods,
   2. vapor content measurements in the wall region,
   3. temperature fluctuation measurements on the heating surface, and
   4. measurement of the intensity and spectral composition of the boiling sound.

The vapor content in the wall region has been measured with regard to the difference in the electric properties of liquid and vapor. Temperature pulsations on the heating surface have been studied with microthermocouples. The representative sensors are shown in Fig. 8.

Information on experimental studies of transition boiling heat transfer and its physical mechanisms is given in Table I [48–54a]. The majority of these works are concerned with pool boiling of saturated liquid at atmospheric pressure and normal gravity. The exceptions are cited under the heading "Variable factors."

B. EXPERIMENTAL STUDIES OF THE TRANSITION BOILING MECHANISM

In the majority of the experimental works dealing with studies of the transition boiling mechanism, the laws of this process have been examined qualitively with regard to such factors as availability of periodic liquid–wall contacts, dry spots appearing on the heating surface, etc. In a number of works, quantitative data were obtained for the frequency and time of liquid and vapor phase contact with the heating surface and their dependence on the governing factors (temperature head, subcooling, etc.).

The following experimental methods were used to study the transition boiling mechanism:

   1. Visual observations, photography, high-speed photography, optical methods,
   2. steam quality measurements in the wall region,
   3. heating surface temperature pulsation measurements,
   4. measurement of the intensity and spectral composition of the sound appearing during boiling.

Based on the analysis of the results of visual observations, photographs,

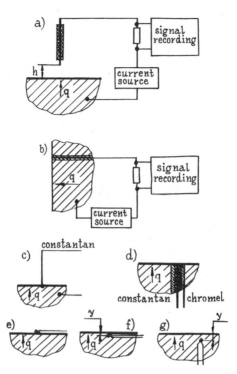

FIG. 8. Schemes for measuring the vapor content and temperature pulsations on the heating surface. (a) and (b): Vapor-content probes with outer (a) and inner (b) electrodes; (c), (d), and (e): surface thermocouples with outer electrode (c), inner electrode (d) with electrode supply to the heating surface (e); (f) and (g): embedded thermocouples with the electrode supply along the heating surface (f) and from the depth of the test section (g).

and high-speed photographs, numerous investigators came to the conclusion that in the transition boiling region, wetted and dry spots exist at each instant on the heating surface, and on rather long heating surfaces each area of the wall appears to be periodically either wet or dry.

In references [9, 10, 19, 33, 34, 55, 56] it is shown that dry spots on the heating surface appear at some temperature head $\Delta T_{DNB} < \Delta T_{crl}$. Farber and Scorah [33] consider that at the point of the departure from stable nucleate boiling (DNB), the heat transfer coefficient α_f is a maximum. Ishigai and Kuno [19] determined the quantity $(\Delta T_{crl} - \Delta T_{DNB}) = 5$ K with saturated water boiling on a copper vertical tube. With subcooling available, this difference may substantially increase. Tolubinsky et al. [43, 55] observed the onset of local unsteady vapor films with strongly subcooled water boiling ($\Delta T_{sub} = 70$ K; $p = 0.29$ MPa) on copper horizontal

TABLE I

EXPERIMENTAL STUDIES OF TRANSITION BOILING

| Actuating medium | Material | Heating surface | | | | Heating technique | Variable factors | Note | Ref. |
		Shape	Dimensions (mm)	Orientation	Roughness				
Water	Nickel, chromel	Cylinder	$D = 1$ $L = 152$	Horizontal		Electroheating, manual control	$p = 0.1 - 0.8$ MPa		[33]
Methanol	Copper	Cylinder (tube)	$D = 9.5, 6.35$ $L = 203$	Horizontal	Polishing	Water-vapor condensation		Effect of large molecular impurities	[48]
n-Pentane	Copper, nickel, In-conel	Flat disk	$D = 50.8$	Horizontal	Polishing, lapping E60, E120, D160, E320	Water-vapor condensation		Effect of heating surface wettability	[7]
Nitrogen Nitrogen	Copper Copper	Cylinder Sphere	$D = 15.9$ $D = 25.4$ $D = 12.7$	Horizontal	Polishing	Nonstationary calorimeter	$g/g_0 = 0.02 - 1$		[49] [37, 46]
Freon-12 Freon-113 Freon-114	Nickel	Cylinder (tube)	$D = 14$ $L = 560$	Horizontal	$R = 0.6$ μm; grooves $h = 0.5$ mm	Auxiliary heat-transfer agent (water)	$p = 0.05 - 3$ MPa		[16]
Water	Copper	Flat rectangle	35×43	Horizontal		Electrical heating of bulky body with nonisothermal bulge		Vapor-content measurements near heating surface	[44]
Water	Copper, brass, stainless steel	Cylinder	$D = 4$ $L = 60$	Horizontal		Electrical heating	$p = 0.1 - 0.5$ MPa, $\Delta T_{sub} = 5 - 120$ K	The region under consideration $\alpha_{max} - q_{max}$	[43]

Freon-113	Stainless steel	Flat rectangle	120 × 12.7 120 × 6.35			Water-vapor condensation		bubble departure and their distribution over surface	
Water, ethanol	Copper	Flat disk	$D = 12–25$	Horizontal, $\gamma = 60°$		Quasistationary heating and cooling	$\Delta T_{sub} = 0–40$ K	Filming, acoustic studies	[34, 35]
Freon-114		Inner surface of tube	$D = 14$	Horizontal		Auxiliary heat-transfer agent, combined heating	$p = 0.5–1.5$ MPa $\rho W = 1200–1400$ kg/m² sec		[17]
Water	Copper	Flat disk	$D = 10;\ 35$	Horizontal, vertical, $\gamma = 180°$		Combined heating	$p = 2–100$ kPa	Filming 1000 frames per second	[9, 51]
Water	Copper, silver	Sphere	$D = 19;\ 25.4$		Polishing	Nonstationary calorimeter	$\Delta T_{sub} = 23$ K, 40 K, 76 K $W = 1.5–6.1$ m/sec	Filming 4000–5000 frames per second	[38, 42]
Freon-113	Copper, coatings with low heat conductivity	Flat disk	$D = 36.6$	Horizontal			Effect of nonisothermicity of heating surface		[50]
n-Pentane	Copper	Flat disk	$D = 50.8$	Horizontal		Water-vapor condensation	Measurements of pulsations and acoustic signal		[8]
Water, Freon-113, nitrogen	Stainless steel, copper	Cylinder, torus	$D = 12.7$ $L_c = 76–152$ $R_t = 67.5$	Horizontal	Lapping E500	Lapping E500	Electroheating, nonstationary calorimeter	Comparison between stationary and nonstationary processes	[39]

(Continued).

TABLE I *(Continued)*

Actuating medium	Heating surface					Heating technique	Variable factors	Note	Ref.
	Material	Shape	Dimensions (mm)	Orientation	Roughness				
Water	Copper	Cylinder (tube)	$D = 40$ $b = 1, 4$ $L = 86, 106$	Vertical	Lapping E60, E320	Water-vapor condensation		Measurements of volume vapor-content and pulsations	[19]
Water	Copper	Flat disk	$D = 36$	Horizontal		Water-vapor condensation			[20]
Water	Copper	Sphere	$D = 59.7$			Nonstationary calorimeter	$\Delta T_{sub} = 5-72$ K		[36]
Nitrogen	Copper, stainless steel, coating $\phi\pi$-3	Inner surface of annular channel	$D = 10$ $b_{ct} = 0.02-0.16$ $b_w = 0.5$ $b_{cup} = 2$	Vertical		Nonstationary calorimeter	$p = 0.17-1.7$ MPa $\Delta T_{sub} = 6-29$ K $W = 0.4-8$ m/sec		[53]
Water	Copper	Flat disk	$D = 8$	Horizontal	Lapping E1000	Combined heating		Measurements of vapor-content pulsations and T_w	[29]
n-Pentane	Gold-plated copper	Flat disk	$D = 25.4$	Horizontal		Water-vapor condensation			[22]
Water	Copper	Cylindrical channel	$D = 12.7$ $L = 104.8$	Vertical		Nonstationary calorimeter	$\Delta T_{sub} = 0-28$ K $W = 68$ kg/m² sec	Vapor-content studies near heating surface	[47, 54]
Water	Platinum	Cylinder	$D = 1-1.4$ $L = 50-100$	Horizontal Vertical		Computer-controlled	$T_H = 0-40$ K		[31, 32]

Fluid	Material	Shape	Dimensions	Orientation	Surface finish	Method	Conditions	Purpose	Ref.
Nitrogen	Copper, nickel, stainless steel	Flat disk, cylinder	$D = 8–36$ $D = 12$	Horizontal Vertical	$R_z = 0.6–14$ μm	Combined heating, nonstationary calorimeter		Comparison between stationary and nonstationary processes	[26, 27]
Water, ethanol, their mixtures	Copper	Cylinder (tube)	$D = 10$ $b = 2.15$ $L = 235$	Horizontal		Water-vapor condensation			[21]
Water	Copper	Flat disk	$D = 29$	Horizontal	Polishing			Study of volume vapor content near heating surface	[4, 30]
Nitrogen	Copper, copper with Teflon coating	Sphere	$D = 25.4, 19, 12.7$		Shot blasting	Nonstationary calorimeter		Effect of vertical oscillations of sphere ($f = 0–12$ Hz)	[45]
Nitrogen	Stainless steel, coating teflon	Cylinder	$D = 10, 20$ $b_w = 0.5–2$ $b_{ct} = 0.01–0.02$	Vertical		Unsteady calorimeter			[52]
Water, Freon-113	Copper	Flat circle	$D = 20$	Horizontal	$R_z = 0.5–6.4$	Combined heating	$\Delta T_{sub} =$ 0–24 K 0–18 K 0–54 K	Measurement of T_w pulsations	[54a]
Nitrogen	Copper, teflon-covered copper	Plate, disc, cylinder, tube (inner surface, sphere)	90×150 $D = 150$ $D = 50$ $d = 48$ $D = 50$	Horizontal Vertical Inclined ($\gamma = 0–180$)		Unsteady calorimeter	$\Delta T_{sub} = 0–10$ K; $p = 0.025–2.1$ MPa		[54a]

tubes at a point ($\Delta T_{DNB} = 15$ K, $q_{DNB} = 1.76$ MW/m^2), i.e., long before the nucleate boiling crisis occurs ($\Delta T_{cr1} = 75$ K $q_{cr1} = 4$ MW/m^2).

All the authors consider that the vapor film in the transition boiling region is hydrodynamically unstable, as a result of which the liquid, from time to time, approaches the heating surface. Further behavior of the liquid depends on the temperature head. In the low-temperature zone of the transition boiling region in the vicinity of the first boiling crisis at the liquid–wall contact place, there arises nucleate boiling characterized by a microfilm burnout under a growing bubble [9]. At high temperature heads, the liquid wets the wall for a short period of time and boils up explosively [34, 42]. Finally, at still higher temperature heads, the liquid rushes toward the heating surface but is thrown back due to intense boiling [34, 57, 58]. In the last case, unstable film boiling occurs and is characterized by elevated heat transfer as compared with stable film boiling. At a further increase in ΔT_W, the vapor film collapses and stable film boiling [33, 34] takes place. When the temperature head is decreased, these processes proceed in the opposite direction. The temperature boundaries at which these different mechanisms of the process in the transition boiling region come into play are unknown. Indirect indications point to the idea that the liquid–wall contacts are possible in the vicinity of the film boiling crisis (e.g., ΔT_{cr2} is decreased with artificial deterioration of the heating surface wettability; see Fig. 9).

The above mechanism of this process is supported by the interesting observation made by Ishigai and Kuno [19]: insufficient purification of the water results in more intense contamination of the heating surface in transition boiling than in nucleate boiling. This fact suggests that there are liquid–wall contacts in the transition boiling region as well as complete liquid film burnout leaving scale deposits on the heating surface, whereas in nucleate boiling conditions the wall layer is impregnated with liquid, as a result of which a greater amount of admixture remains in the liquid.

As our observations have shown, the behavior of the transition boiling process strongly depends on the liquid subcooling up to the saturation temperature. A relatively thick vapor film with local breaks is formed during boiling of saturated and weakly subcooled liquid. With increasing subcooling, the film thickness is decreased and the number of breaks is increased. The nature of the boiling process sharply changes when passing through some threshold value of subcooling. There appear on the heating surface regions occupied by extremely small vapor spots (0.2–0.3 mm) forming a dull surface. In this regime, called "cavitation," transition boiling is accompanied by the specific high-intensity noise. The threshold subcooling, under which there occurs a transition to cavitation boiling, is decreased with increasing temperature head and decreasing heating surface roughness.

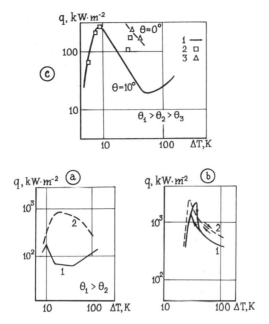

FIG. 9. Effect of the heating surface wettability on transition boiling heat transfer: (a) water–copper surface (from Nishikawa *et al.* [29]): (1) artificial bad wettability, (2) pure surface: (b) water–platinum surface (from Sakurai and Shiotsu [32]): (1) without admixture, (2) with surfactant added; (c) *n*-pentane–copper surface (from Berenson [7]); (1) without admixture, (2) with surfactant added, (3) maximum concentration of surfactant.

Denisov [34], Tolubinsky *et al.* [43, 55] investigated the acoustic signal in the transition boiling region with water boiling while Aoki and Welty [59] carried out a similar study with pool *n*-pentane boiling. The sound signal under different boiling conditions was received by a hydrophone and was either recorded on magnetic tape with a subsequent spectral analysis [34, 59] or immediately passed via the amplifier into a spectrum analyzer. In Denisov [34], high-speed photography was carried out while in reference [43] photographs of the process were made.

Denisov [34] found that transition boiling is characterized by a specific pulse acoustic signal. With increasing ΔT_{w}, this signal appears while approaching the first boiling crisis, achieves maximum loudness and frequency in the transition region, and ceases in stable film boiling. These pulses probably appeared under rapid evaporation of the liquid impinging on the hot surface at the moment that the vapor film breaks. The shape of the sound pulse and pulse frequency and duration regularly varied with changing temperature head. The pulse nature of the sound signal in the transition region is also supported by the data of Ishigai and Kuno [19] who observed the initiation of low-explosive sounds in transition boiling and classified it as one of its important differences from nucleate boiling.

Tolubinsky *et al.* [43, 55] studied the acoustic signal with $\Delta T_W < \Delta T_{crl}$ and emphasized that the pulse signal formed between the points α_{max} and q_{max}. In this case, the pulse frequency corresponded to one that would be characteristic of a forming vapor film (50–70 Hz). Nucleate boiling is characterized by continuous noise with a wide high-frequency spectrum (1–20 kHz) having one or several maxima. An increase in subcooling, heat flux and roughness resulted in a widening of the higher-frequency range. For high heat fluxes, when vapor bubbles coalesced and local unsteady vapor films were formed, low-frequency components (200–800 Hz) appeared in the acoustic signal spectrum and the amplitude of the high-frequency vibrations decreased. At α_{max} the spectrum became completely low frequency (f < 1000 Hz) and, as a rule, high-frequency components vanished. A further increase in the heat flux was accompanied by the formation of a pulse signal with a discrete spectrum in a wide frequency range. Thus, the analysis of the sound signal suggests that on the boiling curve (between α_{max} and q_{max}) the mechanism of the process is more close to transient than to nucleate boiling. Taking into account the differences in the mechanism of this process, the authors of references [43, 55] proposed that the heat-transfer crisis occurs when individual bubbles start to coalesce, thus forming unstable vapor films, i.e., the condition α_{max} but not q_{max} is satisfied.

Aoki and Welty [59] conducted acoustic studies in all regions of the boiling curve and found that in the transition region, the sound pressure achieves its maximum and the loudest sound was observed within the frequency range of 250 to 500 Hz.

It should be noted that the present acoustic methods used to study transition boiling yield little information on the mechanism of this process. First, this is due to the fact that undirected microphones are used, receiving sound waves from different points of the heating surface and the boiling medium as well as from the waves reflected from the vessel walls, the phase interface, and the free liquid surface. As a result, the analyzed signal is a superposition of numerous waves arriving from different elementary emitters and having different direction, frequency, phase, and amplitude. Second, even in the case of a signal captured by a narrow-beam microphone, it is difficult to interpret, since the sound generation mechanism is as complex as the heat-transfer mechanism (sound waves appear with generation, growth, coalescence, departure, and collapse of vapor bubbles, with phase interface vibrations, etc.).

Apparently, acoustic methods might be very useful to establish the boundaries of the different mechanisms, since a qualitative change of the mechanism results in sharp changes in the sound signal. These are a transition from a continuous noise to a pulse noise at α_{max}, with noise ceasing in the vicinity of q_{cr2}, a transition through the maximum of the sound intensity in the middle of the transient region, and the appearance of a specific audible signal during "cavitation" boiling of the subcooled liquid. Unfortunately, in the majority of the acoustic studies, the main attention was directed to the determination of the quantitative characteristics of the sound but not of the boundaries where sharp changes occurred.

Ishigai and Kuno [19] and Iida and Kobayasi [4] studied the transition boiling mechanism for water by measuring the vapor content near the heating surface on vertical tubes. Similar measurements were carried out by Nishikawa et al. [29] and Rubin and Roizen [44] on horizontal spherical surfaces in pool boiling and by Ragheb et al. [47] and Stevens and Witte [60] in ascending vertical channel flow. Measurements of the vapor content were based on the substantial difference in the electric conductivity of water vapor and liquid water.

In Ragheb et al. [47], the electric circuit had a gap between the heating surface and the zirconium 1.02-mm-diameter electrode embedded flush with this surface and electrically insulated from the test section with a zirconium oxide layer. In the remaining works, the investigated region of the boiling medium was between the heating surface and the electrode (probe) located normal to this surface and at some distance from it.

In references [19, 29, 47], the time variation of the dc signal (voltage on the resistance switched into the circuit) was recorded, and in this case increasing current (voltage) corresponded to decreasing volumetric vapor content in the gap. In references [4, 44], an ac current of 3 to 5 kHz passed through the discriminator pulses that corresponded to large [4] or small [44] electric resistance in the gap. The discriminated signal was recorded on the oscillograph and was accumulated in the pulse counter, allowing determination of the mean residence time of the vapor (or liquid) on the noninsulated point of the probe, i.e., φ in Iida and Kobayasi [4] or $(1 - \varphi)$ in Rybin and Roizen [44].

Experimental results are presented in Fig. 10. For convenient comparison, the data of some authors are rearranged to have, on the ordinate axis, the same quantity

$$\tau_L = \frac{t_L}{t_L + t_V} = 1 - \varphi$$

the relative residence time of the liquid near the heating surface. In all cases, the fraction of time spent for wall–liquid contacts decreases with increasing temperature head. According to Iida and Kobayasi [4], the dry spots on the heating surface appear in the nucleate boiling region at a heat

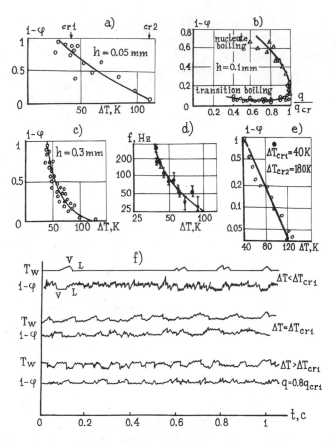

FIG. 10. Vapor content probes used for studying the transition boiling mechanism. $(1 - \varphi)$ is the relative liquid residence time near the heating surface, h the gap between the probe and heating surface. (a) (from Iida and Kobayasi [4]); (b) (from Nishikawa *et al.* [29]). (c) and (d) (from Rybin and Roizen [44]), (e) subcooled water flow in the channel [47], (f) comparison of wall temperature pulsations and vapor content (from Ishigai and Kuno [19]).

flux, $q_{\mathrm{DNB}} = 0.9q_{\mathrm{cr1}}$; at the point of the first boiling crisis, $1 - \varphi \approx 0.8$ (i.e., dry spots occupied about 20% of the area), and at the point of the second boiling crisis $1 - \varphi \approx 0.04$ (i.e., 4% of time was spent with liquid–wall contacts). At the same time Nishikawa *et al.* [29] obtained substantially different results: even at $q = 0.6q_{\mathrm{cr1}}$ in the nucleate boiling region, the fraction of liquid $(1 - \varphi)$ greatly differs from one and sharply falls to $1 - \varphi \approx 0.05$ near the point of the first crisis. Such a great difference in the reported data is apparently attributed to the strong effect of the probe height h between the probe and the heating surface due to large vapor content gradients near the heating surface [4] as well as due to the possible influence of the probe on the boiling process at small h.

Ragheb et al. [47] used a probe with the electrode supplied through the wall (Fig. 8b) to measure the vapor content on the heating surface. The location of the probe inside the wall has certain advantages over one in the boiling medium: first, the probe location does not affect hydrodynamic processes near the wall; second, the electric circuit is closed when the liquid contacts the wall on the electrically insulated layer between the electrode and the heating surface. A gradual change in the signal of the vapor content probe was observed under unsteady cooling of a bulky copper unit with the flow of saturated or subcooled water through a cylindrical channel in this unit. In the film boiling region, the "dry" impingements of the liquid on the wall (without wetting), the point at which the signal pulses, is close to the initial one (before the liquid gets into the channel). In the transition region, the nature of the signal changed, which was responsible for some wetting of the wall. As long as the wall was cooled, the wetting lasted longer, and in the nucleate region, the signal stabilized about the level showing that there was a certain amount of the liquid in the electric circuit gap. The discontinuous wall wetting started near ΔT_{cr2}, whereas the continuous one occurred near ΔT_{cr1}. However, the accuracy of determining the initiation of the second and first boiling crises and ΔT_{cr1} and ΔT_{cr2} [47] was not high because the experimental heat-transfer data were determined through the indications of one thermocouple embedded at the channel middle and the heat fluxes over the bulky high heat-conducting block were not taken into account.

Ishigai and Kuno [19] made simultaneous measurements of the signals of the vapor content probe and the surface thermocouple, the electrodes of which were shaped as a copper heating surface and a constantan 0.2-mm-diameter wire, normal to the surface from the boiling medium side and soldered into a 0.4-mm-diameter boring and at a 0.3-mm depth (Fig. 8c). The water boiling was studied in a vertical 40-mm-diameter tube heated inside by vapor injected from the inner tube through a set of holes at a velocity of 40 m/sec. The vapor content probe was located near the thermocouple electrode and at a distance of 0.2 mm from the wall. The measurement results are shown in Fig. 10f. Under nucleate boiling, far from ΔT_{cr1} the thermocouple did not experience vibrations and the probe recorded a high-frequency signal corresponding to the nucleation conditions. Near ΔT_{cr1}, the high-frequency probe signal was occasionally and irregularly interrupted by a signal typical of the vapor phase and followed by a heating-surface temperature splash. This points to a brief onset of a dry wall section at the location of the thermocouple and probe. At maximum heat flux, the liquid drying became more frequent and, on the average, lasted longer. In the transition region, the picture did not change qualitatively: the vapor content probe recorded a high-frequency signal typical either of nucleate boiling or of an area corresponding to a vapor phase; the wall temperature varied simultaneously with the vapor signal (it decreased when nucleate boiling occurred and increased when dry spots were formed. The reciprocal coincidence of the signals of the vapor content probe and the surface thermocouple allowed the thermocouple to be used for studying the phase alternation on the heating surface, which is of importance in the understanding of the transition boiling mechanism and in the establishment of quantitative heat-transfer laws.

Comparison of the vapor content probe and the thermocouple signals (Fig. 10f) shows that the proper processing and interpretation of the vapor contact measurements involve great difficulties, since vapor content vibrations are far from being consistent with the phase alternation on the heating surface. A formal (and moreover mechanical) account of the vapor content signal vibrations will lead to a systematic overestimation of the phase alternation frequency. This evidently depends on the high frequencies of the liquid–wall contacts (Fig. 10d) near the first boiling crisis. Indeed, as is seen from Fig. 10f, the high frequency of a change in the electric resistance of the circuit between the probe and the heating surface has nothing in common with the low-frequency process of the formation and disappearance of dry spots. Hence, it follows that the transition boiling mechanism results obtained by a vapor content probe should be treated only as qualitative information. The heating surface temperature pulsation thermogram may be used to determine the quantitative characteristics of the process (frequency of the phase alternation and the time of surface heating for each phase contact), but in this case, it should be borne in mind that the wall temperature pulsation does not correspond to the phase alternation.

Heating-surface temperature pulsations in transition pool-boiling heat transfer were investigated by Ishigai and Kuno [19] as well as by Kesselring et al. [18], Nishikawa et al. [29], Aoki and Welty [8], Kalinin et al. [52]. Water [29, 29], freon-113 [18], n-pentane [7] and liquid nitrogen [52] served as working liquids. Test sections, copper (in the majority of cases) or stainless steel [18] included vertical cylinders [9, 52] and bodies with horizontal rectangular [18] or circular [8, 29] heating surfaces. Low-inertia thermocouples were embedded directly on the heating surface or near it with thermoelectrodes through the wall or boiling medium (Fig. 8c–f) and used to measure temperature pulsations. A different combination of liquids, boiling conditions, probe locations, processing methods, and data presentation hampers the comparison of the data presented in the above-mentioned works.

As the measurement of surface temperature pulsations involves substantial errors, the majority of the authors analyzed the pulsation frequency and fraction of time associated with decreasing and increasing temperature. The difficulties associated with processing T_W pulsation measurements are connected with the irregularity or disordering of the process as well as with the possibility that very small pulsations are not taken into account (starting wih a certain level, the measuring devices cannot record them).

Numerous authors considered the interval between the start of decreasing wall temperature and the beginning of the subsequent temperature rise to be the liquid–wall contact time t_L. Only in Aoki and Welty [8], it was assumed that the liquid contacted the wall only during the decreasing temperature period, and then the wall temperature was kept at the low level due to spheroid boiling and started increasing only after complete evaporation.

Figure 11 illustrates different experimental frequencies of the liquid–wall contact f, the mean contact time t_L, and the interval between contacts t_V as functions of the temperature head ΔT_W. Comparison of the data [19, 29] shows that irrespective of the heating surface orientation, the contact time t_L decreases with increasing temperature head from several tens of milliseconds to several milliseconds (the authors [17] did not allow for small temperature pulsations T_W in processing the obtained data and this, probably, led to overestimated values of t_L at high ΔT_W. In Aoki and Welty [8] it was found that $t_L = 2$ msec, which did not depend on ΔT_W; this result, which contradicts the data of other authors, apparently, points to the inaccuracy of the processing methods (i.e., interpretation of t_L as a period of a decreasing wall temperature without regard for the time interval during which T_W remained at a low level).

The interval t_V between contacts on vertical surfaces is somewhat reduced as the temperature head grows. This may be attributed to the increasing hydrodynamic instability of the interface with increasing vapor velocity. The analysis of Fig. 11b shows that on horizontal surfaces as ΔT_W increases, t_V increases as well, perhaps because of vapor film thickening and insufficient change of the factors responsible for the hydrodynamic stability of the interface.

The frequency of liquid–wall contacts at the first boiling crisis rapidly increases with the temperature head; it then decelerates, showing a slow decrease of frequency with temperature head in some cases. In Nishikawa et al. [29], which details the measurement of temperature pulsations by thermocouples of two kinds (with an external electrode and one under the heating surface at $y = 0.2$ mm deep), a discrepancy in the experimental data was observed at high ΔT_W values. The analysis of primary thermo-

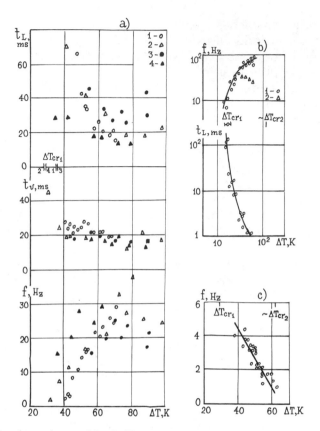

FIG. 11. Results on the transition boiling mechanism obtained from the measured temperature pulsations on the heating surface. (a) Water–copper surface, vertical tubes (from Ishigai and Kuno [19]): 1 and 2, tube thickness $b_w = 1$ mm; 3, 4, $b_w = 4$ mm, 1–3, smooth chrome-plated surface; 2, 4, rough surface (E-60); (b) water–copper surface, horizontal disk (from Nishikawa et al. [29]): 1, thermocouple with outer electrode, 2, embedded thermocouple, (c) n-pentane–copper, horizontal disk (from Aoki and Welty [8]).

grams of this work shows that when determining f the authors [29] take into account all the pulsations including small T_w fluctuations during vapor insulation of the heating surface rather than considering only pulsations typical of the liquid–wall contact (with a sharp temperature drop). It is therefore more reasonable to consider the frequencies otained on the thermocouples with the internal electrode, which only responded to sharp wall temperature fluctuations. The data on the liquid–wall contact frequencies obtained in Aoki and Welty [8] differ from the results of other investigators both in the dependence on the temperature head and in the order of magnitude. The reasons for such a discrepancy lies in the interpretation of the thermograms and in their processing.

Ishigai and Kuno [19] studied temperature pulsations of the heating surfaces during transition boiling on the test sections with different wall thicknesses and roughness (Fig. 11a). For a wall 4 mm thick, the surface roughness results in shorter liquid–wall contacts, which may be accounted for by the increasing density of active nucleation sites. The duration of the interval between contacts is independent of the wall roughness or thickness since it is determined by the hydrodynamic instability of the vapor film. Therefore, with increasing height of the microroughnesses, the time fraction for high-intensity enhanced heat transfer to liquid decreases and the mean heat flux falls in the transition region as well. In the nucleate boiling region at $\Delta T_W < \Delta T_{DNB}$ with continuous wetting of the wall, an increase in the number of the nucleation sites provides a higher heat-transfer rate. As a result, with increasing heating-surface roughness both the nucleate and transition sections of the boiling curve $q(\Delta T_W)$ are shifted to lower temperature drops (Fig. 12). However, no roughness effect on t_L, t_V, and f is observed with test sections having a wall 1 mm thick, though the data of the same study point to equal dependence of heat transfer on the surface roughness for sections of different thickness.

It is worth noting the following relationship [19]: Assuming a linear approximation for $q(\Delta T_W)$ in the nucleate boiling region and extrapolating it to temperature heads $\Delta T_W > \Delta T_{DNB}$, and then neglecting the heat transfer to the vapor, one may determine with sufficient accuracy the heat flux for the transition boiling section adjacent to the first crisis:

$$q_{tr} \approx \alpha_f(\Delta T_W - \Delta T_0)t_L/(t_L + t_V) \qquad (36)$$

This is important qualitative evidence that the dominant contribution is nucleate boiling on the wetted heating surface areas as compared to the mechanism of heat transfer in the low-temperature transition region. At high temperature-head values, the experimental values of q_{tr} are less than those predicted by relation (36), which indicates a variation in the heat-transfer mechanism. The duration of the liquid–wall contact becomes negligible. So, nucleate boiling fails to develop and heat removal is only due to liquid heat conduction, and as a result the heat transfer intensity decreases.

The authors of this review have studied transition boiling of water, ethanol, freon-113 and nitrogen on horizontal and vertical heating surfaces.

An analysis of the thermograms for transition boiling of saturated and subcooled liquids has exposed the fact that the heating-surface temperature fluctuations are independent of subcooling. The representative oscillograms for the horizontal heating surface are given in Figs. 13–15.

The transition boiling region adjacent to the first crisis is characterized by the same temperature fluctuations as those near the critical nucleate

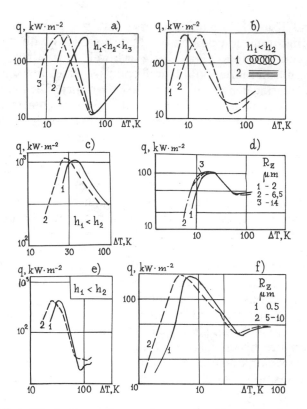

FIG. 12. Effect of the heating-surface roughness on transition boiling heat transfer. (a) and (b) *n*-pentane–copper surface (from Berenson [7]); (c) water–copper surface (from Ishigai and Kuno [19]); (d) ethanol–copper surface (from Nishikawa *et al.* [61]); (e) liquid nitrogen–steel surface (from Grigor'ev *et al.* [27]); (f) liquid nitrogen–copper (from Grigor'ev *et al.* [26]).

boiling region where the abrupt wall-temperature drop is followed by small-amplitude fluctuations. As the temperature head increases, the duration of these situations decreases sharply. For small subcoolings this process involves an essential rise of the duration and amplitude of sharp heating-surface temperature jumps with insignificant increases in their frequency, whereas large subcoolings are characterized by an appreciable growth of the frequency of high-T_W pulsations associated with small increases of dry-spot duration. Thus, the physical mechanism of the process in the near-critical nucleate boiling region is similar to that in the adjacent transition region. The transition of the heat flux through the maximum stems from the gradual decrease of the relative wall-wetting time

$$\tau_L = t_L/(t_L + t_V)$$

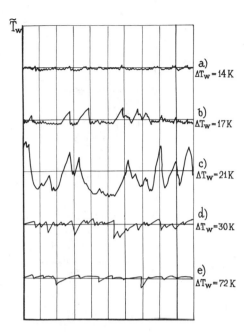

FIG. 13. Typical oscillations of the heating surface temperature with saturated liquid boiling and small subcooling (water, $\Delta T_{sub} = 0$ K).

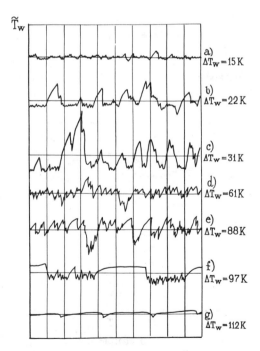

FIG. 14. Typical oscillations of the heating surface temperature with mean subcooling (water, $\Delta T_{sub} = 6$ K).

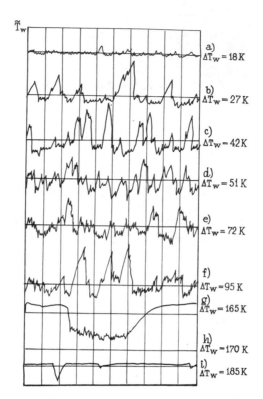

FIG. 15. Typical oscillations of the heating-surface temperature with strongly subcooled liquid boiling (water, $\Delta T_{sub} = 24$ K).

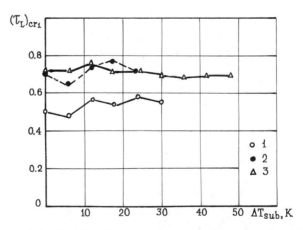

FIG. 16. Mean relative liquid–wall contact time at a point q_{max} under different subcoolings, 1, ethanol; 2, water; 3, Freon-113.

with increasing wall superheat. In this case, the value of τ_L for each of the tested liquids at point q_{max} is practically independent of subcooling (Fig. 16).

At a certain temperature drop in the middle of the transition boiling, the thermograms typical of nucleate boiling disappear and are replaced by rather stable wall-temperature pulsations, in which a sharp drop is followed by an abrupt rise. With increasing liquid subcooling, the boundary of the transition boiling with some responses typical of nucleate boiling shifts to higher temperature heads.

As the second boiling crisis is approached, some reduction in the frequency and amplitude of the wall temperature fluctuations is observed at small subcoolings. A set of large subcoolings is characterized by high-frequency fluctuations involving essentially temperature drops and jumps together with dominant fluctuations moderate in their amplitude and duration.

A comparison of the response of surface thermocouples at different distances from the test center do not suggest qualitative differences in the boiling mechanism.

Distribution histograms are presented for the liquid–wall contact time t_L in Fig. 17 for several temperature-head and subcooling values. The statistical hypothesis of the coincidence of the random t_L distributions at

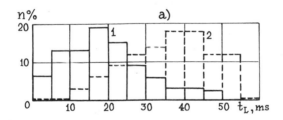

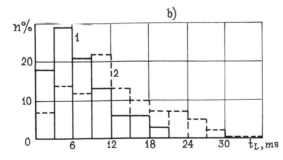

FIG. 17. Histograms of the liquid–wall contact time distribution with ethanol boiling. (a) $\Delta T_w = 19$ K; (b) $\Delta T_w = 29$ K. 1, $\Delta T_{sub} = 0$ K; 2, $\Delta T_{sub} = 30$ K.

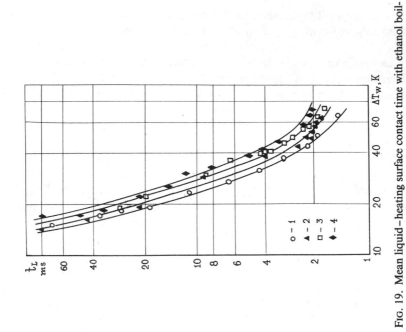

FIG. 19. Mean liquid–heating surface contact time with ethanol boiling. 1, $\Delta T_{sub} = 0$ K; 2, $\Delta T_{sub} = 6$ K; 3, $\Delta T_{sub} = 12$ K, 4, $\Delta T_{sub} = 18$ K.

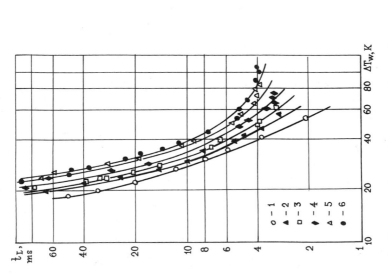

FIG. 18. Mean liquid–heating surface contact time with Freon-113 boiling. 1, $\Delta T_{sub} = 0$ K; 2, $\Delta T_{sub} = 12$ K; 3, $\Delta T_{sub} = 24$ K; 4, $\Delta T_{sub} = 36$ K; 5, $\Delta T_{sub} = 48$ K; 6, $\Delta T_{sub} = 54$ K.

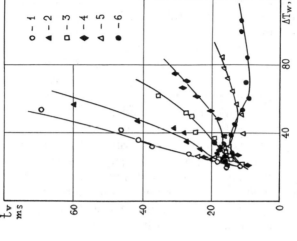

FIG. 21. Mean vapor phase–heating surface contact time with Freon-113 boiling. 1, $\Delta T_{sub} = 0$ K; 2, $\Delta T_{sub} = 12$ K; 3, $\Delta T_{sub} = 24$ K; 4, $\Delta T_{sub} = 36$ K; 5, $\Delta T_{sub} = 48$ K; 6, $\Delta T_{sub} = 54$ K.

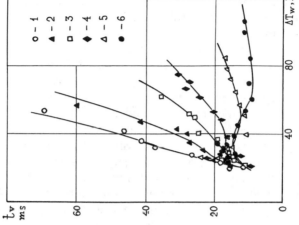

FIG. 20. Mean liquid heating surface contact time with water boiling. 1, $\Delta T_{sub} = 0$ K; 2, $\Delta T_{sub} = 6$ K; 3, $\Delta T_{sub} = 12$ K; 4, $\Delta T_{sub} = 18$ K; 5, $\Delta T_{sub} = 24$ K.

279

different heating-surface points for similar boiling regimes has been tested, yielding χ^2 with a significance of 0.05 to show that the experimental data are in agreement with the hypothesis. As the temperature head increases, the distribution curve t_L shifts to lower values and attains a more explicit maximum. An increase of subcooling, on the contrary, accounts for the flat-top distribution and shifts to higher time values.

The mean parameters of the physical transition boiling mechanism are presented in Figs. 18–29. For all the test liquids, the mean contact time t_L drops with increasing temperature head and increases with subcooling. The mean interruption time τ_L between liquid–wall contacts generally grows with increasing temperature head and falls with increasing subcooling. At high subcoolings only, t_V attains a maximum at the center of the transition region, then decreases slowly with rising ΔT_W. The liquid–wall contact frequency f (which may also be considered as the dry-spot frequency) with low-temperature heads increases at any subcooling, $\partial f/\partial T_W$ being practically independent of the subcooling. But for saturated or slightly subcooled liquids at the center of the transient region the frequency achieves a maximum and then gradually decreases. In the case of subcooled liquids,

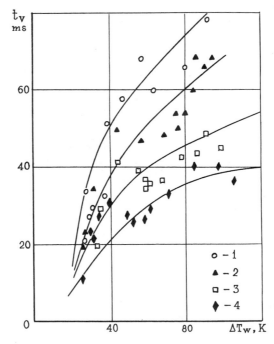

FIG. 22. Mean vapor phase–heating surface contact time with ethanol boiling. 1, $\Delta T_{sub} = 0$ K; 2, $\Delta T_{sub} = 6$ K; 3, $\Delta T_{sub} = 12$ K; 4, $\Delta T_{sub} = 18$ K.

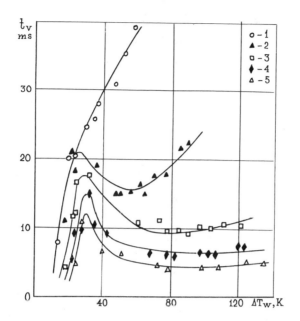

FIG. 23. Mean vapor phase–heating surface contact time with water boiling. 1, $\Delta T_{sub} = 0$ K; 2, $\Delta T_{sub} = 6$ K; 3, $\Delta T_{sub} = 12$ K; 4, $\Delta T_{sub} = 18$ K; 5, $\Delta T_{sub} = 24$ K.

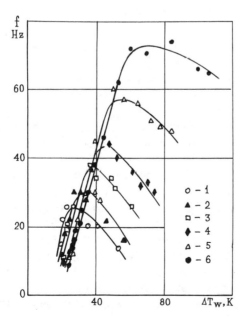

FIG. 24. Frequency of liquid–wall contacts with Freon-113 boiling. 1, $\Delta T_{sub} = 0$ K; 2, $\Delta T_{sub} = 12$ K; 3, $\Delta T_{sub} = 24$ K; 4, $\Delta T_{sub} = 36$ K; 5, $\Delta T_{sub} = 48$ K; 6, $\Delta T_{sub} = 54$ K.

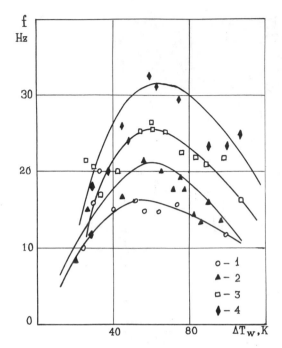

FIG. 25. Frequency of liquid–wall contacts with ethanol boiling. 1, $\Delta T_{sub} = 0$ K; 2, $\Delta T_{sub} = 6$ K; 3, $\Delta T_{sub} = 12$ K; 4, $\Delta T_{sub} = 18$ K.

f increases only to the second boiling crisis and only a slight decrease in the derivative $\partial f / \partial T_W$ is observed.

Changing the heating-surface roughness does not alter either the physical mechanism of the process or its characteristics. As the height of the microroughnesses decreases, some shift of t_L values to higher temperature heads is observed, the difference decreasing with the growth of subcooling. No roughness affect on the liquid–wall contact frequency is observed (Fig. 30).

The impact of the heating-surface orientation on the physical mechanism of transition boiling has been studied with the installation turned 90° to make the heating surface vertical. In this case, the liquid–wall contact time changes insignificantly, and the pulsation frequency for the vertical orientation is about twice as high as the values for the horizontal surface (Figs. 31, 32). However, with the vertical heater, a significant inhomogeneity of the process is observed with different physical behavior at the lower, central, upper, and side heating-surface areas.

The physical mechanism of the film boiling crisis essentially depends on liquid subcooling. In the case of a saturated liquid with low subcooling, a

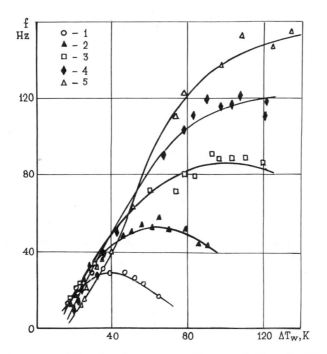

FIG. 26. Frequency of liquid–wall contacts with water boiling. 1, $\Delta T_{sub} = 0$ K; 2, $\Delta T_{sub} = 6$ K; 3, $\Delta T_{sub} = 12$ K; 4, $\Delta T_{sub} = 18$ K; 5, $\Delta T_{sub} = 24$ K.

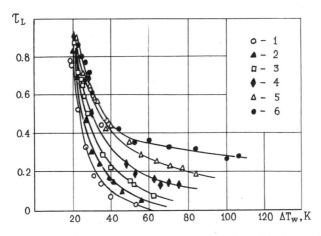

FIG. 27. Mean relative liquid–heating surface contact time with Freon-113 boiling. 1, $\Delta T_{sub} = 0$ K; 2, $\Delta T_{sub} = 12$ K; 3, $\Delta T_{sub} = 24$ K; 4, $\Delta T_{sub} = 36$ K; 5, $\Delta T_{sub} = 48$ K; 6, $\Delta T_{sub} = 54$ K.

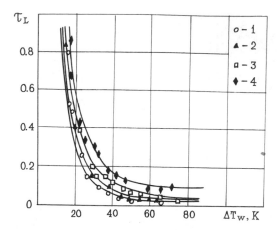

FIG. 28. Mean relative liquid–heating surface contact time with ethanol boiling. 1, $\Delta T_{sub} = 0$ K; 2, $\Delta T_{sub} = 6$ K; 3, $\Delta T_{sub} = 12$ K; 4, $\Delta T_{sub} = 18$ K.

smooth transition to film boiling is observed due to the gradual cessation of vapor film breaking around ΔT_{cr2}. During the departure of the vapor bubbles from the film in the nucleate boiling region, the heating surface thermocouple yields a 10–20-Hz frequency of temperature fluctuations. For growing ΔT_W the frequency essentially decreases, although large-amplitude temperature drops are observed occasionally. This presumably

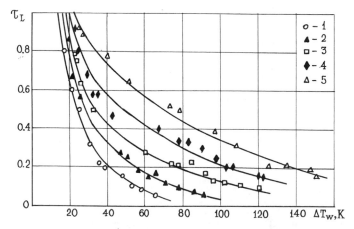

FIG. 29. Mean relative liquid–heating surface contact time with water boiling. 1, $\Delta T_{sub} = 0$ K; 2, $\Delta T_{sub} = 6$ K; 3, $\Delta T_{sub} = 12$ K; 4, $\Delta T_{sub} = 18$ K; 5, $\Delta T_{sub} = 24$ K.

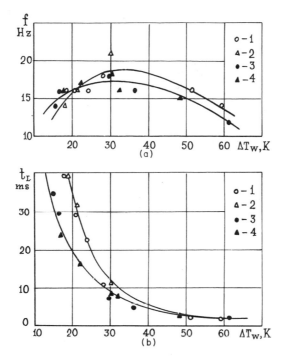

FIG. 30. Effect of the heating surface roughness on the frequency (a) and the liquid–wall contact time (b) with ethanol boiling. 1 and 2, mean microroughness height $R_Z = 0.54 \; \mu$m; 3 and 4, $R_Z = 6.4 \; \mu$m; 1 and 3, saturated liquid; 2 and 4, $\Delta T_{sub} = 6$ K.

means that with increasing temperature head and vapor film thickness, a still smaller portion of the interface pulsation causes a short-time liquid–wall contact due to bubble departure.

For moderate and large subcoolings, the onset of the second boiling crisis induces a multiple change in film and transition boiling. At a certain temperature head, the entire heating surface is spasmodically covered with a thin vapor film, which is also spasmodically changed by transition boiling over the entire surface (at large subcoolings) or on part of the surface (at moderate subcoolings). This interchange of boiling regimes proceeds at 2–10 Hz and involves spasmodic variation of the heating surface temperature by 10–15 K. This phenomenon covers the range of temperature heads between 10 and 20 K somewhat increasing with growing subcooling. Further rise of the wall superheating results in a stable thin vapor film. Still greater growth of ΔT_W leads to film thickening and roughness. During bubble departure from the film, the thermogram shows sharp large-amplitude temperature drops typical of short-time liquid–wall contacts.

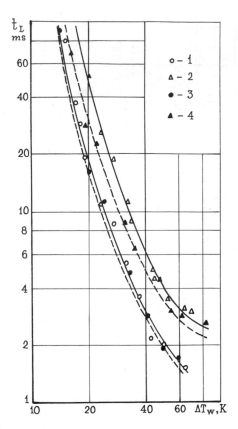

FIG. 31. Effect of the heating-surface orientation on the mean liquid–wall contact time with ethanol boiling. 1 and 2, Horizontal orientation; 3 and 4, vertical orientation; 1 and 3, saturated liquid; 2 and 4, $\Delta T_{sub} = 24$ K.

C. CORRELATION OF THE EXPERIMENTAL DATA ON THE TRANSITION BOILING MECHANISM

The characteristics obtained for the transition boiling mechanism as a function of the process parameters coincide well with the theoretical results. As a result, mathematical models may be used to correlate the obtained experimental data.

From Figs. 18–20, 30, and 31 it is seen that irrespective of the heating surface orientation, the liquid–wall contact time is sharply increased at low temperature heads (approximately according to the law $t_L \propto \Delta T_W^{-4}$) and with increasing ΔT_W the rate of change decreases and tends to $t_L \sim \Delta T_W^{-2}$. As shown in Section III, the relationship $t_L(\Delta T_W)$ must be of this

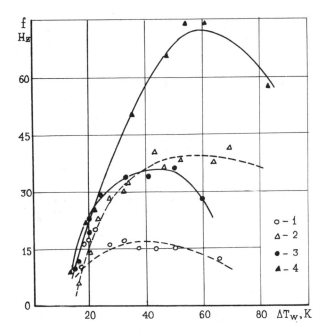

FIG. 32. Effect of the heating surface orientation on the frequency of liquid–wall contacts with ethanol boiling. 1 and 2, Horizontal orientation; 3 and 4, vertical orientation; 1 and 3, saturated liquid; 2 and 3, $\Delta T_{sub} = 24$ K.

form since at small ΔT_W, the greatest part of the contact time is spent for bubble growth and coalescence (duration of this process is proportional to ΔT_W^{-4}) and at large ΔT_W the contact time is limited by the nucleation, which is less dependent on the temperature head ($\propto \Delta T_W^{-2}$). Thus, the empirical constants in (27) and (28) must be determined to correlate the experimental data on liquid–wall contact time. The use of the experimental values of t_L for three liquids: water, ethanol, and Freon-113 (data of the present authors) yields

$$\frac{a_L t_L \rho_{vs}^2 r^2 \, \Delta T_W^2}{4\sigma^2(1 + \cos\theta)^2 T_S^2} = \left(490 + \frac{1.34 \times 10^6}{Ja^2}\right)\left(1 + 0.025\frac{\rho_L C_{pL} \, \Delta T_S}{\rho_{vs} r}\right) \quad (37)$$

Relationship (37) is shown in Fig. 33, and the correlation error does not exceed 30% (in this case, the range of dimensional values of t_L covers two orders of magnitude).

The right hand side (RHS) of expression (37) may be divided into two components, one of which, similar to Eq. (22) does not depend on the Jacob number and the other, similar to Eq. (23) is proportional to Ja^{-2}.

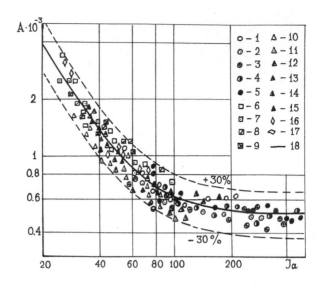

FIG. 33. Correlation of the experimental data on the mean liquid–heating surface time.
1–15, test sections with roughness $R_Z = 4.1\ \mu m$; 16 and 17, $R_Z = 0.54\ \mu m$. 1–5, Water,
$\Delta T_{sub} = 0$–24 K; 6–9, ΔT_{sub} = ethanol, $\Delta T_{sub} = 0$–18 K; 10–15, Freon-113, $\Delta T_{sub} = 0$–
54 K; 16 and 17, ethanol, $\Delta T_{sub} = 0$ K; 18, by Eq. (37),

$$A = \frac{a_L t_L \rho_{vs}^2 r^2\, \Delta T_w^2}{4\sigma^2 (1 + \cos\theta)^2 T_s^2}\left(1 + 0.025\frac{\rho_L C_{pL}\, \Delta T_{sub}}{\rho_{vs} r}\right)^{-1}$$

Hence, the dimensionless formulas are obtained for the components of t_L:
the time of unsteady liquid heating due to heat conduction t_T and the time
of growing bubbles up to their coalescence into a continuous vapor film
(t_{col})

$$\frac{a_L t_T \rho_{vs}^2 r^2\, \Delta T_w^2}{4\sigma^2 (1 + \cos\theta)^2 T_s^2} = 490\left(1 + 0.025\frac{\rho_L C_{pL}\, \Delta T_s}{\rho_{vs} r}\right); \qquad (38)$$

$$\frac{a_L t_{col} \rho_{vs}^2 r^2\, \Delta T_w^2}{4\sigma^2 (1 + \cos\theta)^2\, \Delta T_s^2} = \frac{1.34 \times 10^6}{Ja^2}\left(1 + 0.025\frac{\rho_L C_{pL}\, \Delta T_s}{\rho_{vs} r}\right) \qquad (39)$$

The experimental data on the heating-surface – vapor-phase contact time
are correlated within 30% by the dimensionless formula

$$t_V\left(\frac{\rho_L g^3}{\sigma}\right)^{1/4} = 21 \times 10^6 \frac{C_{pV}\, \Delta T_w}{r}\frac{\mu_V a_V}{\rho_L g l_0^3}\left(1 + 0.04\frac{\rho_L C_{pL}\, \Delta T_s}{\rho_{vs} r}\right)^{-1} \qquad (40)$$

Formula (40) is shown in Fig. 34.

When the transition boiling mechanism is studied on a high heat-conducting heating surface, the amplitude of the temperature pulsations is negligibly small, as compared to the mean temperature head, i.e., $\Delta T_{\text{w}} \approx \Delta T_b \approx \Delta T_{\text{w0}}$. For this case, a comparison of the correlating formulas (38), (40) and relationships (27) and (29) gives the empirical parameters entering into expressions (27) and (29) as

$$K_1 = 1.96 \times 10^3; \qquad K_2 = 5.36 \times 10^6 (1 + \cos\theta)^2; \qquad K_3 = 21 \times 10^6 \frac{\mu_{\text{v}} a_{\text{v}}}{\rho_{\text{L}} g l_0^3}$$

$$K_{\text{S}} = 0.025\, \text{Ja}; \qquad K_{\text{S}}' = 0.025; \qquad\qquad\qquad K_{\text{S}}'' = 0.04$$

$$(41)$$

Coefficients K_{T}, K_{col}, K_{V} are determined by formulas (27) and (29) with regard for (41).

An effect of surface roughness on the coefficients is not found, since it is comparable in value to the point scatter in dimensionless coordinates.

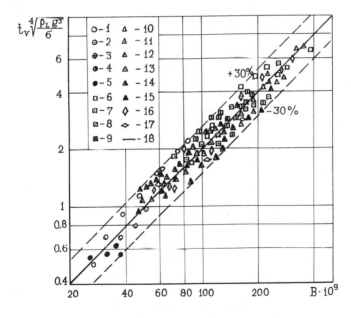

FIG. 34. Correlation of the experimental data on the mean vapor phase–heating surface contact time. 1–15, Test sections with roughness $R_Z = 4.1\,\mu\text{m}$; 16 and 17, $R_Z = 0.54\,\mu\text{m}$. 1–5, Water; $\Delta T_{\text{sub}} = 0$–24 K; 6–9, ethanol, $\Delta T_{\text{sub}} = 0$–18 K; 10–15, Freon-113, $\Delta T_{\text{sub}} = 0$–54 K; 16 and 17, ethanol, $\Delta T_{\text{sub}} = 0$ K; 18, by Eq. (40)

$$B = \frac{C_{p\text{v}} \Delta T_{\text{w}}}{r} \frac{\mu_{\text{vs}} a_{\text{vs}}}{\rho_{\text{L}} g l_0} \left(1 + 0.04 \frac{\rho_{\text{L}} C_{p\text{L}} \Delta T_{\text{sub}}}{\rho_{\text{vs}} r}\right)^{-1}$$

Formulas (37)–(41) may be used for pool transition boiling on horizontal high heat-conducting ($\varepsilon_W \gg \varepsilon_L$) surfaces with roughnesses of $R_Z = 0.5–6$ μm for liquids that are thermodynamically similar to water, ethanol, and Freon-113 (i.e., having a dimensionless saturated-vapor pressure within the range $P_S/P_K = 0.0046–0.0168$ at a dimensionless temperature $T/T_K = 0.625$).

D. Experimental Results for Transition Boiling Heat Transfer

Transition boiling is a complex physical phenomenon involving unsteady processes in the liquid and vapor phases of the test fluid, in the wall, and at the boundaries of these media. This results in a great number of factors that affect transition boiling heat transfer: parameters of the test section (shape, size, and roughness of the heating surface, thermophysical properties of material, boiling medium properties (hydrodynamic, thermodynamic, thermophysical), process conditions, characteristics of the force fields and the test section orientation relative to these fields, free or finite liquid volume velocity of the channel liquid, or external flow past a test section, pressure, and subcooling of the boiling liquid.

The great number of governing factors pose considerable difficulties in studying this process and analyzing and interpreting the data obtained, since it is almost impossible to vary only one factor while keeping constant the remaining parameters of the process.

Therefore, the results of different authors, even under similar conditions, diverge greatly. It is expedient to analyze the reasons for this divergence.

1. *Effect of Heating Surface Characteristics*

The difference in the heat-transfer data obtained by different authors on one and the same liquid under similar external conditions is, first of all, the result of the difference in test section characteristics. A knowledge of the effect, on the heat transfer of such factors as roughness and wettability of the heating surface, the shape and size of the test section, and the thermophysical properties of the wall material are of paramount importance.

The effect of the heating-surface roughness on the shape of the boiling curve was studied by Berenson [7] on n-pentane, Ishigai and Kuno [19] on water, Nishikawa *et al.* [61] on ethanol, Hesse [16] on Freon-114, and Grigorev *et al.* [26–28] on liquid nitrogen. Saturated liquid pool boiling was investigated on heating surfaces of relatively high heat-conducting materials. The results obtained are shown in Fig. 12.

As a rule, with decreasing height of the microroughnesses on the heating surface, the transition boiling region contracts due to an increase in ΔT_{crl}

for a constant value of ΔT_{cr2}. When the effect of roughness on heat transfer in the transition boiling region is analyzed, it should be kept in mind that film boiling heat transfer does not depend on the microroughness height R_Z (if this height does not exceed the vapor film thickness δ_v) and in nucleate boiling, the roughening of the heating surface considerably enhances the heat-transfer rate due to the increasing density of nucleation sites at a given temperature head. The nucleate boiling crisis related to the hydrodynamic instability of the two-phase flow is determined by the nucleation rate, i.e., by the heat-flux density, and the quantity q_{cr1} (which practically does not depend on the surface roughness) and ΔT_{cr1} as shown on the nucleate boiling curve. Thus on smooth surfaces the heat flux attains its critical value at higher values of ΔT_{cr1}. The film boiling crisis is governed by the thermodynamic liquid instability and does not depend on the heating-surface roughness (at $R_Z < \delta_v$). Thus, the effect of the microroughness height on heat transfer in the transition region depends on the boiling crisis parameters. Under the same temperature heads, the heat flux in transition boiling is higher on the smoother surfaces (in nucleate boiling, vice versa, the heat transfer is higher on rough surfaces). It is obvious that at equal ΔT_W in the transition region, a smaller part of the area of smooth surfaces is occupied by a vapor film, which results in increasing heat removal. With increasing ΔT_W, the roughness effect in the transition region is decreased, which causes less of a contribution of nucleate boiling (which is sensitive to roughness) to the total heat flux.

For boiling of cryogenic liquids, which are characterized by high wettability, an interesting and important specific behavior is observed. As is shown in references [26, 28], the effect of roughness on transition and nucleate boiling of cryogenic liquids is found only on very smooth surfaces, and heat transfer does not depend on the roughness on more rough surfaces (Fig. 12e, f). In transition boiling, this similarity zone (where heat transfer does not depend on roughness) starts approximately at a microroughness height $R_Z = 2\,\mu m$ and in nucleate boiling, at $R_Z = 6-7\,\mu m$. The similarity boundary shifts to the side of decreasing R_Z at transition boiling, as compared to nucleate boiling. This is explained by the relative contribution of nucleate boiling to the total heat flux [26, 28]. As the value of this contribution in the vicinity of q_{cr1} is large, it may be expected that in this region the similarity boundary of the heat transfer with respect to roughness must approach the one typical of nucleate boiling. However, from Fig. 12f it is seen that, up to q_{cr1}, the roughness from 2 μm up does not affect the heat fluxes in the transient region. Probably, this is attributed to the different mean time of nucleate boiling at the liquid–wall contact for different-roughness surfaces. On surfaces with $R_Z = 6\,\mu m$, where α_n is lower than with $R_Z = 2\,\mu m$ at q_{cr1} a relatively larger contact time occurs during which the same amount of vapor necessary to repel the liquid from the wall is formed.

The existence of a similarity zone in transition boiling heat transfer of cryogenic liquids with respect to the roughness is of importance in the study of transition boiling and the interpretation of the obtained data. It is therefore possible to obtain the dependence of transition boiling heat transfer of cryogenic liquids on such factors as the shape and size of the test section, the heating-surface material, etc. as well as to compare the data of different authors if the roughness of their test sections is in the similarity zone.

For highly boiling liquids, the similarity zone of heat transfer which is independent of the roughness, is not observed, and therefore it should be borne in mind in the analysis of the data of different authors that the difference in the heating surface roughness causes an uncertainty and hampers the comparison of experimental data.

As Berenson [7] has shown, the boiling process is affected not only by the miroroughness height but also by the depth and shape of recesses depending on the method of finishing a surface with a given grinding composition. In particular, circular grinding promotes recesses shaped as semiclosed cavities, which are more effective nucleation sites than are scratches obtained with longitudinal grinding even if these scratches have a larger depth (Fig. 12b). Owing to this fact, large difficulties appear when comparing the data of different authors for same-roughness surfaces.

The effect of the boiling liquid wetting the heating surface was studied by Berenson [7] on n-pentane, Nishikawa et al. [29] on saturated water, and Sakurai and Shiotsu [32] on subcooled water. All pool boiling experiments were made at atmospheric pressure on flat surfaces and horizontal high heat-conducting (copper, platinum) cylinders.

In references [7, 32], the wettability was improved by adding a small amount of a surfactant (oleinic acid for n-pentane, oleinic acid sodium for water) that decreases the contact angle θ (for n-pentane from $10°$ to $0°$) without changing the thermophysical and thermodynamic properties of the boiling medium. In Nishikawa et al. [29], on the other hand, deterioration of the surface wettability was produced artificially. Experimental data are shown in Fig. 9. As is seen, the results are not single valued. According to the data of reference [29] the values of q_{cr1} and T_{cr1} are increased with decreasing θ (Fig. 9a) and according to the data of reference [32], the value of q_{cr1} does not vary even while the value of ΔT_{cr1} decreases with decreasing θ (Fig. 9b). The results of reference [7] do not throw any light on this problem since there are no points on the boiling curve at $\theta = 0°$ in the temperature-head range $\Delta T_w = 10-25$ K (Fig. 9c). However, in all cases, the heat transfer is substantially increased in the transition region with decreasing contact angle, and the wettability does not affect heat transfer in the nucleate region. Berenson [7] has pointed out that transition boiling heat-transfer augmentation at $\theta = 0°$ gives evidence that there are liquid–wall contacts in this temperature-head range. In view of this, it should be noted that for the lower boiling curve in Fig. 9a under film boiling conditions ($\Delta T_W = 50-110$ K), the liquid may contact the wall hydrodynamically and only the bad wettability does not promote transition boiling. Transition boiling (upper curve, Fig. 9a) is implemented at the same temperature heads and smaller contact angle. Thus, the hydrodynamic conditions that promote liquid–wall contacts may originate even in the film boiling region.

The mechanism of the effect of the wetting angle θ on transition boiling heat transfer is, apparently, related to expanding the liquid–wall contact area with decreasing θ, other conditions being equal.

There are no data for the wetting effect on the film boiling crisis in the above-mentioned works. If film boiling heat transfer is assumed not to depend on θ, then it may be concluded (Fig. 9a) that q_{cr2} and ΔT_{cr2} are increased with decreasing θ.

It should be borne in mind that a change in the contact angle or surface tension is not the only mechanism of the influence of admixtures on the boiling curve. Lowery and Westwater [23] have investigated the effect on methanol boiling of introducing admixtures that do not vary either wetting or thermophysical properties of the liquid. High-molecular-weight compounds such as Hyamine 1622 (relative molecular mass $M = 466$) giving cations, aerosol OT ($M = 444$) giving anions, and nonionizing Span 20 ($M = 340$) served as additives. Experimental results show that the introduction of even 0.01% admixture noticeably deforms the boiling curve, whereas 0.1% admixture substantially enhances transition boiling heat transfer and q_{cr1} and ΔT_{cr1}. The addition of 1% Hyamine admixture increases heat fluxes by an order of magnitude in the transition region. In this case, the process on the descending branch of the boiling curve is represented as intensive nucleate boiling. The authors [23] explain the admixture effect by the fact that their relatively coarse molecules serve as nucleation sites. Moreover, the change in the heat-transfer rate is caused by the heating surface contamination (in the above-mentioned work, it is shown that for large admixture concentrations deposits are observed on the test section surface after the experiment). Thus, the presence of admixtures in the boiling medium may be one of the reasons for the bad reproducibility of the experimental data. Hence, high demands for the purity of the working liquid must be satisfied.

The effect of the heating surface material on transition boiling was studied by Farber and Scorah [33] for water, Berenson [7] for n-pentane, and Rhea and Nevins [45], Grigoriev *et al.* [27, 28] and the authors of the present review [3] for liquid nitrogen. Saturated liquid pool boiling experiments were made on flat horizontal surfaces, horizontal wire, spheres, and on vertical tubes. Experimental data are given in Fig. 35. The comparison of the data for liquid nitrogen boiling on different-material surfaces is the most reliable, since the wetting angle for nitrogen is almost equal to zero and the surface roughness is either constant or in the similarity region. The analysis of the experimental data supports the conclusion that with decreasing thermal conductivity of the surface material λ_w, the nucleate and transition boiling curves $q = f(\Delta T)$ shift to the right into the region of

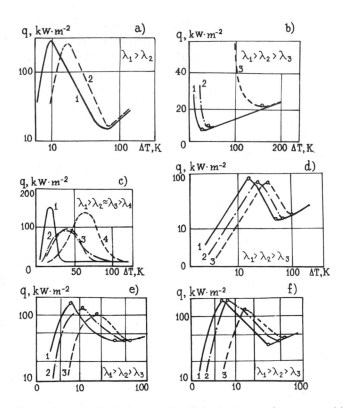

FIG. 35. Effect of thermophysical properties of the heating surface on transition boiling heat transfer. (a) *n*-pentane, horizontal surfaces (from Berenson [7]): 1, Nickel; 2, inconel. (b) Nitrogen, spheres (from Rhea and Nevins [45]): 1, pure copper; 2, oxidized copper; 3, teflon-covered copper. (c) Nitrogen, vertical cylinders (from Koshkin *et al.* [3]): 1, copper; 2, titan; 3, steel; 4, teflon. (d) Nitrogen, vertical cylinders, $R_Z = 7\ \mu$m, unsteady-state process (from Klinenko [28]): 1, copper; 2, nickel; 3, steel. (e) and (f) Nitrogen, horizontal surfaces, $R_Z = 7\ \mu$m, quasi-stationary. (e) Steady-state processes. (f) Processes (from Grigor'ev *et al.* [27]: 1, copper; 2, nickel; 3, steel.

higher temperature gradients, i.e., the nucleate boiling heat-transfer rate is decreased while the transition heat-transfer rate is increased. Film boiling heat transfer is not affected by the surface material. Temperature heads ΔT_{cr1} and ΔT_{cr2} are increased with decreasing λ_w; the values of q_{cr2} are slightly increased in accordance with the film boiling curve, and those of q_{cr1} are slightly decreased. (The more complex change in q_{cr1} shown in Fig. 35c may be attributed to the effect of process unsteadiness, which will be considered below).

The analysis of the mathematical models for transition and nucleate boiling heat transfer shows that the thermal activity coefficient $\varepsilon_w = \sqrt{(\rho c \lambda)_w}$ characterizes the effect of the heating surface material and determines the wall temperature behavior under the bubble or at the liquid–wall contact place. Different mean heating-surface temperatures correspond to equal temperatures at the liquid–surface contact location for different surface materials, which is responsible for the shift of the boiling curves and values of ΔT_{cr1}, ΔT_{cr2} to the right with decreasing ε_w.

It should be noted that in nucleate boiling, the heat transfer is higher on surfaces with larger thermal activity and vice versa in the transition region. It is obvious that in the transition boiling region at the same temperature heads on the surfaces with high thermal activity, a larger part of the area is occupied by a vapor film, which results in decreasing wall heat flux.

The strong effects of roughness, wettability, and heating-surface thermal activity on transition boiling is the reason for the high process sensitivity to the contamination and oxidation of the heating surface. This phenomenon was observed by Berenson [7] in the case of n-pentane boiling, by Nishikawa et al. [61] in the case of ethanol boiling, by Denisov [34] and Sakurai and Shiotsu [32] in the cases of saturated and subcooled water, by Veres and Florschietz [62] and Bergles and Thompson [39] in the case of Freon-113 boiling, and by the authors [40] with liquid nitrogen boiling.

The reasons for the contamination effect on transition boiling may be different. For liquid nitrogen, at $\theta = 0°$, with the heating-surface roughness being in the similarity region, the main reason is associated with the varying thermal activity of the surface layer. For n-pentane, Berenson [7] considers that this is caused by varying the contact angle from $10°$ to $0°$ due to the surface contamination and due to the surfactant addition. The effect of surfactant addition and contamination in the transient boiling region is the same, whereas in the nucleate boiling region, it is different (a change in the wettability does not involve a variation of q_n, whereas surface contamination decreases q_n by as much as 1.5 to 2 times). This may be the result of a decrease in the surface roughness due to the filling of recesses with the contaminating substance.

Nishikawa et al. [61] also report that the effect of the test section contamination with ethanol boiling may be attributed to varying wettability, since on the completely contaminated surface the contact angle was almost equal to zero and on the pure surface $\theta \approx 50°$. The boiling curve variation occurs gradually from experiment to experiment until a completely contaminated surface is attained. A similar check of the scale effect on a water–surface interface has given a smaller effect since the water contact angles on a contaminated surface and on a pure surface are close. The authors [61] note that the heating-surface contamination occurs and develops from experiment to experiment not only in the nucleate and transition regions of the boiling curve but also in the film boiling region. The test section contamination occurred because of silicon rubber dissolution in ethanol and the subsequent deposition of the soft organic scale on the wall when the liquid boiled at the place of its contact with the wall. The scale deposition under steady-state film boiling at $\Delta T_W > \Delta T_{cr2}$ supports the contention that a brief liquid–wall contact also takes place in this temperature-head region.

In the general case, the contamination of heating surfaces may, in a complicated manner, simultaneously affect roughness, wettability, and thermal activity of the heating surface, thus causing the nonreproducibility of experimental data. In addition, the deposition of the contaminating layer on the heating surface promotes an additional thermal resistance, which must be specially taken into account in the treatment of experimental data. If this is not done, this results in a systematic error in determining the temperature head.

The substantial effect of the heating surface contamination on transition boiling heat transfer requires that special care be taken in providing for the purity of the heating surface and the test liquid. When this is not done, it becomes impossible to compare the data obtained even on the same test section. This fact brings out the great difficulties involved in analyzing experimental data of a process under natural operating conditions when it is sometimes impossible to clean the scale from the surface. (In this case, a knowledge of the dynamics of the growing contaminating layer and its properties is necessary.)

On the other hand, the control of a heat-transfer process involves high sensitivity of the boiling curve to the deposition of a thin substance layer onto the heating surface. In particular, when a low heat-conducting coating is deposited on a metal surface, the nucleate and transition boiling regions shift to the high temperature-head region, placing heavy demands on body cooling even at cryogenic temperatures.

The effective thermal activity of the heating surface has been varied by covering the metal surface with a different-thickness Teflon layer (see the works of the present authors). On the heating surfaces with such a coating,

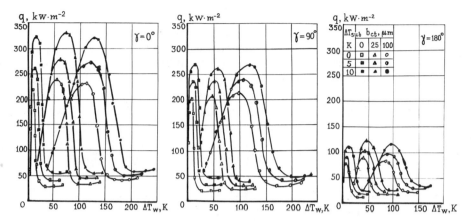

FIG. 36. Heat-flux density versus the temperature heat for various subcooling, heating-surface orientation, and low heat-conducting coating thicknesses with nitrogen boiling on the plate (90 × 150 mm).

the transition boiling section shifts to the region of higher temperature heads (Fig. 36). As is seen from Fig. 37, the above shift is increased with growing coating thickness within $b_{ct} < 100$ μm, and a further coating increase does not affect heat transfer in the transition boiling region and its temperature boundaries. This is supported by the fact that during short

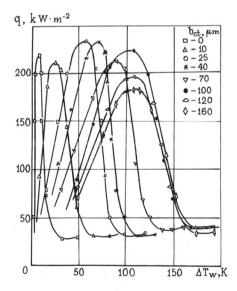

FIG. 37. Effect of the low heat-conducting coating thickness on nitrogen boiling heat transfer on the horizontal disk.

liquid–wall contacts typical of the transition boiling region on low heat-conducting surfaces there occurs a greater temperature fall than on the high heat-conducting metal surfaces without such a coating. However, at a small coating thickness, the effect of the main wall material promotes a more rapid temperature increase on the coating surface and interrupts the liquid–wall contact, i.e., in this case, the heat-transfer enhancement due to contacts is relatively small. At a rather large thickness of the low heat-conducting coating, the thermal disturbance front at the local place of the liquid–heating surface contact cannot approach the coating boundary during the contact time t_L. In this case, the main wall material cannot affect the temperature field at the contact place, and the process occurs in the same manner as it would proceed on a thick wall fabricated from the coating material.

The effective thermal activity of the heating surface is an integral characteristic of the heat exchange between the liquid and the two-layer wall at the place of their short contact. The laws for the effect of the low heat-conducting coating thickness on the temperature transition boiling range agree well with the results found for varying effective thermal activity of the two-layer copper–Teflon wall, which is decreased with increasing coating thickness at $b_{ct} < 2\sqrt{a_{ct}t_L}$ (Fig. 6) and attains a constant value $\varepsilon_{ef} = \varepsilon_{ct}$ at $b_{ct} > 2\sqrt{a_{ct}t_L}$. This law is of practical importance in determining the optimum coating thickness that allows minimization of the cooling time and coolant losses. To a first approximation, the quantity $b_{ct} = 2\sqrt{a_{ct}t_L}$ may be taken as an optimum coating thickness, providing the highest heat-transfer enhancement at high temperature heads and has the lowest thermal resistance against other coating thicknesses, resulting in the same augmentation.

2. Effect of Process Unsteadiness and Heating Surface Nonisothermality

Since transition boiling heat transfer is investigated using both steady and unsteady methods, and in engineering practice transition boiling often occurs as an unsteady process, the effect of process unsteadiness is of importance. It is necessary to know when laws found under steady conditions may be applied to unsteady processes (and vice versa) and how to account, quantitatively, for the effect of the process unsteadiness beyond the ranges of applicability.

The works of Ruzicka [41], Bergles and Thompson [39], Veres [62], Grigorev et al. [27, 28], and the authors [3] are devoted to studies of the presence or the absence of process unsteadiness on the boiling curve. The works of Berenson [7] under steady conditions, Nishikawa et al. [61], Denisov [34] under quasi-stationary conditions, and Sakurai and Shiotsu [31] deal with the investigation of the presence or absence of a hysteresis in the transition boiling region. Unfortunately, the data are rather contradictory, although a general tendency is clear: the slower the unsteady process is, the less the boiling curves differ from the steady-state curve obtained under the same conditions.

Merte and Clark [37] used an unsteady method to study liquid nitrogen boiling heat transfer and obtained a complete coincidence of the boiling curves for spheres of 12.7-mm and 25.4-mm diameter. As the values of the diameters are far beyond the heating surface sizes in the similarity region, it may be concluded that under these conditions the cooling velocity change has affected neither the position of the boiling curve extrema nor heat transfer in each of its three regions. Ruzicka [41] has found good agreement of boiling curves obtained with liquid nitrogen unsteady cooling of vertical copper tubes 10–20 mm in diameter with a 1-mm-thick wall with stationary experimental data with film boiling and in the crisis part of the nucleate region (involving the points of both boiling crises).

In the remaining works, a substantial difference in the steady and unsteady results is observed in the transition and nucleate boiling regions (Fig. 38). In the film boiling region, the unsteadiness effect is not manifested since unsteady processes proceed rather slowly. The largest difference in the steady and unsteady data is, as a rule, observed in the q_{cr1} region where the unsteady process rate is maximum.

A comparison of the boiling curves for vertical and horizontal surfaces (Fig. 38c,d) shows that the unsteadiness effect is stronger on the vertical heating surface. Klimenko [28] accounts for this by noting that for vertical tubes the removal of the vapor formed in the transition boiling regime is hindered in the unsteady process in the vicinity of q_{cr1}, thus deteriorating the nucleate boiling heat-transfer conditions, and α_n is decreased. In this work it is emphasized that in unsteady processes in the transition boiling region, the heat flux is more sharply increased with decreasing ΔT_w close to q_{cr2}. This is supported by the fact that in the steady process, liquid–wall contacts are limited by the heat removal from the deep body layers, whereas under unsteady cooling these contacts result in an avalanche-type increase of heat fluxes with decreasing wall temperature. Therefore, the steady-state data are governed by the concave transition boiling curve in the coordinates $\log q = F(\log \Delta T_w)$ and the unsteady-state ones, by the convex curve.

The effect of the process unsteadiness on the instantaneous characteristics may take the form of hysteresis in the boiling curve, i.e., of a difference in the relationships $q(\Delta T_w)$ obtained with increasing or decreasing heating surface temperature.

For saturated water in a slow (quasi-stationary) process, hysteresis was not observed [34] but at higher rates of temperature change (e.g., under

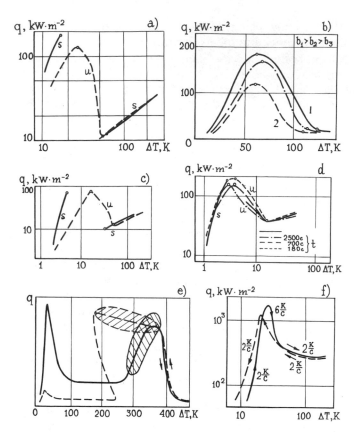

FIG. 38. Effect of the process unsteadiness on boiling heat transfer: (a) Freon-113–copper (from Veres and Florschietz) [62]): u, unsteady cooling; s, steady process. (b) Nitrogen (from Koshkin *et al.* [3]), unsteady cooling of steel tubes of various thicknesses with teflon coating of thickness $b_{ct} = 40\ \mu$m. 1, $b_w = 2$ mm; 2, $b_w = 0.5$ mm. (c) and (d), nitrogen–copper surface (from Klimenko [28]), boiling on the vertical (c) and horizontal (d) surfaces. (e) Water–copper surface, $\Delta T_{sub} = 20$ K (from Denisov [34]). (f) Water–platinum surface, $\Delta T_{sub} = 10$ K (from Sakurai and Shiotsu [31]).

subcooling conditions), a different course of the process was found under heating and cooling conditions (Fig. 38e,f). Denisov [34] reports that the hysteresis is practically eliminated if the copper heating surface is covered with a 3-μm-thick palladium layer. Hence, it may be concluded that hysteresis has appeared due to some variation of heating-surface properties (e.g., formation and clearing of the scale deposit or roughness change). In Sakurai and Shiotsu [31, 32], the effect of the process direction in the vicinity of q_{cr1} and in the nucleate boiling region may be attributed to the delay of the activation and deactivation of vaporization sites. Moreover, hysteresis in the vicinity of q_{cr1} may be a reason for the different nature of the wall nonisothermality under heating and cooling conditions.

An analysis of the data for rapid unsteady boiling cooling must also account for the possibility of a systematic methodological error, since in treating the experimental data, all the authors have assumed that the heat-flux density q is proportional to the test section cooling rate dT/dt. However, in the transition and nucleate boiling regions, where the heat-flux density strongly depends on the temperature head, high rates of varying heat flux dq/dt are observed under conditions of rapid cooling. It is therefore necessary to take into account at least the derivative d^2T/dt^2 and possibly higher-order derivatives. Moreover, in an unsteady-state process, an essential effect of the heating surface nonisothermality is possible (heat flux from the low heat-transfer-rate regions to the q_{max} region), which have not been accounted for in processing the experimental data.

The effect of the heating surface nonisothermality on boiling heat transfer was studied in the works of Denisov [34] and Petukhov et al. [63]. Experiments were conducted in pool boiling of saturated liquid at atmospheric pressure using water, ethanol, and Freon-113. In Petukhov et al. [63], the boiling occurred on the surface of a foil 0.1 mm thick and 60 mm in dia separated by a liquid gallium layer from the horizontal steel plate whose central part was electrically heated. The microthermocouple was put into the gallium layer; there it measured a temperature field used to calculate the local heat transfer at different points on the heating surface. The heating-surface temperature decreased from the center to the periphery. Temperature gradients were varied by changing the supplied power, the thickness of the gallium layer, or the steel plate. With Freon-113 boiling, temperature gradients attained 4.5 K/mm in the nucleate region and 15 K/mm in the transition or film regions; with water boiling, these values were 5 K/mm and 50 K/mm, respectively. Comparison of the boiling curves for isothermal and nonisothermal surfaces shows that both for Freon-113 and water q_{cr1} on the nonisothermal surface is decreased by 15%, q_{cr2} is increased (for Freon-113 weakly and for water as much as twice), and the whole transition boiling curve shifts to the right by 5 to 8 K for Freon-113 and by 15 to 60 K for water.

The author of reference [63] considers the surface nonisothermality to be a function of q_{cr1} due to the reciprocal hydrodynamic effect of adjacent boiling regimes. Because of the powerful liquid flow above the surface section on which q_{cr1} is implemented, there occurs a vapor-phase efflux along the surface toward this section, and the increasing vapor content yields a decreasing q_{cr1}. The smoothed form of the boiling curve is attributed to the difference in the operating conditions of nucleation sites (averaging of the nucleation conditions under a growing bubble on the nonisothermal surface). Enhancement of the heat transfer in the transition region is due to the improvement of wall temperature pulsations on the nonisothermal surface, which causes periodic chifts of the boiling boundaries. With Freon-113 boiling on the nonisothermal surface, an interesting fact is revealed: the local heat transfer depends on the presence or absence of the film boiling section. This supports the reciprocal effect of the adjacent boiling regimes on the nonisothermal surface.

It should be noted that the above results are obtained at very high values of the longitudinal temperature gradient along the heating surface when the entire length of the transition boiling section was several millimeters. It is obvious that the reciprocal effect of these various mechanisms will diminish with decreasing temperature gradient.

3. Effect of Ambient Conditions

As noted in the previous section, the overwhelming majority of experimental studies were made for transition pool boiling of saturated liquid at atmospheric pressure and normal gravity, i.e., under conditions that do not require special arrangements. Some of these investigators studied the dependence of transition boiling heat transfer on the process conditions.

The effect of the force field strength on the boiling curve behavior was examined by Merte and Clark [37] under free fall (zero gravity) and counterweight fall (lower gravity).

The curves of nitrogen surface boiling on a polished copper sphere 25.4 mm in diameter obtained in the course of unsteady cooling are presented in Fig. 39. As the strength of mass forces decreases, the heat transfer in the nucleate and transition regions does not change while it decreases in the film region. The values of ΔT_{cr1} remain constant, while ΔT_{cr2} grows

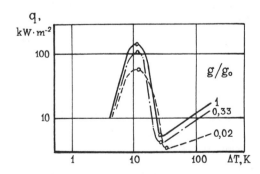

FIG. 39. Effect of the body force-field strength on boiling heat transfer (from Merte and Clark [37]).

slightly, and q_{cr1} and q_{cr2} fall (by a power of $\frac{1}{4}$). This is consistent with the concept of the hydrodynamic nature of the first boiling crises and the thermodynamic character of the second boiling crises.

Variations in the orientation of the heating surface from horizontal facing upwards ($\gamma = 0°$) to vertical ($\gamma = 90°$) and then to horizontal facing downwards ($\gamma = 180°$) involves some decrease of the heat-transfer intensity and some shift of transition boiling to lower temperature heads. This is mainly attributed to the changing conditions of hydrodynamic instability at the liquid–vapor interface, which causes the contacts between the liquid and superheated wall and to the conditions of vapor removal responsible for the vapor film thickness.

At $\gamma = 0°$ the interface is most unstable while the conditions for vapor removal are most favorable. Hence, the liquid–wall contacts appear at higher temperature drops and the interruption between contacts is short. This leads to high heat-transfer intensity in transition boiling.

With increasing γ the role of the Taylor instability due to phase-interface orientation in the gravity field decreases. Meanwhile the effect of the Helmholtz instability due to the relative phase motion along the interface increases at γ, goes from $0°$ to $90°$, and then decreases with growing γ beyond γ equal to $90°$. Thus, as the horizontal upward-facing surface changes its orientation to vertical, the two different mechanisms of hydrodynamic instability compensate each other, whereas the conditions of vapor removal are somewhat deteriorated if the height of the vertical heating surface is much in excess of the dominant wavelength. This accounts for some decrease in the heat-transfer intensity and in the transition-boiling temperature range with increasing inclination angle γ of the heating surface from 0 to $90°$.

At $\gamma > 90°$ the effect of two mechanisms of hydrodynamic instability is no longer compensated: the orientation of the phases in the gravity field becomes increasingly stable with increasing γ; the vapor velocity relative to the liquid decreases with increasing γ, thus lowering the Helmholtz instability. The conditions for vapor removal continue to deteriorate as γ increases. This results in a drastic decrease of the heat-transfer rate and a narrowing of the temperature range in the transition boiling region as γ increases from $90°$ to $180°$.

Boiling on a flat horizontal downward-facing surface ($\gamma = 180°$) demonstrates the greatest interface stability and the worst conditions for vapor removal. Both factors increase the interruptions between liquid–wall contacts and, consequently, reduce the heat-transfer rate in the transition region to a minimum.

The dependence of the transition boiling temperature range on the heating-surface orientation is most explicit on surfaces of low thermal activity. Here the thermodynamic conditions for liquid–wall contact are realized at higher temperature drops and the hydrodynamic conditions turn out to be restricting (Fig. 36).

Comparison of experimental data obtained on surfaces of different shapes and equal orientation of the test surface area points to the dominant role of the local heating-surface orientation rather than its shape. Boiling inside a cylindrical tube is an exception due to the specific conditions of vapor removal depending on the useful channel section. As compared to boiling on flat surfaces and on external cylindrical and spherical surfaces, the process of vapor removal during boiling inside pipes, is restricted, which increases the mean vapor-film thickness and the interruptions between liquid–wall contacts, thus causing lower heat removal in the transition region (Fig. 40).

The effect of liquid subcooling on transition boiling heat transfer has been studied for water by Bradfield [36], Denisov [34], Sakurai and Shiotsu [31], and Stevens and Witte [38], for ethanol by Nishikawa et al. [61], and for liquid nitrogen by the present authors. The experimental results are presented in Figs. 36 and 41.

Liquid subcooling affects transition boiling heat transfer irrespective of the thermophysical properties and the orientation of the heating surface. With increasing ΔT_{sub} the transition boiling region becomes wider and shifts to higher temperature heads, and the heat transfer is enhanced at all ΔT_W values. In contrast, heat removal in the nucleate region does not depend on subcooling, which may be attributed to two compensating

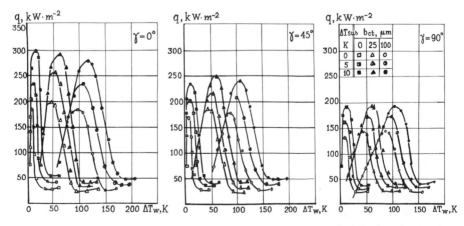

FIG. 40. Heat-flux density versus the temperature head, subcooling, heating-surface orientation, and low heat-conducting coating thickness with nitrogen boiling inside the cylindrical 45-mm-diameter tube.

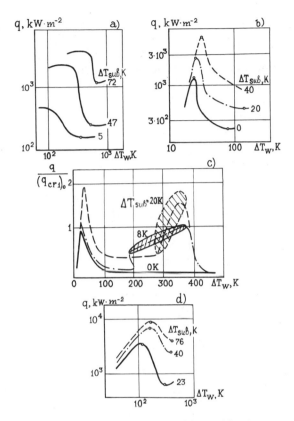

FIG. 41. Effect of the liquid subcooling on transition boiling heat transfer: (a) Water–copper surface (from Bradfield [36]), (b) water–platinum (from Sakurai and Shiotsu [31]), (c) water–copper surface (from Denisov) [34]), (d) water–silver surface, external flow (from Stevens and Witte [38]).

factors such as the increasing temperature drop between the wall and liquid and the decreasing rate of bubble growth due to subcooled liquid condensation at their caps. In the transition boiling region these very factors act in the same direction since the increasing temperature difference $(T_{\mathrm{w}} - T_{\mathrm{L}})$ increases the heat release during the liquid–wall contact while the falling bubble growth increases the duration of this contact. The decreasing rate of bubble growth in subcooled liquid also accounts for the fact that the bubbles collapse at higher temperature drop than in saturated liquid to form a solid film in some sections of the heating-surface area (constituting a departure from pure nucleate boiling). This fact explains the increase of ΔT_{DNB} with increasing subcooling. The decreasing vapor formation with increasing liquid subcooling increases the maximum heat flux that is determined by the hydrodynamic conditions of vapor removal [65]. The enhancement of heat transfer with growing superheating in the high-temperature transition boiling region adjacent to the film boiling region has the following explanation. Heat is transferred to the subcooled liquid both during liquid–wall contacts and during interruptions between contacts when the temperature of the liquid–vapor interface exceeds the temperature in the bulk of the liquid.

The pressure effect on transition boiling heat transfer has been studied for water by Farber and Scorah [33], for Freons-12, 113, and 114 by Hesse [16], and for liquid nitrogen by the present authors. Representative results are given in Figs. 42 and 43. As the pressure increases, the transition boiling region shifts to lower temperature drops and the heat transfer at a prescribed temperature drop is enhanced during nucleate and film boiling but is decreased during transient boiling. This may be attributed to shorter liquid–wall contacts since nucleation and bubble growth are favored by increasing pressure. This agrees with relationships (37) and (40).

In general, studies of the effect of various factors on transition pool boiling heat transfer confirm the dominant role of the processes taking place at the liquid–heating surface contacts.

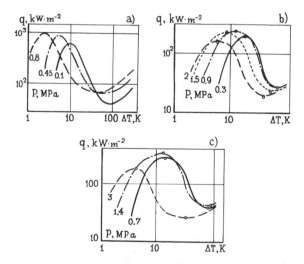

FIG. 42. Effect of the pressure on transition boiling heat transfer: (a) water–nickel surface (from Farber and Scorah [33]), (b) Freon-114–nickel surface (from Hesse [16]), (c) Freon-12–nickel surface (from Hesse [16]).

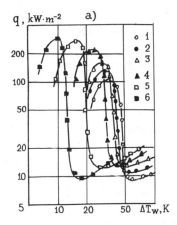 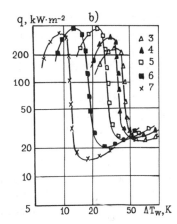

FIG. 43. Effect of the pressure on transition liquid-nitrogen boiling heat transfer. (a) $\Delta T_{sub} = 0$, (b) $\Delta T_{sub} = 10$ K. 1, $p = 25$ kPa; 2, 50 kPa; 3, 100 kPa; 4, 200 kPa; 5, 600 kPa; 6, 1200 kPa; 7, 2100 kPa.

E. GENERALIZATION OF EXPERIMENTAL DATA ON TRANSITION BOILING HEAT TRANSFER

The available formulas and methods that generalize the experimental data on transition boiling heat transfer may be divided into two groups:

1. Empirical relations that are mathematical approximations of the transition boiling curve,

2. semiempirical relations that are based on the analysis of the physical mechanism of the process.

The empirical methods of generalizing the transition boiling heat-transfer data have found wide application. Petukhov, Kovalev and others [20, 63, 66] have suggested the power-law dimensional formulas

$$q = A \, \Delta T^{-m} \qquad (42)$$

These formulas generalize the heat-transfer data for the middle part of the transition boiling region; i.e., the experimental points around ΔT_{cr1} and ΔT_{cr2} are neglected. Each series of experimental data has its own values of the parameters A and m. In the above relations the parameter m varies approximately from 2 to 4. Their simple form and ease of use are the obvious advantages of these formulas. As relations (42) do not cover the entire transition boiling region, they may be recommended only for approximate calculations. The values of the constants for Freon-113 boiling on horizontal copper surface within $27 \text{ K} \leq \Delta T \leq 50°\text{K}$ are $A = 76 \times 10^6$

kW/m^2 and $m = 3.85$ (in this case, $\Delta T_{cr1} = 21$ K, $\Delta T_{cr2} = 76$ K) for water boiling on horizontal cylinders in the same $A = 1.16 \cdot 10^6$ kW/m^2 and $m = 2.12$.

In the works of a number of authors, the proposed correlations are based on the parameters of the first and second boiling crises. Unlike formulas of the type of Eq. (42), these correlations describe the entire descending part of the boiling curve involving the regions near the crisis points. Moreover, these formulas based on the boiling crisis parameters have much in common because their use takes into account the factors affecting transition boiling heat transfer and the specifying critical heat fluxes and temperature heads. (Of course, this accounting may be incomplete, and such factors not only change the position of the points of the transition boiling curve but also its configuration.) The values of q_{cr1}, ΔT_{cr1}, q_{cr2}, and ΔT_{cr2} are taken from the experimental data for each series of experiments obtained under the same conditions to construct the correlating formulas. It should be noted that when these formulas are used in engineering practice the values of these quantities are calculated by empirical relations that are still imperfect. Therefore the calculation error may be much larger than that associated with the correlation equation and the original data.

Berenson [7] has proposed to connect the points of the first and second boiling crises in the coordinates $\log q - \log \Delta T$ by a straight line. The resulting formula may be considered to be a dimensionless version of relation (42) with the difference that it covers the crisis points and, hence, it less exactly describes the experimental curve in the middle of the transition region.

It should be noted that in reference [7], the descending part of the experimental boiling curves in the majority of cases has no points around ΔT_{cr1}, and a rather large scatter of the data is observed around ΔT_{cr2}. Nevertheless, the proposed approximation [7] is convenient for approximate calculations.

Adiutori [66a] thinks that the points of the first and second crises must be correlated by a straight line in the coordinates $q - \Delta T$.

Klimenko [28] has introduced a dimensionless parameter to account for the unsteadiness of the process in the transition boiling region to correlate numerous experimental data of different authors:

$$w = \frac{\overline{dT}}{dt} \frac{t_L}{\Delta T_{cr2} - \Delta T_{cr1}} \sqrt{\frac{(\rho C \lambda)_W}{(\rho C \lambda)_L}} = \frac{t_L}{t_{tr}} \sqrt{\frac{(\rho C \lambda)_W}{(\rho C \lambda)_L}} \tag{43}$$

where

$$\frac{\overline{dT}}{dt} = \frac{\Delta T_{cr2} - \Delta T_{cr1}}{t_{tr}}$$

is the mean cooling rate for the transition boiling time t_{tr}, t_L is the liquid–wall contact time in transition boiling and is estimated by the formula

$$t_L = \rho_L r \delta_{cr2}^2 / 2\lambda_L \Delta T_{cr2} \qquad (44)$$

Relation (44) is obtained from solving the problem of the evaporation of a liquid layer from a semi-infinite body surface when the vapor film thickness at the film boiling crisis δ_{cr2} was taken as the initial liquid film thickness δ_{L0} [28].

The mean rate of decrease of temperature was estimated by analyzing the cooling of bodies under transition boiling

$$\frac{\overline{dT}}{dt} = (m+1)\frac{q_{cr2}S}{\rho_w C_w V}\left(1 - \frac{\Delta T_{cr1}}{\Delta T_{cr2}}\right)F(\overline{Bi}, E),$$

where

$$F(\overline{Bi}, E) = \begin{cases} 1 & \text{at} \quad \overline{Bi} \leq 1 \\ 1 + 0.25\sqrt{E}(\overline{Bi} - 1) & \text{at} \quad 1 < \overline{Bi} < 100 \\ 1 + 25\sqrt{E} & \text{at} \quad \overline{Bi} \geq 100 \end{cases}$$

where

$$\overline{Bi} = \frac{V}{mS\lambda_w}\frac{q_{cr1} - q_{cr2}}{\Delta T_{cr2} - \Delta T_{cr1}}; \qquad E = \sqrt{\frac{(\rho C\lambda)_L}{(\rho C\lambda)_w}}; \qquad \text{and}$$

$$m = \frac{\log(q_{cr1}/q_{cr2})}{\log(\Delta T_{cr2}/\Delta T_{cr1})}$$

Here, V and S are the volume and heat-transfer surface of the cooled body, respectively.

Correlation of the experimental data has allowed three regions of w to be found and the configuration of the boiling curve to be described for each value of w:

$$y = 1 - x - 0.1\sin[2\pi(1-x)^2] \qquad \text{at} \quad w < 2 \times 10^{-4} \qquad (45)$$

$$y = 1 - x^{4/3} \qquad \text{at} \quad 2 \times 10^{-4} \leq w < 2 \times 10^{-3} \quad (46)$$

and

$$y = 1 - x^2 \qquad \text{at} \quad w \geq 2 \times 10^{-3} \qquad (47)$$

where

$$x = \frac{\log(\Delta T/\Delta T_{cr1})}{\log(\Delta T_{cr2}/\Delta T_{cr1})} \qquad \text{and} \qquad y = \frac{\log(q/q_{cr2})}{\log(q_{cr1}/q_{cr2})}.$$

It should be noted that we have somewhat changed the determination of the coordinate x, as compared to the one made in reference [28] so that increasing x corresponds to growing ΔT and the location of dimensionless transition boiling curves is as usual.

As is seen in Fig. 44a, the deviation of the experimental data from relations (45) through (47) is 0.2 over the logarithmic coordinate y. At a value of $q_{cr1}/q_{cr2} \approx 10$ this is consistent with an error of up to 60% in determining the heat flux density in the most favorable case when the exact values of q_{cr1}, q_{cr2}, ΔT_{cr1} and ΔT_{cr2} are known. Moreover, a certain discrepancy between the unsteadiness parameter w and its effect is observed. The scatter of the experimental data on w around the second boiling crisis is much more substantial than near the first crisis (Fig. 44c), although this is just the region of larger heat fluxes where the rate of decreasing temperature is a maximum. In addition, the boundary values of w that distinguish

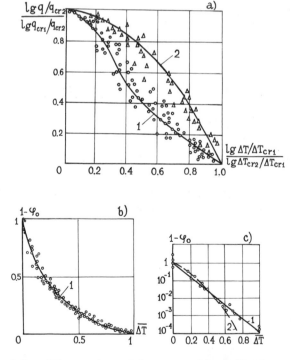

Fig. 44. Correlation of the experimental data on transition boiling heat transfer. (a) V. A. Grigoriev and V. V. Klimenko's correlation of the data (from Grigor'ev [2]): 1, by Eq. (45); 2, by Eq. (47). (b) dimensionless contact area $(1 - \varphi_0)$ versus dimensionless temperature head ΔT (from Kalinin et al. [40]). (c) $(1 - \varphi_0)$ versus ΔT (from Kalinin et al. [67]): 1, by Eq. (52); 2, by Eq. (53).

one region from another are very low. From Eq. (43) it follows that at $w = 2 \times 10^{-3}$ the cooling time in the transient boiling regime t_{tr} is very high, as compared to the time characteristic of the physical mechanism of the process, and as a result, the boiling must be of a quasi-stationary nature. In our opinion, one possible reason for the transition boiling curve configuration as a function of the parameter is the effect of the heating-surface nonisothermality, which was not taken into account when processing the unsteady experimental data. It is quite probable that the effect of this factor increases with growing mean cooling rate, and the heat overflux along the heating surface is most pronounced in the minimum heat-transfer region.

The correlation of experimental data using semi-empirical theories based on the laws of the transition boiling mechanism is most promising but it remains a rather difficult approach since it requires a deep understanding of the physics of the process. This approach has been mainly developed in references [3, 67, 68] and is being improved as new experimental data have become available.

At first, the semi-empirical correlation was based on the equality:

$$q_{tr} = [\bar{q}_T + \alpha_n(T_b - T_s)](1 - \varphi_0) + \alpha_f(T_{wo} - T_s)\varphi_0 \qquad (48)$$

where φ_0 is the time-mean area of the heating surface in contact with the vapor and $(1 - \varphi_0)$ that with the liquid. It was assumed that at the liquid–wall contact, heat is transferred by unsteady heat conduction and nucleate boiling with a heat-transfer coefficient α_n, and on the dry spots of the heating surface there occurs film boiling with a heat-transfer coefficient α_f. The values of α_n and α_f are determined by extrapolating the relations for the developed nucleate and film boiling regions to the transition region.

The value of φ_0 must be known to make calculations by formula (48). Also a knowledge of t_L is needed to calculate $\bar{q}_T$ by formulas (18) or (2). Based on the measured temperature pulsations of heating surface and design estimates of transition pool boiling of cryogenic liquids, an approximate relation is proposed

$$t_L \approx \begin{cases} 0.01s & \text{at } \Delta T_{cr1} \leq \Delta T \leq \Delta T_{cr2}; \\ \left(\dfrac{\Delta T - \Delta T_{DNB}}{\Delta T_{cr1} - \Delta T_{DNB}}\right)^{-2} \times 0.01s & \text{at } \Delta T_{DNB} < \Delta T < \Delta T_{cr1} \end{cases} \qquad (49)$$

By using the experimental values of q_{tr} and solving (48) with respect to $(1 - \varphi_0)$ we have constructed a correlating function (Fig. 44b)

$$1 - \varphi_0 = \frac{1 - \overline{\Delta T}}{1 + 3\,\overline{\Delta T}} \qquad (50)$$

where

$$\overline{\Delta T} = \frac{\Delta T - \Delta T_{DNB}}{\Delta T_{cr2} - \Delta T_{DNB}}$$

is the dimensionless temperature head. The use of $(\Delta T_{cr2} - \Delta T_{DNB})$ as the temperature scale has made it possible to correctly describe the boiling curve in the region adjacent to and on both sides of the first crisis.

However, the practical use of this correlation and its comparison with the data of other authors are hindered by the absence of correlating relationships for ΔT_{DNB} and t_L. In Kalinin et al. [67, 68], an attempt was made to simplify this approach. The assumption of unsteady heat conduction at the liquid–wall contact place with nucleate and transition boiling was made, allowing the heat flux density in transition boiling to be given as

$$q_{tr} = q'_n(1 - \varphi_0) + q'_f \varphi_0 \tag{51}$$

The function $(1 - \varphi)$ of ΔT was determined through the experimental data from (51) without additional assumptions regarding t_L. Such an approach has allowed correlation of the experimental data of different authors (Fig. 44c) by

$$1 - \varphi_0 = e^{-9.2\overline{\Delta T}} \tag{52}$$

or

$$1 - \varphi_0 = (1 - \overline{\Delta T})^7 \tag{53}$$

where

$$\overline{\Delta T} = \frac{\Delta T - \Delta T_{cr1}}{\Delta T_{cr2} - \Delta T_{cr1}}$$

Rubin and Roizen [44] correlated the data of a number of authors on water transition boiling heat-transfer by the formula

$$q = \frac{2\lambda_L \Delta T}{(1 + \sqrt{(\rho C \lambda)_L/(\rho C \lambda)_w}) \sqrt{\pi a_L}} \sqrt{(1 - \varphi_0)f} + \alpha_f \Delta T \varphi_0 \tag{54}$$

Equation (54) differs from (48) only by the absence of the component related to nucleate boiling. To determine φ_0 and the pulsation frequency $f = (1 - \varphi_0)/t_L$ Rubin and Roizen used their measurements of vapor content pulsations near the heating surface. As noted in Section IV,B, the formal processing of vapor content pulsations leads to an overestimated frequency f of dry spot formation at low temperature heads (near ΔT_{cr1}), and this has probably compensated for neglecting the nucleate boiling contribution.

During recent years the authors of the present review have performed a new correlation of transition boiling heat-transfer using a mathematical model for transition boiling heat transfer (Section III). Two versions of the correlation were made, one based on the physical transition boiling mechanism and the other on the parameters of the boiling curve points.

For a thoroughly studied specific case, pool boiling on a horizontal high heat-conducting surface, governed by the empirical constants of the mathematical model Eq. (41), the heat-flux density was calculated by Eqs. (26)–(29). When calculating each curve $q_{tr} = F(\Delta T_W)$ the values of q_f and q_n were determined by the experimental data, assuming a linear extrapolation of $q_n = F(\Delta T_W)$ to $\Delta T_W > \Delta T_{DNB}$ [19].

Comparison of the prediction with the experimental data of the present work and of the other authors is shown in Fig. 45.

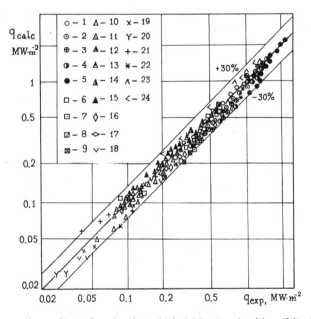

FIG. 45. Comparison of heat flux density calculated by the algorithm (26)–(29) with the data of the present work and other authors. 1–17, Data of the present authors: 1–5, water, $\Delta T_{sub} = 0$–24 K; 6–9, ethanol, $\Delta T_{sub} = 0$–18 K; 10–15, Freon-113, $\Delta T_{sub} = 0$–54 K, heating surface, copper, $R_Z = 4 \, \mu m$; 16 and 17, ethanol–copper surface, $R_Z = 0.5 \, \mu m$, $\Delta T_{sub} = 0$ K; 6 K; 18, nitrogen–copper surface (from Grigor'ev et al. [2]); 19, nitrogen–steel surface (from Lowery and Westwater [23]); 20, n-pentane–copper surface (from Berenson [7]); 21, Freon-113–steel surface (from Kesselring et al. [18]); 22–24, water–copper surface (data from Nishikawa et al. [29] and Katto and Yokoya [51]).

The predominance of the experimental points where $q_{exp} > q_{calc}$ results from the fact that the empirical parameters (4) entering into formulas (26)–(29) were determined only through temperature pulsations on the heating surface, and heat-transfer data were not used for this purpose. A maximum divergence of approximately 30% is observed in the film boiling crisis region where the determination of the experimental heat-flux density involves the largest errors. Thus, the correlation accuracy should be considered high in view of the fact that the values of the heat-flux density vary by two orders of magnitude.

Since the empirical constants given by Eq. (41) have limited applicability, the correlation of the data for transition boiling heat transfer on vertical and inclined surfaces with forced flow, on low heat-conducting surfaces, etc. cannot be made directly by formulas (26)–(29). In the general case, the correlation is therefore based on Eq. (26), and the constants of Eq. (41) are replaced by the parameters of the boiling curve characteristic points. By analogy with Eq. (38), (39), and (40), the following laws for varying characteristic intervals as a function of a temperature head are assumed:

$$t_T \propto \Delta T_W^{-2}; \qquad t_{col} \propto \Delta T_W^{-4}; \qquad t_V \propto \Delta T_W$$

The heat-flux density under film and nucleate boiling as a function of a temperature head is extrapolated to the transition boiling region by the linear functions

$$q_n = a_1 + b_1 \Delta T_W \tag{55}$$

$$q_f = a_2 + b_2 \Delta T_W \tag{56}$$

Reducing Eq. (26) to a dimensionless form, consistent with the assumptions, gives a structural form of the adopted formula for correlating experimental transition boiling heat transfer data, given by

$$Q_{tr} = \frac{B_2 \vartheta + A_2 + E\vartheta^{-1} + B_1\vartheta^{-4} + A_1\vartheta^{-5}}{1 + G\vartheta^{-3} + H\vartheta^{-5}}, \tag{57}$$

where

$$Q_{tr} = q_{tr}/q_{cr1}; \qquad \vartheta = \Delta T_W/\Delta T_{cr1}$$

$$A_2 = a_2/q_{cr1}; \qquad B_2 = b_2 \Delta T_{cr1}/q_{cr1}$$

Five unknown dimensionless coefficients (E, G, H, A_1, B_1) are determined from the conditions, at which the transition boiling curve passes through the point of departure from stable nucleate boiling (DNB) and through the boiling crisis points and from the extrema conditions at the crisis points

$$q_{tr}(\Delta T_{DNB}) = q_{DNB}$$

$$q_{tr}(\Delta T_{cr1}) = q_{cr1}$$

$$q'_{tr}(\Delta T_{cr1}) = dq_{tr}/d(\Delta T_w) = 0 \qquad (58)$$

$$q_{tr}(\Delta T_{cr2}) = q_{cr2}$$

$$q'_{tr}(\Delta T_{cr2}) = \frac{dq_{tr}}{d(\Delta T_w)} = 0$$

Substituting conditions (58) into formula (57) and passing to dimensionless temperature heads $\vartheta_i = \Delta T_i/\Delta T_{cr1}$ and dimensionless heat fluxes $Q_i = q_i/q_{cr1}$ we obtain a system of linear algebraic equations for the unknown coefficients

$$\mathbf{Ax = b} \qquad (59)$$

where the matrix **A** and vectors **b**, **x** assume the form

$$\mathbf{A} = \begin{bmatrix} 1 & \vartheta_{DNB} & \vartheta_{DNB}^4 & -Q_{DNB} & -Q_{DNB}\vartheta_{DNB}^2 \\ 1 & 1 & 1 & -1 & -1 \\ 5 & 4 & 1 & -5 & -3 \\ 1 & \vartheta_{cr2} & \vartheta_{cr2}^4 & -Q_{cr2} & -Q_{cr2}\vartheta_{cr2}^2 \\ 5 & 4\vartheta_{cr2} & \vartheta_{cr2}^4 & -5Q_{cr2} & -3Q_{cr2}\vartheta_{cr2}^2 \end{bmatrix}$$

$$\mathbf{b} = \begin{bmatrix} \vartheta_{DNB}^5(Q_{DNB} - A_2 - B_2\vartheta_{DNB}) \\ 1 - A_2 - B_2 \\ B_2 \\ \vartheta_{cr2}^5(Q_{cr2} - A_2 - B_2\vartheta_{cr2}) \\ B_2\vartheta_{cr2}^6 \end{bmatrix}; \qquad \mathbf{x} = \begin{bmatrix} A_1 \\ B_1 \\ E \\ H \\ G \end{bmatrix}$$

The Computer EC-1033 was used to solve the system (59) for each experimental boiling curve followed by calculation of $Q_{tr}(\vartheta)$ by formula (57) using the values of ΔT_{DNB}, ΔT_{cr1}, ΔT_{cr2}, q_{DNB}, q_{cr1}, q_{cr2}, a_2, b_2.

Typical examples of the experimental and predicted $q_{tr}(\Delta T_w)$ are shown in Figs. 46 and 47 [69]. Similar calculations are made for all of the experimental data obtained in the present work and in the known works of different authors. Comparison of the predicted and experimental heat-flux density is shown in Fig. 48. As is seen, the divergence does not exceed $\pm 20\%$, with the heat-flux density varying by more than by two orders of magnitude. Such a relatively high accuracy may be explained by the fact that experimental values of the parameters ΔT_{DNB}, ΔT_{cr1}, ΔT_{cr2}, q_{DNB}, q_{cr1}, q_{cr2} and of the function coefficients $q_f = a_2 + b_2 \Delta T_w$ were used as the

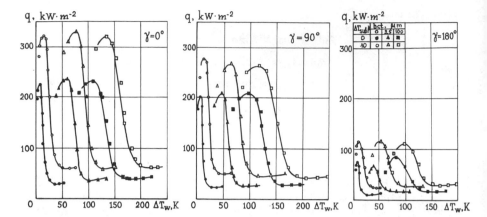

FIG. 46. Typical examples of the comparison of transition boiling heat transfer calculated by the algorithm (57)–(59) with the experimental data of the present authors. Lines stand for calculations, points for experimental values.

initial data in the calculations. Formulas or algorithms are required to calculate these quantities for *a priori* calculation of transition boiling heat transfer. The following recommendations may be made proceeding both from the comparison of the experimental values of the above parameters and the available formulas and from the correlation of experimental data. Methods for calculating the function $q_f = F(\Delta T_w)$ in the form of Eq. (56) are cited in Kalinin *et al.* [54a, 70]. The maximum heat-flux density can be calculated by Kutateladze's formula [65] with a correction for the heating surface orientation in the mass force field:

$$q_{cr1} = 0.137 \sqrt{\rho_V} \sqrt[4]{(\rho_L - \rho_V)g\sigma} \sqrt{1 - \gamma/190°}$$
$$\times \left(1 + 0.065 \left(\frac{\rho_L}{\rho_V}\right)^{0.8} \frac{C_{pL} \Delta T_{sub}}{r}\right] \tag{60}$$

The minimum heat-flux density may be calculated by the approximate relation being a modification of Eq. (31) for the subcooled liquid:

$$q_{cr2} \approx a_2 + 2b_2 \Delta T_{cr2} \tag{61}$$

It is recommended that the film boiling crisis temperature head ΔT_{cr2} be calculated by the formula [71]

$$\frac{\Delta T_{cr2}}{T_{crit} - T_L} = \left[0.16 + 2.5 \sqrt{\frac{\varepsilon_L}{\varepsilon_W}} + \left(\frac{\varepsilon_L}{\varepsilon_W}\right)^2\right](1 + 0.13 \cos \gamma)\frac{1 + \cos \theta}{2} \tag{62}$$

The effective thermal activity determined by the nomogram given in Fig.

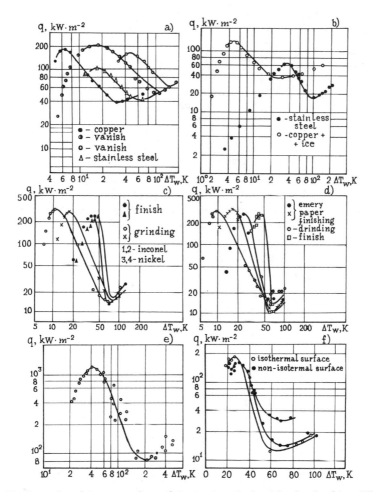

FIG. 47. Examples of the comparison of the results calculated by the algorithm (57)–(59) with the experimental data of other authors: (a) and (b) (from Klimenko [28]), nitrogen; (c) and (d) (from Berenson [7]), n-pentane; (e) and (f) (from Kovalev et al. [69]), water, Freon-113. Lines stand for calculations, points for experimental values.

6 must be used as ε_w for two-layer heating surfaces. The nucleate boiling crisis temperature head is determined by the relation [3]:

$$\Delta T_{cr1} \approx 0.6 \, \Delta T_{cr2} \tag{63}$$

The histograms of dimensionless temperature head and heat-flux density distributions are built to correlate the experimental data at the point of departure from stable nucleate boiling. It may be assumed within $\pm 15\%$

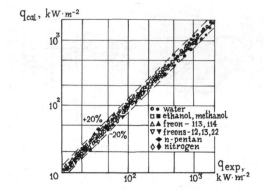

FIG. 48. Comparison of the heat flux density calculated by the algorithm (57)–(59) with transition pool boiling and the experimental data of the present review (○, □, △, ▽, ◇) and other authors (●, ■, ▲, ▼, ◆).

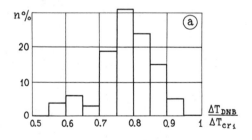

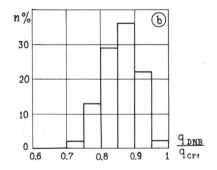

FIG. 49. Histograms of (a) dimensionless temperature head and (b) heat-flux density distribution at the point of the deviation from stable nucleate boiling.

(Fig. 49) that

$$\vartheta_{DNB} = \Delta T_{DNB}/\Delta T_{cr1} = 0.78; \qquad Q_{DNB} = \frac{q_{DNB}}{q_{cr1}} = 0.86 \qquad (64)$$

In such a determination of parameters, the heat-flux density error calculated by algorithm (57)–(59) is increased up to 30%.

NOMENCLATURE

a	thermal diffusivity, $\lambda/\rho c$	$\bar{q}_V$	volumetric power of heat release
Bi	Biot number	R_3	minimum radius of an equilibrium vapor nucleus
b	thickness of the wall, coating		
C	specific heat (C_p, at $p =$ const)	R_Z	mean height of microroughnesses of the heating surface
D	diameter		
f	frequency of liquid–wall contacts	r	specific evaporation heat
F	function of	T	temperature
g	acceleration of gravity	t	time
h	maximum height of microroughness	t_{col}	period of the growth of bubbles before their coalescence
Ja	Jacob number, $\rho_L C_{pL} \Delta T_w/\bar{\rho}_V r$		
k	wave number	t_L	wall–liquid contact time
L	characteristic length	t_V	wall–vapor contact time
l	wavelength	t_T	period of liquid heating due to heat conduction
l_0	capillary constant $\sqrt{\sigma/(\rho_L - \rho_V)g}$		
p	pressure	x, y	coordinates
q	heat flux density		

Greek Symbols

α	heat transfer coefficient	θ	boundary wetting angle
β	coefficient for increasing amplitude	λ	thermal conductivity
γ	angle of inclination of the heating surface to the horizon	μ	dynamic viscosity coefficient
		ρ	density
ΔT	temperature head $\Delta T_w = T_w - T_s$	σ	surface tension coefficient
ΔT_{sub}	liquid subcooling up to the saturation temperature $T_s - T_L$	τ_L	fraction of time per liquid–wall contact
ΔT_b	liquid superheating at the wall boundary $T_b - T_s$	φ	volumetric vapor content
		φ_0	heating surface section occupied by the vapor
δ	thickness of the vapor film or liquid layer		
ε	thermal activity coefficient, $\sqrt{\rho C \lambda}$	ω	angular frequency

Subscripts

b	interphase boundary	DFB	point of deviation from stable film boiling
cal	calculated		
cr1	nucleate boiling crisis (q_{max})	DNB	point of deviation from developed nucleate boiling laws
cr2	film boiling crisis (q_{min})		
crit	critical	d	dominating wave
ct	coating	ef	effective

Subscripts (cont'd)

exp	experimental	S	saturation
f	film boiling	tr	transition boiling
L	liquid	v	vapor
lim	limiting	w	wall
n	nucleate boiling		

REFERENCES

1. S. S. Kutateladze, "Heat Transfer During Condensation and Boiling." Mashgiz, Leningrad, 1952 (in Russian).
2. V. A. Grigor'ev, Yu. M. Pavlov, and E. V. Ametistov, "Boiling of Griogenic Liquids." Energia, Moscow, 1977 (in Russian).
3. V. K. Koshkin *et al.*, "Unsteady Heat Transfer." Mashinostroenie, Moscow, 1973 (in Russian).
4. Y. Iida and K. Kobayasi, An experimental investigation of the mechanism of pool boiling phenomena by a probe method. *Proc. Int. Heat Transfer Conf., 4th, 1970,* Vol. 5, B 1. 3 (1970).
5. S. Bankoff and V. Mehra, A quenching theory for transition boiling. *Ind. Eng. Chem. Fundam.* **1,** No. 1, 38–40 (1962).
6. N. Zuber, On the stability of boiling heat transfer. *J. Heat Transfer* **80,** No. 3, 711–720 (1958).
7. P. J. Berenson, Experiments on pool boiling heat transfer. *Int. J. Heat Mass Transfer* **5,** 985–999 (1962).
8. T. Aoki and J. R. Welty, Energy transfer mechanisms in transition pool boiling. *Int. J. Heat Mass Transfer* **13,** 1237–1240 (1970).
9. Y. Katto *et al.*, Mechanism of boiling crisis and transition boiling on pool boiling. *Proc. Int. Heat Transfer Conf., 4th, 1970,* Vol. 5, B 3. 2 (1970).
10. R. F. Gaertner, Photographic study of nucleate pool boiling on a horizontal surface. *Int. J. Heat Transfer* **86,** No. 2 (1964).
11. R. F. Gaertner and J. W. Westwater, Population of active sites in nucleate boiling heat transfer. *Chem. Eng. Prog., Symp. Ser.* **46,** No. 30, 39–48 (1960).
12. Y. P. Chang, Wave theory of heat transfer in film boiling. *J. Heat Transfer* **81,** No. 1, 1–12 (1959).
13. P. J. Berenson, Film boiling heat transfer from a horizontal surface. *Int. J. Heat Transfer* **83,** No. 3 (1961).
14. J. H. Lienhard and P. T. Y. Wong, The dominant unstable wavelength and minimum heat flux during film boiling on a horizontal cylinder. *Int. J. Heat Transfer* **86,** No. 2 (1964)
15. B. S. Petukhov and S. A. Kovalev, Methodics and some results of measuring critical load during transition from film to nucleate boiling. *Teploenergetika* **5,** 65 (1962) (in Russian).
16. G. Hesse, Heat transfer in nucleate boiling, maximum heat flux and transition boiling. *Int. J. Heat Mass Transfer* **16,** 1611–1627 (1973).
17. K. Stephan and E. G. Hoffmann, Transition- and flow-boiling heat transfer inside a horizontal tube. *Int. J. Heat Mass Transfer* **20,** 1381–1387 (1977).
18. R. C. Kesselring *et al.*, Transition and film boiling from horizontal strips. *AIChE J.* **13,** No. 4, 669–675 (1967).
19. S. Ishigai and T. Kuno, Experimental study of transition boiling on a vertical wall in open vessel. *Bull JSME* **9,** No. 34, 361–368 (1966).

20. S. A. Kovalev, On methods of studying heat transfer in transition boiling. *Int. J. Heat Mass Transfer* **11**, No. 2, 279–283 (1968); **13**, No. 11, 1505–1506 (1970).
21. N. Yu. Tobilevich *et al.*, On the regime transitional from nucleate to film boiling for water and water-alcohol mixtures during vapor heating of heatingchange surface. *Inzh.-Fiz. Zh.* **16**, No. 4, 610 (1969) (in Russian).
22. R. M. Canon and E. L. Park, Transition boiling of normal pentane from a horizontal flat gold surface at one atmosphere pressure. *Int. J. Heat Mass Transfer* **19**, No. 6, 696–698 (1976).
23. A. J. Lowery and J. H. Westwater, Heat transfer to boiling methanol—Effect of added agents. *Ind. Eng. Chem.* **49**, No. 9, 1445–1448 (1957).
24. K. Stephan, On methods of studying heat transfer in transition boiling. *Int. J. Heat Mass Transfer* **11**, 1735–1736 (1968).
25. P. G. Poletavkin *et al.*, A new method of transition boiling study. *Dokl. Akad. Nauk SSSR* **90**, No. 5, 775 (1953) (in Russian).
26. V. A. Grigor'ev *et al.*, Experimental study of heat transfer during nitrogen boiling. *Izv. Vuzov. Energ.* **3**, 72 (1975) (in Russian).
27. V. A. Grigor'ev *et al.*, Investigation of heat transfer during nitrogen boiling. *Tr. MEI,* **268**, 43 (1975) (in Russian).
28. V. V. Klimenko, Investigation of transition and film boiling of criogenic liquids. Thesis for Candidate's Degree, MEI, Moscow, 1975 (in Russian).
29. K. Nishikawa *et al.*, Experimental study on the mechanism of transition boiling heat transfer. *Bull. JSME* **15**, No. 79, 93–103 (1972).
30. Y. Iida and K. Kobayasi, Distribution of void fraction above and adjacent to a horizontal heating surface in pool boiling. *Technol. Rep. Tohoku Univ.* **34**, No. 1, 1–18 (1969).
31. A. Sakurai and M. Shiotsu, Temperature-controlled pool-boiling heat transfer. *Heat Transfer, Proc. Int. Heat Transfer Conf., 5th, 1974,* Vol. 4, pp. 81–85 (1974).
32. A. Sakurai and M. Shiotsu, Studies on temperature-controlled pool-boiling heat transfer. *Tech. Rep. Inst. At. Energy, Kyoto Univ.* **175**, 1–12, (1978).
33. E. A. Farber and R. L. Scorah, Heat transfer in water boiling under pressure. *Int. J. Heat Transfer* **70**, No. 3, 369–384 (1948).
34. Yu. P. Denisov, Processes and characteristics of transition boiling on massive heat-conductive bodies. *Proc. Sci.-Technol. 8th, Conf., 1968,* p. 156 (1968).
35. Yu. P. Denisov, Experimental investigation of transition boiling on the surface of massive heat-conductive bodies. *Tr. Vses. Nauchno-Issled. Inst. Giprostal., 1969,* Vols. 11–12, p. 236 (in Russian).
36. W. S. Bradfield, On the effect of subcooling on wall superheat in pool boiling. *Int. J. Heat Transfer* **89**, No. 2 (1967).
37. H. Merte and J. A. Clark, Boiling heat transfer with cryogenic fluids at standard, fractional and near-zero gravity. *Int. J. Heat Transfer* **86**, No. 3 (1964).
38. J. W. Stevens and L. C. Witte, Loss of stability of vapor film during boiling on spheres. *Inzh.-Fiz. Zh.* **16**, No. 3, 669 (1973) (in Russian).
39. A. E. Bergles and W. G. Thompson, The relationship of quench data to steady-state pool boiling data. *Int. J. Heat Mass Transfer* **13**, No. 1, 55–68 (1970).
40. E. K. Kalinin *et al.*, The study of film boiling crisis and transition boiling of cryogenic liquids. *Proc. Int. Cryog. Eng. Conf.* **4**, 287–290 (1972).
41. J. Ruzicka, "Heat Transfer to Boiling Nitrogen. Problems of Low-Temperature Physics and Thermodynamics," pp. 323–329. Pergamon, Oxford, 1959.
42. J. W. Stevens and L. C. Witte, Transient film and transition boiling from a sphere. *Int. J. Heat Mass Transfer* **14**, 443–450 (1971).
43. V. I. Tolubinsky, A. M. Kichigin, and S. G. Povsten, Spectral analysis of noise during

boiling in transition region. *In* "Heat Transfer, 1974. Soviet Investigations," p. 243. Nauka, Moscow, 1975 (in Russian).

44. I. R. Rybin and L. I. Roizen, Investigation of boiling mechanism of water on nonisothermal surface. *In* "Heat Transfer, 1974. Soviet Investigations," p. 207. Nauka, Moscow, 1975 (in Russian).

45. L. G. Rhea and R. G. Nevins, Film boiling heat transfer from an oscillating sphere. *Int. J. Heat Transfer* **91,** No. 2 (1969).

46. H. Merte and J. Clark, Boiling heat transfer data for liquid nitrogen at standard and near-zero gravity. *Adv. Cryog. Eng.* **7,** 546–550 (1962).

47. H. S. Ragheb, S. Cheng, and D. Groeneveld, Measurement of transition boiling boundaries in forced convective flow. *Int. J. Heat Mass Transfer* **21,** No. 12, 1621–1624 (1978).

48. J. W. Westwater and J. G. Santangelo, Photographic study of boiling. *Ind. Eng. Chem.* **47,** No. 8, 1605–1610 (1955).

49. T. M. Flynn *et al.,* The nucleate and film boiling curve of liquid nitrogen at one atmosphere. *Adv. Cryog. Eng.* **7,** 535–545 (1962).

50. S. A. Kovalev *et al.,* Effect of coating with low thermal conductivity upon boiling heat transfer of liquid on isothermal and non-isothermal surfaces. *Proc. Int. Heat Transfer Conf., 4th, 1970,* Vol. 5, B 1.4 (1970).

51. Y. Katto and S. Yokoya, Principal mechanism of boiling crisis in pool boiling. *Int. J. Heat Mass Transfer* **11,** No. 6, 993–1002 (1968).

52. E. K. Kalinin *et al.,* Investigation of transition pool boiling of criogenic liquids. *Izv. Akad. Nauk B SSR,* **4,** 24 (1971) (in Russian).

53. E. K. Kalinin, I. I. Berlin *et al.,* Investigation of the crisis of film boiling in channels. *Proc. 90th Winter Annu. Meet. Am. Soc. Mech. Eng.,* pp. 89–94 (1969).

54. H. S. Ragheb *et al.,* Measurements of transition boiling data under forced convective conditions. *Int. J. Heat Transfer* 1621, No. 12 (1978).

54a. E. K. Kalinin *et al.,* "Methods of Calculation of Adjoint Problems of Heat Transfer." Mashinostroenie, Moscow, 1983.

55. A. M. Kichigin and S. G. Povsten, Investigation of acoustic phenomena during the change of pool boiling regimes. *Teplofiz. Teplotekh.,* **19** (1971) (in Russian).

56. D. Kirby and J. Westwater, Bubble and vapor behavior on heated horizontal plate during pool boiling near burnout. *Chem. Eng. Prog., Symp. Ser.* **61,** No. 57, 238–248 (1965).

57. V. G. Pron'ko and L. B. Bulanova, Thermodynamic film boiling crisis. *Inzh.-Fiz. Zh.* **35,** No. 1, 62 (1978) (in Russian).

58. S. A. Kovalev, V. M. Zhukov, and Yu. A. Kuzma-Kichta, Methodics of investigation of interfacial oscillations during liquid film boiling with the aid of laser. *Inzh.-Fiz. Zh.* **25,** No. 1, 20 (1973) (in Russian).

59. T. Aoki and J. R. Welty, Frequency distribution of boiling-generated sound. *Int. J. Heat Transfer* **92** (1970).

60. J. W. Stevens and L. C. Witte, Transient film and transition boiling from a sphere. *Int. J. Heat Mass Transfer* **14,** 443–450 (1971).

61. K. Nishikawa *et al.,* Studies of boiling characteristic curve (relation between the surface temperature fluctuation and heat transfer). *Trans. Jpn. Soc. Mech. Eng.* **34,** 152 (1967).

62. D. R. Veres and L. W. Florschietz, A comparison of transient and steady state pool-boiling data obtained using the same heating surface. *Int. J. Heat Transfer* **98,** No. 2 (1971).

63. B. S. Petukhov *et al.,* The study of heat transfer during liquid boiling on a nonisothermal surface. *In* "Heat and Mass Transfer During Phase Transitions," Part 1. Minsk, 1974 (in Russian).

64. S. A. Zhukov *et al.,* Heat transfer during liquid boiling on a surface with a coating with

low heat conductivity. *In* "Heat Transfer and Physical Gas Dynamics," pp. 116–129 Nauka, 1974 (in Russian).

65. S. S. Kutateladze, "Fundamentals of Heat-Novosibirsk." Nauka, Moscow, 1970 (in Russian).

66. Yu. A. Kuzma-Kichta, Investigation of heat transfer and the boiling mechanism on a metal surface without coating and with a coating with low heat conductivity. Thesis for Candidate's Degree, IVT Akad. Nauk SSR, 1974 (in Russian).

66a. F. Eugene Adiutori, "Stability Consultants. The New Heat Transfer." Ventuno Press, 1974.

67. E. K. Kalinin *et al.,* heat transfer during transition boiling of cryogenic liquids. *Teplofiz. Vys. Temp.* **14,** No. 2, 410 (1976) (in Russian).

68. E. K. Kalinin, I. I. Berlin *et al.,* Heat transfer in transition boiling of cryogenic liquids. *Adv. Cryog. Eng.,* pp. 273–277 (1976).

69. S. A. Kovalev *et al.,* Effect of coating with low thermal conductivity upon boiling heat transfer of liquid on isothermal and non-isothermal surfaces. *Heat Transfer, Proc. Int. Conf. Heat Transfer, 4th, 1970,* Vol. 5, B 1.4 (1970).

70. E. K. Kalinin, I. I. Berlin, and V. V. Kostiouk, Film boiling heat transfer. *Adv. Heat Transfer* **11** (1975).

71. I. I. Berlin *et al.,* Transition boiling of cryogenic liquids mechanism, mathematic model and generalization. *Supercond. Electroenerg. Electrotech.* **1,** 292–308 (1982).

Thermodynamic and Transport Properties of Pure and Saline Water

DAVID J. KUKULKA

Department of Technology
State University College at Buffalo
Buffalo, New York 14222

BENJAMIN GEBHART

Department of Mechanical Engineering
University of Pennsylvania
Philadelphia, Pennsylvania 19104

JOSEPH C. MOLLENDORF

Department of Mechanical and Aerospace Engineering
State University of New York at Buffalo
Buffalo, New York 14260

I. Introduction

Thermodynamic and transport properties of water enter into many analyses and interpretations of diverse aspects of the environment and in technology, such as life in the ocean, industrial processes that use seawater, underwater communication, defense measures, and the natural processes that occur in terrestrial surface waters. Most of the present knowledge of seawater properties has been obtained from thermodynamic and transport relationships. The newly calculated properties that are presented here were made using various thermodynamic and transport relations and the equation of state of Gebhart and Mollendorf. Often the solutes in saline water are treated as a single chemical component. For example, Richardson *et al.* [2], like many other investigators, considered a 3.5 wt % solution of NaCl as equivalent to seawater in the description of various transport properties. Many measurements of thermal diffusivity and thermal conductivity of seawater have actually been made using NaCl solutions. In the absence of other information one must rely on these results for seawater.

This study has surveyed the available fundamental data and correlations of both the molecular transport and thermodynamic properties of saline water for a wide range of temperature t, salinity s, and pressure p. The objective was to establish a data base that could be used to achieve systemic representations of as many as possible of the basic properties required in many areas of analysis and in calculations concerning terrestrial surface water transport processes and circulations.

The important molecular transport quantities are the thermal conductivity k, the viscosity μ, and the chemical-species or mass diffusivity of salinity components D. These, along with the specific heat C_p and the density ρ, determine the Prandtl (Pr), Schmidt (Sc), and Lewis (Le = Sc/Pr) numbers.

The thermodynamic quantities were determined by using thermodynamic relations and an equation of state, $\rho(t, s, p)$, of seawater. For given values of t and s, calculations were made of the change with pressure of specific heat ΔC_p, enthalpy Δh, and entropy ΔS from the standard pressure 1 atm abs, to 1000 bars. The thermal and saline volumetric expansion coefficients β and β^* are calculated directly from a density function $\rho(t, s, p)$.

The following section summarizes the information sources and presents information and results concerning the various properties studied. New results are presented graphically with corresponding tabular results presented in appendix A. The first subsection summarizes the molecular transport quantities μ, k, and D. The second subsection presents background information concerning the various thermodynamic properties (β, β^*, ΔC_p, Δh, ΔS, and ρ). Prandtl number and Schmidt number quantities are summarized in the third subsection. The calculation techniques are given in Appendix B.

II. Property Information

Previous studies of the thermodynamic and transport properties of water have contributed a wealth of background information. The portion of that material that is of specific interest here is that which concerns the dependence of such properties on the temperature, salinity and pressure levels.

According to the standards commonly applied in dealing with the states of fluids, both t and p are well defined and definite. However, the historical and perhaps inevitable imprecisions in the definition and determination of salinity remains. This effect, as it relates to the determination of properties, is generally an undetermined uncertainty in property predictions. That uncertainty remains in the background information and in the new results here.

A. MOLECULAR TRANSPORT PROPERTIES

A collection of information concerning μ, k, and D is summarized below over a wide range of t, s, and p.

1. Viscosity

At atmospheric pressure, data were obtained by Krummel and Ruppin (as reported by Dorsey [3]), Isdale et al. [4], and Chen et al. [5]. At elevated pressures, data were obtained by Stanley and Batten [6] and Horne and Johnson [7] for 19.37‰ chlorinity (normal seawater), pressures to 1406 bars and temperatures to 30°C. Comparing this data we see that there is fair agreement at 20°C. However, at 10°C, 6°C, and 0°C and for pressures above 350 bars there are larger differences. Stanley and Batten [6] suggested the difference is due to experimental error in the work of Horne and Johnson [7] and conclude that they have eliminated those sources of error in their experiment. An expression for viscosity $\mu(t, s, p)$ was developed by Matthaus [8], (see Eq. 1). It is said to be accurate to within 0.7% in the range $0°C \leq t \leq 30°C$, $0‰ \leq s \leq 36‰$ and 1 to 1000 bars abs, but for larger ranges of t, s, and p the error is slightly greater. The viscosity $\mu(t, s, p)$, in centipoise is given by Matthaus [8] as:

$$
\begin{aligned}
\mu(t, s, p) = {} & 1.7910 - (6.144 \times 10^{-2})t + (1.4510 \times 10^{-3})t^2 \\
& - (1.6826 \times 10^{-5})t^3 - (1.5290 \times 10^{-4})p \\
& + (8.3885 \times 10^{-8})p^2 + (2.4727 \times 10^{-3})s \\
& + t(6.0574 \times 10^{-6}p - 2.6760 \times 10^{-9}p^2) \\
& + s(4.8429 \times 10^{-5}t - 4.7172 \times 10^{-6}t^2 \\
& + 7.5986 \times 10^{-8}t^3)
\end{aligned}
\tag{1}
$$

where t is in degrees Celsius, p is in bars absolute and s is in ‰.

2. Thermal Conductivity

Limited data at atmospheric pressure were obtained by Krummel [9] and Riedel [10] (as reported by Richardson et al. [2]). Additionally,

Krummel [9] assumed that the thermal diffusivity of seawater is equal to that of pure water. More complete sets of data are given by Jamieson and Tudhope [11], Castelli and Stanley [12], Castelli *et al.* [13], and Caldwell [14]. Castelli and Stanley [12] present data for temperatures to 30°C and pressure to 1400 bars abs, at a salinity of 34.944‰. Caldwell [14] developed an equation for the thermal conductivity of seawater for oceanic conditions (see Eq. 2). It is a fit of his data and is presented as a function of temperature, salinity, and presure. Comparisons of Caldwell [14] to Jamieson and Tudhope [11] indicated small differences.

Thermal conductivity $k(t, s, p)$ in cal/cm °C sec is given by Caldwell [14] as

$$k(t, s, p') = k(0, t, s)[1 + p'(f(t, s, p'))] \tag{2}$$

$$f(t, s, p') = C_1 + C_2 t^2 + C_3 p' + C_4 s \tag{3}$$

where $C_1 = 0.0690$, $C_2 = -8 \times 10^{-5}$, $C_3 = -0.0020$, and $C_4 = -0.00010$ and t is in °C, p' is in kilobars and s is in ‰. It is said to be accurate to within 0.3% for ocean water.

A simpler formula for $k(t, s, p')$, which is "useful for ocean waters and probably accurate to 0.5%," is also presented by Caldwell [14]:

$$k(t, s, p') = 0.001365(1 + A_1 t - A_2 t^2 + A_3 p' - A_4 s) \tag{4}$$

where $A_1 = 0.003$, $A_2 = 1.025 \times 10^{-5}$, $A_3 = 0.0653$ and $A_4 = 0.00029$.

3. *Mass Diffusivity*

The value of D has been determined for only a limited range of oceanic conditions. Measurements of diffusivity have been reported for temperatures of 18.5°C and 25°C, for various salinities at atmospheric pressure by Clack [15], Fuoss [16], Onsager and Fuoss [17], Gordon [18], Stokes [19], Vitagliano and Lyons [20], O'Donnel and Gosting [21], Dunlop and Gosting [22], Richardson *et al.* [2, 23], and Caldwell [24, 25].

A comparison of reported values of D at atmospheric pressure for various salinities and the temperatures 18.5°C and 25°C is given in Table I. Most of the data is in good agreement.

In a complex solution such as seawater, the movements of the various and diverse constituents cannot be easily traced accurately. Caldwell's [24] experiment used a single salt solution (NaCl) and depended on conductivity measurements to determine solute movement. Because these movements cannot be easily traced in seawater, all values reported to date for D are for single salt solutions. Therefore, we must rely on these results in the absence of more accurate data.

TABLE I

COMPARISON OF DIFFUSIVITY D OF AN NaCl SOLUTION AT
ATMOSPHERIC PRESSURE FOR VARIOUS SALINITIES TAKEN FROM
RICHARDSON et al. [2]

Salinity (‰)	Diffusivity (10^{-3} cm²/sec) with reference[a]	
	At 18.5°C	At 25°C
5	1.235(a); 1.25(d)	
5.5		1.505(e)
6.0	1.23(b); 1.245(c)	1.485(e)
10.0	1.22(a); 1.225(d)	1.508(f)
11.5	1.215(b)	
12.0		1.478(g)
12.5	1.22(c)	1.475(e)
15.0	1.215(a); 1.215(d)	1.52(f); 1.475(h)
18.0		1.475(e)
20.0	1.215(a); 1.205(d)	1.48(e); 1.535(f)
23.0	1.20(c)	
25.0	1.217(a); 1.205(d)	1.545(f)
26.0		1.475(i)
27.5	1.22(b)	
29.5		1.475(e); 1.48(g)
30.0	1.22(a); 1.208(d); 1.208(c)	1.555(f)
35.0	1.225(a); 1.21(d)	1.575(f); 1.440(j); 1.448(j); 1.486(j)
40.0	1.23(a); 1.214(d)	1.48(e); 1.59(f); 1.475(g)

[a] Reported by (a) Gordon [18] (semitheoretical), (b) Gordon [18] (data points plotted on curve), (c) Clack [15] (values taken from a curve), (d) Clack [15] (experimental data), (e) Vitagliano and Lyons [20] (experimental data), (f) Onsager and Fuoss [17] (theoretical), (g) Stokes [19] (experimental data), (h) O'Donnel and Gosting [21] (experimental data), (i) Dunlop and Gosting [22] (experimental data), (j) Richardson et al. [2] (theoretical integral coefficients).

Caldwell [25] gave the following equation for the temperature dependence of diffusivity, $D(t, 28.5, 1$ atm abs$)$, for a sodium chloride concentration of 28.5 g/kg, in square centimeters per sec:

$$D(t, 28.5, 1 \text{ atm abs}) = 10^{-7}[62.5 + 3.63t] \qquad (5)$$

where t is in degrees Celsius. Equation 5 is a least-squares fit of 51 data points over a temperature range of 0 to 40°C. The standard deviation of the fit was reported to be 8.76×10^{-7}.

Measurements of D for elevated pressures have been attempted by Richardson et al. [23], Caldwell [25], and Caldwell and Eide [26]. Richardson et al. [23] attempted to measure the pressure dependence of D in an aqueous solution of 3.5 wt % NaCl solution. The data gives a 7% increase of D across a range of pressure of 1 to 1000 bars. However, Horne [27] has observed that the experimental uncertainty was so great that the pressure effect could not be accurately determined. Likewise, there is too much scatter in Caldwell's [25] data to allow an accurate determination of a pressure effect. For this data there was perhaps, a slight increase in D with increasing pressure.

Furthermore, Caldwell and Eide [26] made measurements of the diffusivity of seawater ($s = 28.5‰$) for temperatures to 40°C and pressures to 620 bars. They present the most complete and useful information published to date on diffusivity. Specifically, they measured the thermal diffusion coefficients and isothermal diffusivity coefficients for several aqueous solutions (sodium chloride, sodium sulfate, magnesium chloride, magnesium sulfate, and calcium chloride). They also present equations for both the thermal diffusion coefficient and the isothermal diffusivity coefficient as functions of temperature and pressure for each of the five electrolytes.

For sodium chloride with a concentration of 28.5 g/kg the equation from reference [26] for isothermal diffusivity $D(T, 28.5, P)$, in 10^{-5} cm^2/sec is given as

$$D(T, 28.5, P) = A_0 T \exp(B_0(1 + C_0 P)/(T - T_0)) \tag{6}$$

where $A_0 = 0.148571(\pm 0.000073)$, $B_0 = -546.453(\pm 0.080)$, $C_0 = -2.414(\pm 0.031) \times 10^{-5}$, and $T_0 = 139.085(\pm 0.023)$, T is absolute temperature in kelvins, and P is gauge pressure in bars. The standard error of the fit is reported to be 0.041.

Equation 6 indicates that at constant temperature and salinity diffusivity will increase slightly with increasing pressure. While at constant salinity and pressure, diffusivity increases with increasing temperature.

B. THERMODYNAMIC PROPERTIES

Next, we summarize information concerning density, the expansion coefficients, specific heat, enthalpy, and entropy of pure and saline water.

1. Density

Several equations of state for seawater have been developed in recent investigations by Chen and Millero [28], Gebhart and Mollendorf [1], Millero et al. [29], Poisson et al. [30], and Millero and Poisson [31], all of apparent comparable accuracy for most practical purposes. The main differences between the studies are the forms in which the equations are presented, data sets used to create the equations, and the actual accuracy.

Subsequent calculations here of both volumetric expansion coefficients and specific-heat changes require a number of derivatives of the density relation, $\rho(t, s, p)$. However, most of the modern density relations that give high accuracy over a wide range of t, s, and p are analytically unmanageable in this respect. Because of this, when using such density relations, one must instead rely on numerical differencing at very fine scale to retain accuracy in all t, s, and p ranges. This poses additional difficulties.

On the other hand, the relation given by Gebhart and Mollendorf [1] was developed for relative simplicity in such manipulations. It contains only one temperature term and the s and p effects are included as systematic polynomials. This permits taking density derivatives with relative ease. The relation and the polynomials are

$$\rho(t, s, p) = \rho_m(s, p)[1 - \alpha(s, p)|t - t_m(s, p)|^{q(s,p)}] \tag{7}$$

$$\rho_m(s, p) = \rho_m(0, 1)[1 + f_1(p) + sg_1(p) + s^2h_1(p)] \tag{8}$$

$$\alpha(s, p) = \alpha(0, 1)[1 + f_2(p) + sg_2(p) + s^2h_2(p)] \tag{9}$$

$$t_m(s, p) = t_m(0, 1)[1 + f_3(p) + sg_3(p) + s^2h_3(p)] \tag{10}$$

$$q(s, p) = q(0, 1)[1 + f_4(p) + sg_4(p) + s^2h_4(p)] \tag{11}$$

$$f_i(p) = \sum_{j=1}^{n_f} f_{ij}(p-1)^j \tag{12}$$

$$g_i(p) = \sum_{j=0}^{n_g} g_{ij}(p-1)^j \tag{13}$$

$$h_i(p) = \sum_{j=0}^{n_h} h_{ij}(p-1)^j \tag{14}$$

where $i = 1, 2, 3, 4$. The pressure p is in bars abs; the salinity s in ppt, the temperature t in degrees Celsius; and the density ρ in kilograms per cubic meter. The basic quantities $\rho_m(0, 1)$, $\alpha(0, 1)$, $t_m(0, 1)$, and $q(0, 1)$ correlate the density of pure water at one bar abs. The quantity $f_i(p)$ is necessary to include the effect of pressure on the density of pure water, while $g_i(p)$ and $h_i(p)$ are additional pressure effects associated with salinity.

An additional benefit of the above density relation is that it incorporates directly the temperature at which the density extremum occurs, $t_m(s, p)$ (Eq. 10). Gebhart and Mollendorf [1] discuss the basis for their determination of $t_m(s, p)$ and point out that there are significant differences among investigators regarding the actual temperature at which the density extremum occurs. For many purposes these differences are not important.

However, for calculations of properties, especially those involving derivatives of density, very small changes in $t_m(s, p)$ have very large effects on the
results. Indeed even the sign of the result can be reversed as would happen,
for example, with the expansion coefficients. For other properties, the
effect is much more subtle and complicated. In most equations of state, the
effect of $t_m(s, p)$ is somewhat buried as the net result of the interactions of
the many terms of the relation. Recognizing its importance for accurately
modeling certain phenomena, Gebhart and Mollendorf [1] include $t_m(s, p)$
as an integral part of their relation. Furthermore, they explain in considerable detail why their formulation of $t_m(s, p)$ gives the best overall quantitative representation of the perceived best experimental evidence. All of this
suggests that our calculated properties, particularly in the vicinity of
$t_m(s, p)$, are perhaps the most accurate reflection of the actual physics that
determines thermodynamic properties there.

2. The Expansion Coefficients

Values of the thermal expansion coefficient have been given by various
investigators. Bradshaw and Schleicher [32] measured β over a range of

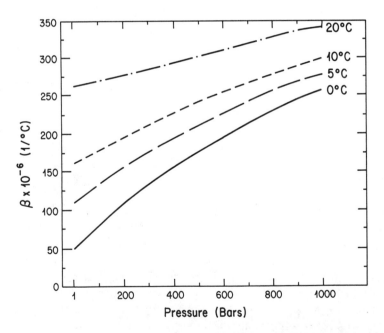

FIG. 1. The temperature and pressure effects on the thermal expansion coefficient $\beta(t, s, p)$
for seawater ($s = 35\%_o$), from Table AI.

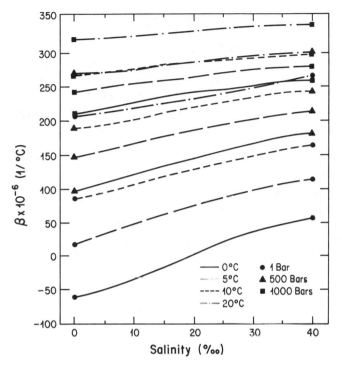

FIG. 2. The temperature and salinity effects on the thermal expansion coefficient $\beta(t, s, p)$ for water at pressures of 1, 500, and 1000 bars, from Table AI.

temperature from $-2°C$ to $30°C$, pressures from 8 to 1001 bars, but for only two salinity levels, 30.5‰, 35.0‰. Wilson and Bradley [33] give values of β for a wide range of temperatures, salinities, and pressures. Fofonoff [34] compared values of β at 35‰ resulting from the equations of state of Knudsen [35] and of Eckart [36]. Sturges [37] compares values of β computed from the density equation of Fofonoff [34] with the experimental results of Wilson and Bradley [38]. Caldwell and Tucker [39] have inferred β by observing the onset of convection in a Rayleigh–Bénard device. Bryden [40] developed a polynomial based on a least-squares fit of the experimental data of Bradshaw and Schleicher [32]. Hydrographic Tables (LaFond [41]), Crease [42], Wilson and Bradley [33, 38], Bradshaw and Schleicher [32], Millero and Lepple [43], Wang and Millero [44], Fine et al. [45], Emmet and Millero [46], Cox et al. [47], and Chen and Millero [28] give values of β over a wide range of conditions. The results of all of these studies are generally in good agreement except at low temperatures and pressures, where very large differences are seen.

Calculations of thermal expansion coefficients (see Figs. 1 and 2 whose data is taken from Table AI in Appendix A) were performed for a range of temperatures to 20°C, salinities to 40‰ and pressures to 1000 bars abs. Figure 1 shows that our calculated β increases with both increasing temperature and pressure for constant salinity. Figure 2 shows that our calculated thermal expansion coefficients for seawater are greater than those of pure water and that they increase with increasing pressure. Negative values can be seen, which indicates contraction with increasing temperature.

A comparison is made in Tables II and III, of presently calculated values of β with the results of various other investigators, for water at various temperatures and pressures. A similar comparison of β at atmospheric pressure is made in Tables IV and V. Good agreement is seen throughout except at low temperatures and pressures. As discussed earlier, such calculated results are extremely sensitive to the accuracy of the representation of the temperature at which the density extremum occurs. Consequently, calculation of thermal expansion near $t_m(s, p)$ may differ by several

TABLE II

COMPARISON OF THERMAL EXPANSION $\beta(t, s, p)$ OF PURE WATER ($s = 0‰$) AT
VARIOUS TEMPERATURES AND PRESSURES

Temperature (°C)	Pressures (bars)	Thermal expansion coefficient ($10^{-6}/°C$) from reference given[a]						
		a	b	c	d	e	f	g
0	1	—	−41.	−34.	−68.	−90.	−34.	−61.3
	200	7.5	25.	28.	3.	11.	28.	5.3
	400	66.6	80.	83.	62.	69.	83.	68.7
	600	121.7	130.	132.	116.	104.	132.	122.2
	800	169.3	174.	177.	162.	162.	177.	169.6
	1000	209.8	215.	216.	202.	216.	216.	211.0
10	1	—	86.	90.	87.	98.	90.	87.2
	200	133.9	128.	133.	134.	118.	133.	133.0
	400	175.3	167.	171.	176.	183.	171.	173.9
	600	212.1	201.	206.	212.	224.	206.	211.5
	800	244.5	232.	236.	243.	255.	236.	244.2
	1000	272.5	260.	264.	272.	271.	264.	272.2
20	1	—	197.	200.	206.	206.	200.	210.6
	200	235.1	224.	228.	236.	245.	228.	237.0
	400	261.4	248.	252.	261.	261.	252.	263.1
	600	285.2	269.	275.	284.	286.	275.	287.6
	800	306.4	288.	295.	306.	307.	295.	307.9
	1000	325.1	306.	312.	327.	333.	312.	324.6

[a] (a) From Chen [59], (b) from Eckart [36], (c) from Wilson and Bradley [33], (d) from Crease [42], (e) from Kell and Whalley [57], (f) from Wilson and Bradley [38], and (g) calculated in this study.

TABLE III

COMPARISON OF THERMAL EXPANSION $\beta(t, s, p)$ OF SEAWATER
($s = 35‰$) AT VARIOUS TEMPERATURES AND PRESSURES

Temperature (°C)	Pressure (bars)	Thermal expansion coefficient (10^{-6} °C^{-1}) from reference given[a]					
		a	b	c	d	e	f
0	1	52.	80.	56.	78.	—	50.2
	200	105.	134.	106.	132.	90.7	109.6
	400	154.	182.	150.	180.	139.5	156.3
	600	198.	225.	171.	222.	182.3	196.0
	800	232.	266.	226.	261.	219.7	230.0
	1000	265.	295.	256.	295.	252.3	259.5
10	1	167.	162.	167.	163.	—	164.4
	200	202.	199.	201.	200.	193.1	200.2
	400	233.	231.	232.	233.	225.3	231.4
	600	261.	260.	261.	263.	254.1	259.5
	800	286.	288.	286.	290.	279.7	283.9
	1000	309.	307.	308.	313.	302.5	305.5
20	1	257.	238.	257.	241.	—	266.5
	200	278.	262.	277.	265.	272.0	282.7
	400	298.	282.	297.	286.	292.4	300.5
	600	317.	300.	—	306.	311.0	318.1
	800	335.	317.	—	323.	328.0	334.1
	1000	355.	329.	—	338.	343.2	348.7

[a] (a) From "Hydrographic Tables" (LaFond [41]), (b) from Eckart [36], (c) from Crease [42], (d) from Wilson and Bradley [33], (e) from Chen [59], (f) calculated in this study.

TABLE IV

COMPARISON OF THERMAL EXPANSION, $\beta(t, s, p)$ OF WATER AT ATMOSPHERIC
PRESSURE FOR VARIOUS TEMPERATURES AND SALINITIES

Temperature (°C)	Reference[a]	Coefficient of thermal expansion (10^{-6}/°C)		
		At salinity 0‰	At salinity 20‰	At salinity 35‰
0	a	−68.0	3.0	48.8
	b	−67.95	3.47	51.68
	c	−61.31	5.38	50.22
10	a	88.1	135.5	167.2
	b	88.10	135.81	166.96
	c	87.18	135.17	164.35
20	a	206.7	236.9	257.3
	b	206.65	237.08	257.01
	c	210.59	239.22	266.53

[a] (a) Reported by Poisson et al. [30], (b) reported by Millero et al. [29], (c) calculated in this study.

TABLE V

COMPARISON OF THERMAL EXPANSION $\beta(t, s, p)$ OF
SEAWATER ($s = 35‰$) AT ATMOSPHERIC PRESSURE AND
VARIOUS TEMPERATURES

Temperature (°C)	Thermal expansion coefficient ($10^{-6}/°C$) from reference given[a]				
	a	b	c	d	e
0	51.8	51.9	49.5	52.1	50.22
5	117.2	117.3	115.9	116.8	109.73
10	172.3	172.1	172.7	171.6	164.35
15	220.4	219.8	221.8	219.4	216.31
20	263.5	263.0	264.0	262.8	266.53

[a] (a) Reported by Millero and Lepple [43] using standard
Copenhagen water, (b) reported by Knudsen [35], (c) reported by Cox et al. [47], (d) reported by Millero and
Leppler [43] using artificial seawater, (e) calculated in this
study.

hundred percent. Therefore, care must be exercised near those temperatures when using the previously reported calculations.

Calculations of saline expansion coefficients (see Figs. 3 and 4 with data
from Table AII) were performed for a range of temperatures to 20°C,
salinities to 40‰ and pressures to 1000 bars abs. Figure 3 shows that our

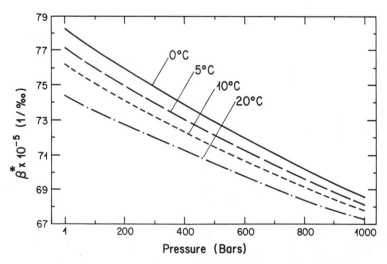

FIG. 3. The temperature and pressure effects on the saline expansion coefficient, $\beta^*(t, s, p)$
for seawater ($s = 35‰$) from Table AII.

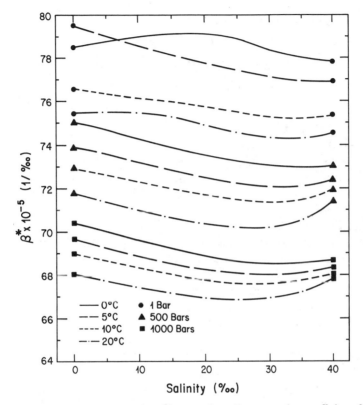

FIG. 4. The temperature and salinity effects on the saline expansion coefficient $\beta^*(t, s, p)$ for water, at pressures of 1, 500, and 1000 bars, from Table AII.

calculated β^* decreases with both increasing temperature and pressure for constant salinity. Figure 4 shows that our calculated saline expansion coefficient decreases with increasing pressure. The variation of β^* with salinity is seen to be much more complicated. No comparisons of the saline expansion coefficients could be made with other investigators since there appears to be no such previously reported information.

3. Specific Heat

There have been several measurements of C_p of seawater at atmospheric pressure. The various investigators were Cox and Smith [48], Bromley et al. [49–51], Kovalenko [52], and Millero et al. (53).

The following equation for $C_p(t, s, 1$ atm abs) in Btu/lbm °F at atmospheric pressure was developed by Bromley et al. [51], as converted by

Horne [27] from salinity in weight percent to ppt

$$C_p(t, s, 1 \text{ atm abs}) = 1.0049 - 0.0016210s + (3.5261 \times 10^{-6}s^2)$$
$$- [(3.2506 - 0.14795s + 7.7765 \times 10^{-4}s^2) \times 10^{-4}t]$$
$$+ [(3.8013 - 0.12084s + 6.121 \times 10^{-4}s^2) \times 10^{-6}t^2]$$
(15)

where t is in °C and s is in ppt.

Calculations were done in the present study (see Figs. 5 and 6 from the data of Table AIII) to determine $\Delta C_p = C_p(t, s, 1 \text{ atm abs}) - C_p(t, s, p)$ for a pressure range to 1000 bars abs, over a salinity range from 0 to 40‰ and a temperature range from 0 to 20°C. The specific heats at elevated pressures $C_p(t, s, p)$ were then calculated from the calculated changes in specific heat and from Eq. 15.

From Fig. 5 it can be seen that ΔC_p increases with increasing pressure and decreases with increasing temperature for constant salinity. Particularly interesting behavior is seen in Fig. 6 at 0°C. At both 500 and 1000 bars abs (at 0°C) an inflection point is noted in the change in specific heat. At 100 bars abs and 0°C, a maximum in the change in specific heat is seen. This behavior is suspected to be a result of the presence of a density

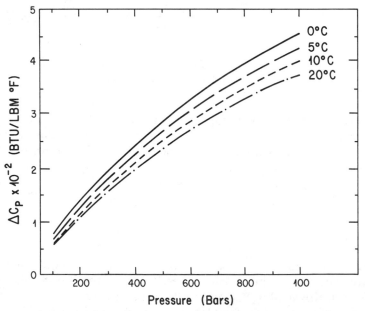

FIG. 5. The temperature and pressure effects on the change in specific heat $\Delta C_p = C_p(t, s, 1 \text{ atm abs}) - C_p(t, s, p)$, for seawater ($s = 35‰$), from Table AIII.

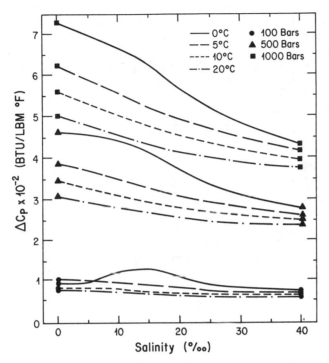

FIG. 6. The temperature and salinity effects on the change in specific heat, $\Delta C_p = C_p(t, s, 1 \text{ atm abs}) - C_p(t, s, p)$, for water at pressures of 100, 500, and 1000 bars, from Table AIII.

extremum. For temperatures less than about 4°C, salinities less than about 19‰ and pressures less than about 190 bars abs, a density extremum spans the range of the calculations. Therefore, only the 0°C curves of Fig. 6 are so influenced. Furthermore, for calculations at 0°C and 100 bars abs, the density extremum effect is felt over the entire pressure range, whereas, for higher pressures, the effect is limited to a progressively smaller percentage of the pressure range. This suggests that the maximum seen at 100 bars abs and 0°C results from the more extensively felt density extremum effects there.

Horne [27] has found no reliable data for the pressure dependence of C_p either for pure water or for seawater. Sverdrup et al. [54] gives a table of differences between C_p at elevated pressure and at atmospheric pressure, for seawater, based on the computation of Ekman [55]. However, Ekman [55] assumed that the "temperature coefficient" of C_p for seawater is the same sign as that of pure water over the same temperature range. Horne [27] pointed out that Ekman's [55] assumption was wrong and therefore the calculated values of ΔC_p that Sverdrup [54] present are in error. Other

TABLE VI

COMPARISON OF THE CHANGE IN SPECIFIC HEAT $\Delta C_p = C_p(t, s, 1 \text{ atm abs}) - C_p(t, s, p)$ OF PURE WATER ($s = 0‰$) AT 0°C FOR VARIOUS PRESSURES

Change in specific heat (Btu/lbm °F) at pressure given

Reference[a]	100 bars	200 bars	300 bars	400 bars	500 bars
a	0.01139	0.02171	0.03105	0.03953	0.04722
b	0.01111	0.02161	0.03141	0.04041	0.04871
c	0.01123	0.02102	0.02914	0.03583	0.04204
d	0.01146	0.02197	0.03201	0.04108	0.04944
e	0.01126	0.02141	0.03053	0.03868	0.04596
f	0.008989	0.02034	0.03044	0.03870	0.04603

Reference[a]	600 bars	700 bars	800 bars	900 bars	1000 bars
a	0.05419	0.06055	0.06633	0.07163	0.07645
b	0.05651	0.06371	0.07041	0.07661	0.08241
c	—	—	—	—	—
d	0.05732	0.06473	0.07141	0.07763	0.08360
e	0.05243	0.05818	0.06328	0.06781	0.07185
f	0.05261	0.05854	0.06385	0.06859	0.07279

[a] (a) Reported by Chen [59], (b) reported by Kell and Whalley [57], (c) using Eq. (3), from Sirota and Shrago [58], (d) using Eq. (1), from Sirota and Shrago [58], (e) reported by Fofonoff and Millard [60], (f) calculated in this study.

TABLE VII

COMPARISON OF THE CHANGE IN SPECIFIC HEAT $\Delta C_p = C_p(t, s, 1 \text{ atm abs}) - C_p(t, s, p)$ OF PURE WATER ($s = 0‰$) AT 10°C FOR VARIOUS PRESSURES

Change in specific heat (Btu/lbm °F) at pressure given

Reference[a]	100 bars	200 bars	300 bars	400 bars	500 bars
a	0.008670	0.01664	0.02391	0.03055	0.03666
b	0.008442	0.01634	0.02354	0.03024	0.03634
c	0.008599	0.01624	0.02317	0.02938	0.03535
d	0.008598	0.01672	0.02388	0.03081	0.03702
e	0.008673	0.01660	0.02384	0.03043	0.03642
f	0.008206	0.01573	0.02261	0.02891	0.03469

Reference[a]	600 bars	700 bars	800 bars	900 bars	1000 bars
a	0.04225	0.04739	0.05209	0.05642	0.06040
b	0.04194	0.04724	0.05194	0.05644	0.06044
c	—	—	—	—	—
d	0.04251	0.04777	0.05278	0.05708	0.06114
e	0.04186	0.04681	0.05130	0.05538	0.05911
f	0.03998	0.04477	0.04910	0.05297	0.05642

[a] (a) Reported by Chen [59], (b) reported by Kell and Whalley [57], (c) using Eq. (3) of Sirota and Shrago [58], (d) using Eq. (1) of Sirota and Shrago [58], (e) reported by Fofonoff and Millard [60], (f) calculated in this study.

TABLE VIII

COMPARISON OF THE CHANGE IN SPECIFIC HEAT $\Delta C_p = C_p(t, s, 1\ \text{atm abs}) - C_p(t, s, p)$ OF PURE WATER ($s = 0‰$) AT 20°C FOR VARIOUS PRESSURES

Reference[a]	Change in specific heat (Btu/lbm °F) at pressure given				
	100 bars	200 bars	300 bars	400 bars	500 bars
a	0.007165	0.01382	0.01994	0.02563	0.03088
b	0.007054	0.01355	0.01955	0.02505	0.03015
c	0.007165	0.01361	0.01982	0.02580	0.03129
d	0.007165	0.01385	0.01982	0.02556	0.03057
e	0.006854	0.01285	0.01985	0.02585	0.03085
f	0.007149	0.01376	0.01988	0.02552	0.03073
g	0.007513	0.01439	0.02064	0.02634	0.03154

Reference[a]	600 bars	700 bars	800 bars	900 bars	1000 bars
a	0.03573	0.04025	0.04443	0.04829	0.05190
b	0.03485	0.03915	0.04325	0.04685	0.05025
c	—	—	—	—	—
d	0.03535	0.03965	0.04371	0.04753	0.05087
e	—	—	—	—	—
f	0.03553	0.03997	0.04406	0.04785	0.05137
g	0.03626	0.04052	0.04434	0.04774	0.05076

[a] (a) Reported by Chen [59], (b) reported by Kell and Whalley [57], (c) using Eq. (3) of Sirota and Shrago [58], (d) using Eq. 1 of Sirota and Shrago [58], (e) VTI experimental values as reported by Sirota and Shrago [58], (f) reported by Fofonoff and Millard [60], (g) calculated in this study.

TABLE IX

COMPARISON OF THE CHANGE IN SPECIFIC HEAT, $\Delta C_p = C_p(t, s, 1\ \text{atm abs}) - C_p(t, s, p)$ OF SEAWATER ($s = 20‰$) AT VARIOUS TEMPERATURES AND PRESSURES

Pressure (bars)	Reference[a]	Change in specific heat (Btu/lbm °F) at temperature given		
		0°C	10°C	20°C
200	a	0.01840	0.01386	0.01159
	b	0.01769	0.01387	0.01176
	c	0.01800	0.01300	0.01182
400	a	0.03365	0.02545	0.02151
	b	0.03207	0.02548	0.02185
	c	0.03097	0.02380	0.02170
600	a	0.04632	0.03523	0.03002
	b	0.04364	0.03514	0.03046
	c	0.04127	0.03271	0.02987
800	a	0.05693	0.04343	0.03731
	b	0.05289	0.04315	0.03782
	c	0.04957	0.04003	0.03659
1000	a	0.06585	0.05036	0.04359
	b	0.06030	0.04982	0.04413
	c	0.05630	0.04602	0.04209

[a] (a) Reported by Chen [59], (b) reported by Fofonoff and Millard [60], (c) calculated in this study.

TABLE X

COMPARISON OF THE CHANGE IN SPECIFIC HEAT,
$\Delta C_p = C_p(t, s, 1 \text{ atm abs}) - C_p(t, s, p)$ OF SEAWATER
$(s = 35‰)$ AT VARIOUS TEMPERATURES AND
PRESSURES

Pressure (bars)	Reference[a]	Change in specific heat (Btu/lbm °F) at temperature given		
		0°C	10°C	20°C
200	a	0.01618	0.01209	0.01037
	b	0.01515	0.01210	0.01056
	c	0.01370	0.01191	0.01121
400	a	0.02964	0.02225	0.01924
	b	0.02752	0.02224	0.01960
	c	0.02436	0.02148	0.02020
600	a	0.04089	0.03078	0.02689
	b	0.03751	0.03068	0.02732
	c	0.03297	0.02931	0.02756
800	a	0.05038	0.03798	0.03346
	b	0.04551	0.03766	0.03390
	c	0.03997	0.03574	0.03361
1000	a	0.05839	0.04407	0.03915
	b	0.05193	0.04345	0.03954
	c	0.04568	0.04103	0.03859

[a] (a) Reported by Chen [59], (b) reported by Fofonoff
and Millard [60], (c) calculated in this study.

investigators that report changes in the specific heat of water include
Fofonoff and Froese [56], Fofonoff [34], Kell and Whalley [57], Sirota and
Shrago [58], Chen [59], and Fofonoff and Millard [60]. However, Caldwell
and Tucker [39] point out that the empirical formula that Foronoff [34]
presents appears to be wrong.

A comparison of ΔC_p of pure water at various temperatures and pressures is made in Tables VI–VIII, while Tables IX and X show a comparison of ΔC_p of saline water at various pressures and temperatures. One sees
agreement and disagreement in these comparisons. The disagreement
among investigators is probably caused by the choice of the temperature at
which the density extremum occurs, density relation used, and also because some calculations are integrals of exact derivatives, while others are
generated from integrals of derivatives obtained by numerical differencing.
It should be noted that the calculated results presented in this study are
integrals of exact derivatives and hence are expected to be more accurate.

However, C_p has never been thoroughly measured at elevated pressures so no further check can be made.

4. *Enthalpy*

Calculations were done in this study (see Figs. 7 and 8 from data of Table AIV) to determine the change in enthalpy, $\Delta h = h(t, s, p) - h(t, s, 1$ atm abs) for a pressure range to 1000 bars abs, over a salinity range of 0 to 40‰ and a temperature range of 0 to 20°C. Figure 7 shows an increase in Δh with increasing pressure for constant salinity, while Fig. 8 shows an increase in Δh with increasing pressure. Furthermore, Figs. 7 and 8 show that the values of Δh is essentially constant for increasing temperature.

Previous investigators that reported values of Δh include Kestin and Whitelaw [61] and Zaworski and Keenan [62]. Table XI shows a comparison of the change of enthalpy of pure water as calculated in this study, to that of Kestin and Whitelaw [61] and good agreement is seen throughout. However, h has never been thoroughly measured or calculated for the complete set of conditions considered here, so no further checks can be made.

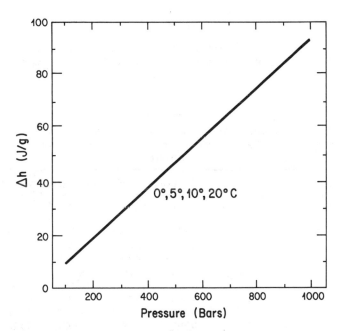

FIG. 7. The temperature and pressure effects on the change in enthalpy, $\Delta h = h(t, s, p) - h(t, s, 1$ atm abs) for seawater ($s = 35$‰), from Table AIV.

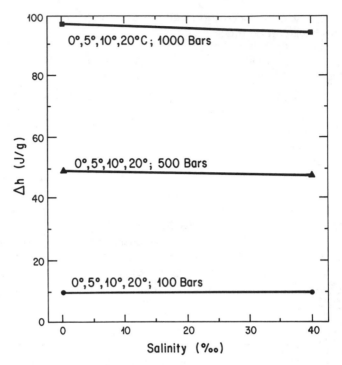

FIG. 8. The temperature and salinity effects on the change in enthalpy, $\Delta h = h(t, s, p) - h(t, s, 1\ \text{atm abs})$ for water at pressures of 100, 500, and 1000 bars, from Table AIV.

TABLE XI

COMPARISON OF THE CHANGE IN ENTHALPY $\Delta h = h(t, s, p) - h(t, s, 1\ \text{atm abs})$ OF PURE WATER ($s = 0‰$) AT VARIOUS TEMPERATURES AND PRESSURES

Temperature (°C)	References[a]	Change in enthalpy (J/g) at pressure given			
		100 bars	200 bars	500 bars	1000 bars
0	a	9.8769	19.8045	49.3075	97.6312
	b	10.114	20.087	49.081	95.258
5	a	9.8749	19.8007	49.2996	97.6184
	b	9.868	19.604	48.114	93.732
10	a	9.8700	19.7914	49.2789	97.5841
	b	9.690	19.26	47.36	92.47
20	a	9.8525	19.7578	49.2042	97.4605
	b	9.392	18.69	46.06	90.22

[a] (a) Calculated in this study, (b) reported by Kestin and Whitelaw [61].

5. *Entropy*

Calculations were done in this study (see Figs. 9 and 10 from data of Table AV) to determine the change in entropy $\Delta S = S(t, s, p) - S(t, s, 1$ atm abs), for a pressure range to 1000 bars abs, over a salinity range of 0 to 40‰ and a temperature range of 0 to 20°C. Figure 9 shows that ΔS increases with increasing temperature and pressure for constant salinity. Figure 10 shows increasing ΔS for increasing salinity and pressure.

Previous investigators reporting values of ΔS include Kestin and Whitelaw [61] and Zaworski and Keenan [62]. Table XII shows a comparison of the change in entropy of pure water as calculated in this study, to that of Kestin and Whitelaw [61] and fair agreement is seen throughout except at low temperatures. However, S has never been thoroughly measured or calculated for the complete set of conditions considered here, so no further check can be made.

C. PRANDTL NUMBER AND SCHMIDT NUMBER

Using both calculated and previously measured properties, we next present a determination of the Prandtl and Schmidt numbers for pure and

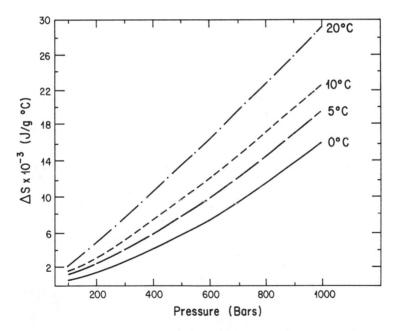

FIG. 9. The temperature and pressure effects on the change in entropy, $\Delta S = S(t, s, p) - S(t, s, 1$ atm abs) for seawater ($s = 35$‰), from Table AV.

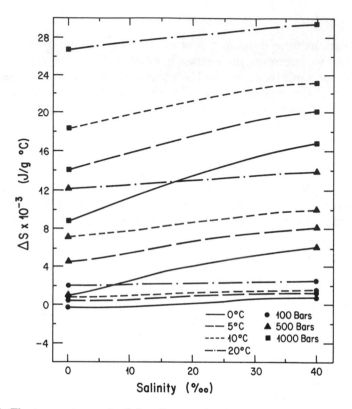

FIG. 10. The temperature and salinity effects on the change in entropy, $\Delta S = S(t, s, p) - S(t, s, 1 \text{ atm abs})$ for water at pressures of 100, 500, and 1000 bars, from Table AV.

TABLE XII

COMPARISON OF THE CHANGE IN ENTROPY, $\Delta S = S(t, s, p) - S(t, s, 1 \text{ atm abs})$
OF PURE WATER ($s = 0‰$) AT VARIOUS TEMPERATURES AND PRESSURES

Temperature (°C)	References[a]	Change in entropy (10^3J/g °C) at pressure given			
		100 bars	200 bars	500 bars	1000 bars
0	a	−0.45591	−0.60245	0.96836	8.53161
	b	−0.5057	−0.6717	1.180	9.023
5	a	0.32677	0.94213	4.2545	13.8224
	b	0.3852	1.078	4.687	14.56
10	a	0.98110	2.1941	7.0268	18.4088
	b	1.021	2.287	7.359	19.06
20	a	2.1508	4.4444	12.0353	26.7014
	b	2.049	4.280	11.90	26.88

[a] (a) Calculated in this study, (b) reported by Kestin and Whitelaw [61].

saline water i.e., $Pr(t, s, p) \equiv \mu C_p/k$ and $Sc(t, s, p) \equiv v/D$. Except for C_p, the properties used to determine the Prandtl and Schmidt numbers are those reported by previous investigators.

The Prandtl number $Pr(t, s, p)$, is the ratio of two molecular transport properties, kinematic viscosity $v \equiv \mu/\rho$, and thermal diffusivity $k/\rho C_p$. The Prandtl number of seawater is used extensively in heat-transfer calculations and until now has been available for a very limited range of temperature, salinity, and pressure.

Calculations were done in this study (see Figs. 11 and 12 from data of Table AVI) to determine Pr for a pressure range to 1000 bars abs, over a salinity range of 0 to 40‰ and a temperature range of 0 to 20°C. In the calculations of the Prandtl number at atmospheric pressure, previously reported data is available for C_p, μ, and k. At higher pressures, Matthaus [8] gives an equation for viscosity (Eq. 1) and Caldwell [14] equations for thermal conductivity, (Eqs. 2 and 4). These equations span the desired temperature, salinity, and pressure ranges of this study. The change in specific heat was calculated in this study and combined with the specific heat at atmospheric pressure (Eq. 15) to give the specific heat at elevated pressures. Figure 11 shows decreasing values of Pr for increasing temperature and pressure for constant salinity. Figure 12 shows decreasing Pr for increasing pressure at constant temperature.

Tabulated values for the Prandtl number at elevated pressures were

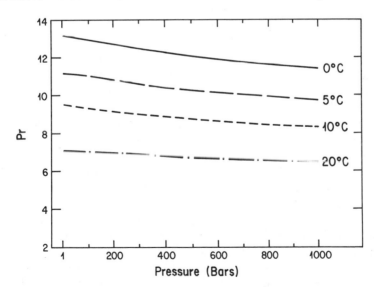

FIG. 11. The temperature and pressure effects on the Prandtl number $Pr(t, s, p)$ for seawater ($s = 35‰$), from Table AVI.

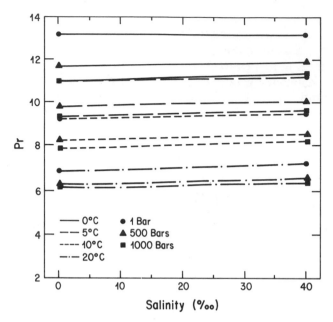

FIG. 12. The temperature and salinity effects on the Prandtl number, Pr(t, s, p), of water at pressures of 1, 500, and 1000 bars, from Table AVI.

reported by Engineering Sciences Data Item (ESDI) [63] and Kestin and Whitelaw [61] for pure water ($s = 0‰$). ESDI [63] also reported Prandtl number values for "normal seawater" ($s = 33‰$) at atmospheric pressure. A comparison of Pr at elevated pressures is made in Table XIII and very good agreement is seen throughout. Table XIV shows a comparison of Pr at 1 bar and good agreement can be seen inasmuch as comparisons are possible at close yet different salinities.

The Schmidt number, Sc(t, s, p) is also used extensively in heat-transfer calculations and until now has not been previously presented for seawater. The calculation of the Schmidt number, is limited by the sparseness of data for D, the diffusivity. The Schmidt number was calculated in this study (see Table XV) at 1 atm abs for various salinities and for temperatures of 18.5°C and 25°C. For a constant salinity of 28.5‰ we calculate Sc (see Fig. 13 from data of Table AVII) for temperatures to 30°C and pressures to 700 bars. Figure 13 shows Sc decreasing for increasing temperature and Sc decreasing slightly for increasing pressure.

In the calculation of Sc, Eq. 1 was used for μ, while for density Eq. 7 was used for temperatures to 20°C; above 20°C a fit by Chen and Millero [28] was used. In the calculation of Sc at atmospheric conditions the diffusivity

TABLE XIII

COMPARISON OF THE PRANDTL NUMBER Pr(t, s, p), OF PURE WATER ($s = 0‰$) AT VARIOUS TEMPERATURES AND PRESSURES

Pressure (bars)	Reference[a]	Prandtl number at temperature given					
		0°C	5°C	10°C	15°C	20°C	50°C
1	a	13.2	11.0	9.31	8.01	6.94	3.56
	b	13.0	—	—	—	—	3.54
	c	13.18	10.99	9.31	8.01	6.99	—
100	a	12.7	10.6	9.02	7.76	6.75	3.50
	b	12.6	—	—	—	—	3.48
	c	12.88	10.74	9.11	7.86	6.88	—
200	a	12.3	10.3	8.76	7.55	6.59	3.44
	b	12.2	—	—	—	—	3.43
	c	12.56	10.51	8.94	7.72	6.77	—
300	a	11.9	9.96	8.52	7.36	6.44	3.40
	b	12.0	—	—	—	—	3.38
	c	12.27	10.30	8.78	7.60	6.68	—
400	a	11.5	9.72	8.32	7.21	6.32	3.36
	b	11.7	—	—	—	—	3.34
	c	12.03	10.12	8.63	7.48	6.59	—
500	a	11.2	9.49	8.15	7.08	6.22	3.33
	b	11.4	—	—	—	—	3.30
	c	11.81	9.95	8.50	7.38	6.52	—

[a] (a) As reported by Engineering Sciences Data Item No. 68011 [63], (b) as reported by Kestin and Whitelaw [61], and (c) Pr calculated in this study using equations of Matthaus [8], for μ, Caldwell [14], for k, and C_p as calculated in this study.

TABLE XIV

COMPARISON OF THE PRANDTL NUMBER Pr(t, s, p) OF STANDARD SEAWATER AT A PRESSURE OF 1 bar FOR VARIOUS TEMPERATURES AND SALINITIES

Temperature (°C)	Reference[a]	Prandtl number at salinity given		
		30‰	33‰	35‰
0	a	—	13.4	—
	b	13.22	—	13.23
5	a	—	11.4	—
	b	11.15	—	11.18
10	a	—	9.7	—
	b	9.52	—	9.56
15	a	—	8.3	—
	b	8.24	—	8.28
20	a	—	7.2	—
	b	7.21	—	7.25

[a] (a) Reported by Engineering Science Data Item No. 68011 [63], (b) Pr calculated in this study using equations of Matthaus [8] for μ, Caldwell [14] for k, and C_p as calculated in this study.

TABLE XV

Schmidt Number Sc(t, s, p) of Seawater at Atmospheric Pressure for Various Salinities and Temperatures[a]

Salinity ‰	Schmidt number at temperature given	
	18.5°C	25°C
5	850.31(a); 840.10(d)	
5.5		601.87(e)
6.0	854.92(b); 844.62(c)	610.40(e)
10.0	866.59(a); 863.05(d)	604.36(f)
11.5	871.90(b)	
12.0		618.28(g)
12.5	869.48(c)	619.96(e)
15.0	875.96(a); 875.96(d)	603.61(f); 622.03(h)
18.0		624.50(e)
20.0	881.71(a); 889.03(d)	624.03(e); 601.67(f)
23.0	896.21(c)	
25.0	885.97(a); 894.79(d)	601.67(f)
26.0		631.04(i)
27.5	886.62(b)	
29.5		633.87(e); 631.73(g)
30.0	889.44(a); 898.27(c); 898.27(d)	601.64(f)
35.0	891.38(a); 902.43(d)	597.76(f); 653.80(j); 650.19(j); 633.56(j)
40.0	893.26(a); 905.03(d)	640.09(e); 595.80(f); 642.26(g)

[a] Calculated using the equations of Matthaus [8] for μ; Gebhart and Mollendorf [1] for ρ; Chen and Millero [28] for ρ; and using the diffusivity D, of the following various investigators taken from Richardson et al. [2]: (a) Gordon [18] (semitheoretical), (b) Gordon [18] (data points plotted on curve), (c) Clack [15] (values taken from a curve), (d) Clack [15] (experimental data) (e) Vitagliano and Lyons [20] (experimental data), (f) Onsager and Fuoss [17] (theoretical), (g) Stokes [19] (experimental data), (h) O'Donnel and Gosting [21] (experimental data), (i) Dunlop and Gosting [22] (experimental data), (j) Richardson et al. [2] (theoretical integral coefficients).

data of previous investigators was used. In calculating Sc at elevated pressures, Eq. 6 developed by Caldwell and Eide [26] was used for D. No comparisons of Sc could be made since there appears to be no previously reported values. Note that all calculations of Sc in this study have used NaCl data for D since there appears to be no diffusivity data for seawater.

As can be seen, there has previously been a great lack of information regarding the Prandtl and Schmidt Numbers of seawater. One of the goals of the present study was to alleviate this deficiency.

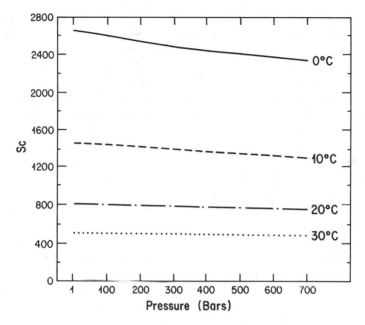

Fig. 13. The temperature and pressure effects on the Schmidt number, Sc(t, s, p), for seawater ($s = 28.5‰$), from Table AVII.

III. Summary and Conclusions

The thermodynamic and transport properties of seawater were not heretofore readily found in one source. The present work can perhaps be considered to be a first step toward a rather complete collection of the properties of seawater. Where it is possible to compare new calculated results to previous investigations we find both agreement and disagreement.

Some of the results here are new, whereas other results are compilations of previously reported measurements. The properties calculated using a density relation presented here are coefficient of thermal expansion, coefficient of saline expansion, change in specific heat, change in enthalpy, and change in entropy. The results for the change in specific heat along with other previously reported data of other properties were used to calculate the Prandtl number. The Schmidt number was compiled solely from previously reported measurements.

Except at low temperatures and pressures, good agreement can be seen in the comparison of thermal expansion results. As discussed earlier, this disagreement is probably caused by the choice of $t_m(s, p)$ of the various investigators. No comparison can be made with the new results presented here for the saline expansion coefficients since there appears to be no such previously reported information. When comparing ΔC_p with various investigators there is some disagreement seen. Again this is probably caused by the choice of $t_m(s, p)$, as well as calculation techniques of the various investigators. When comparing the change in enthalpy, good agreement can be seen throughout, and fair agreement can be seen in a comparison of the change in entropy. However, no complete comparison can be made for β^*, ΔC_p, Δh, or ΔS because there appears to be essentially no previously reported data that spans the conditions considered here.

Similarly, a comparison of the Prandtl number as reported here to that of previously reported data can only be made at a limited number of points. For these points good to very good agreement can be seen. As for the Schmidt number, no previously reported values are available for comparison. Consequently, most of the values presented here for the Prandtl and Schmidt numbers are the only known information.

For thermal conductivity, viscosity, specific heat (at 1 atm) and diffusivity, the present work serves mainly as a collection of previously reported measurements.

ACKNOWLEDGMENT

The authors gratefully acknowledge the expert typing efforts of Ms. Marty Bratton. J.C.M. and B.G. acknowledge support by the National Science Foundation under research grants MEA-8406405 and MEA-8418214, respectively.

Appendix A: Data Tables for Figures 1–13

TABLE AI

THERMAL EXPANSION COEFFICIENT $\beta(t, s, p)$ OF WATER AT VARIOUS TEMPERATURES AND SALINITIES[a]

Temperature (°C)	Pressure (bars)	Thermal expansion coefficient (10^{-6}/°C) at salinity given								
		0‰	5‰	10‰	15‰	20‰	25‰	30‰	35‰	40‰
0	1	−61.31	−47.83	−33.25	−16.96	5.38	23.43	37.86	50.22	60.75
	100	−31.19	−17.24	4.14	24.01	40.80	55.92	69.71	82.17	92.97
	500	96.20	109.73	122.59	134.82	146.44	157.43	167.66	176.86	184.54
	1000	210.97	219.44	227.44	234.97	242.02	248.53	254.41	259.54	263.77
5	1	17.15	34.40	50.13	64.30	77.13	88.88	99.74	109.73	118.49
	100	48.11	62.57	76.29	89.23	101.46	113.10	124.15	134.39	143.16
	500	147.45	157.80	167.82	177.56	186.99	196.06	204.58	212.22	218.37
	1000	243.01	249.73	256.23	262.45	268.33	273.79	278.71	282.95	286.32
10	1	87.18	101.20	113.57	124.75	135.17	145.17	154.94	164.35	172.82
	100	111.00	123.08	134.12	144.49	154.49	164.30	173.93	183.04	190.74
	500	193.20	201.18	209.06	216.89	224.65	232.25	239.45	245.85	250.78
	1000	272.17	277.65	283.04	288.27	293.26	297.89	302.04	305.53	308.17
15	1	150.35	161.47	171.06	179.89	188.57	197.49	206.82	216.31	225.16
	100	168.83	178.56	187.23	195.48	203.74	212.25	220.98	229.48	236.62
	500	235.62	241.58	247.66	253.90	260.28	266.66	272.78	278.18	282.08
	1000	299.12	303.77	308.32	312.79	317.07	321.04	324.55	327.55	329.44
20	1	210.59	218.59	225.40	232.04	239.22	247.36	256.57	266.53	276.18
	100	223.97	231.33	237.57	243.84	250.58	258.05	266.17	274.37	281.28
	500	275.71	279.85	284.33	289.18	294.35	299.70	304.91	309.47	312.50
	1000	324.61	328.47	332.37	336.24	339.96	343.39	346.38	348.73	350.22

[a] At each temperature and salinity entry various pressures are reported.

SALINE EXPANSION COEFFICIENT $\beta^*(t, s, p)$ OF WATER AT VARIOUS TEMPERATURES AND SALINITIES[a]

Temperature (°C)	Pressure (bars)	Saline expansion coefficient ($10^{-5}/‰$) at salinity given								
		0‰	5‰	10‰	15‰	20‰	25‰	30‰	35‰	40‰
0	1	78.518	78.735	78.948	79.179	79.300	78.800	78.397	78.067	77.819
	100	78.479	78.747	79.008	78.535	78.106	77.707	77.346	77.047	76.869
	500	75.091	74.726	74.354	73.980	73.617	73.287	73.025	72.896	73.007
	1000	70.454	70.031	69.631	69.266	68.952	68.709	68.562	68.544	68.696
5	1	79.544	79.036	78.580	78.178	77.824	77.511	77.244	77.044	76.964
	100	78.112	77.760	77.414	77.077	76.750	76.440	76.169	75.990	76.012
	500	73.882	73.567	73.241	72.911	72.593	72.317	72.132	72.121	72.419
	1000	69.683	69.293	68.928	68.601	68.331	68.140	68.057	68.118	68.368
10	1	77.919	77.567	77.261	76.979	76.713	76.461	76.247	76.123	76.196
	100	76.740	76.484	76.222	75.949	75.666	75.391	75.167	75.081	75.311
	500	72.958	72.666	72.358	72.045	71.748	71.506	71.382	71.482	71.971
	1000	69.069	68.693	68.346	68.044	67.807	67.660	67.636	67.773	68.118
15	1	76.559	76.400	76.223	76.012	75.770	75.515	75.294	75.203	75.418
	100	75.576	75.455	75.272	75.034	74.756	74.474	74.260	74.250	74.699
	500	72.261	71.972	71.659	71.340	71.044	70.820	70.746	70.952	71.640
	1000	68.569	68.194	67.856	67.570	67.359	67.252	67.283	67.494	67.936
20	1	75.496	75.528	75.439	75.239	74.953	74.626	74.338	74.234	74.575
	100	74.643	74.669	74.545	74.304	73.985	73.651	73.410	73.454	74.134
	500	71.765	71.458	71.120	70.774	70.460	70.238	70.205	70.514	71.406
	1000	68.154	67.773	67.436	67.161	66.973	66.903	66.987	67.272	67.813

[a] At each temperature and salinity entry various pressures are reported.

TABLE AIII

The Change in Specific Heat $\Delta C_p = C_p(t, s, 1\ \text{atm abs}) - C_p(t, s, p)$ of Water for Various Temperatures and Salinities[a]

Temperature (°C)	Pressure (bars)	Change in specific heat (Btu/lbm °F) at salinity given								
		0‰	5‰	10‰	15‰	20‰	25‰	30‰	35‰	40‰
0	100	0.008989	0.009532	0.011635	0.012323	0.010043	0.008607	0.007843	0.007314	0.006894
	500	0.046026	0.045063	0.043447	0.040961	0.036398	0.033104	0.030780	0.028888	0.027193
	1000	0.072789	0.069665	0.066228	0.062192	0.056297	0.051844	0.048494	0.045678	0.043121
5	100	0.009454	0.008924	0.008382	0.007882	0.007454	0.007108	0.006838	0.006627	0.006443
	500	0.038565	0.036539	0.034464	0.032512	0.030768	0.029261	0.027969	0.026830	0.025730
	1000	0.062257	0.058674	0.055244	0.052107	0.049323	0.046892	0.044770	0.042868	0.041054
10	100	0.008206	0.007822	0.007426	0.007069	0.006773	0.006548	0.006390	0.006284	0.006199
	500	0.034693	0.033013	0.031344	0.029809	0.028470	0.027341	0.026400	0.025588	0.024795
	1000	0.056418	0.053492	0.050724	0.048220	0.046021	0.044124	0.042486	0.041027	0.039613
15	100	0.007765	0.007357	0.006974	0.006648	0.006396	0.006221	0.006116	0.006066	0.006041
	500	0.032782	0.031124	0.029560	0.028178	0.027017	0.026078	0.025332	0.024717	0.024125
	1000	0.053092	0.050379	0.047881	0.045673	0.043777	0.042180	0.040835	0.039661	0.038526
20	100	0.007513	0.007079	0.006691	0.006377	0.006146	0.005998	0.005925	0.005912	0.005930
	500	0.031540	0.029864	0.028340	0.027037	0.025978	0.025156	0.024539	0.024061	0.023615
	1000	0.050761	0.048160	0.045818	0.043791	0.042090	0.040694	0.039553	0.038585	0.037659

[a] At each temperature and salinity entry various pressures are reported.

TABLE AIV

The Change in Enthalpy $\Delta h = h(t, s, p) - h(t, s, 1\ \text{atm abs})$ (J/g) of Water for Various Temperatures and Salinities[a]

Temperature (°C)	Pressure (bars)	Change in enthalpy (J/g) at salinity given								
		0‰	5‰	10‰	15‰	20‰	25‰	30‰	35‰	40‰
0	100	9.8769	9.8373	9.7981	9.7592	9.7208	9.6827	9.6450	9.6076	9.5705
	500	49.3075	49.1162	48.9265	48.7384	48.5521	48.3675	48.1844	48.0027	48.8220
	1000	97.6312	97.2652	96.9023	96.5425	96.1859	95.8323	95.4813	95.1327	94.7858
5	100	9.8749	9.8354	9.7963	9.7576	9.7193	9.6813	9.6436	9.6063	9.5692
	500	49.2996	49.1086	48.9194	48.7318	48.5459	48.3616	48.1787	47.9973	47.8169
	1000	97.6184	97.2530	96.8909	96.5318	96.1758	95.8226	95.4721	95.1239	94.7773
10	100	9.8700	9.8307	9.7919	9.7534	9.7152	9.6774	9.6399	9.6026	9.5655
	500	49.2789	49.0889	48.9006	48.7139	48.5288	48.3451	48.1629	47.9819	47.8020
	1000	97.5841	97.2205	96.8600	96.5023	96.1476	95.7955	95.4460	95.0986	94.7529
15	100	9.8625	9.8236	9.7851	9.7469	9.7090	9.6713	9.6339	9.5966	9.5597
	500	49.2469	49.0584	48.8715	48.6861	48.5020	48.3192	48.1377	47.9573	47.7779
	1000	97.5311	97.1699	96.8116	96.4561	96.1031	95.7525	95.4043	95.0580	94.7133
20	100	9.8525	9.8141	9.7761	9.7383	9.7007	9.6632	9.6258	9.5886	9.5516
	500	49.2042	49.0177	48.8326	48.6488	48.4661	48.2843	48.1036	47.9238	47.7450
	1000	97.4605	97.1025	96.7471	96.3941	96.0433	95.6945	95.3476	95.0026	94.6592

[a] At each temperature and salinity entry various pressures are reported.

356

TABLE AV

The Change in Entropy $\Delta S = S(t, s, p) - S(t, s, 1 \text{ atm abs})$ of Water for Various Temperatures and Salinities[a]

Temperature (°C)	Pressure (bars)	Change in entropy (10^{-3} J/g °C) at salinity given								
		0‰	5‰	10‰	15‰	20‰	25‰	30‰	35‰	40‰
0	100	-0.45591	-0.32328	-0.16367	0.038108	0.23341	0.38865	0.52291	0.64080	0.74152
	500	0.96836	1.71553	2.48655	3.25159	3.97199	4.61664	5.20237	5.72455	6.16088
	1000	8.53161	9.78401	11.0229	12.2232	13.3488	14.3685	15.2963	16.1210	16.8087
5	100	0.32677	0.47956	0.62096	0.75043	0.86938	0.97964	1.08224	1.17617	1.25661
	500	4.2545	4.87101	5.45509	6.00728	6.53054	7.02632	7.49054	7.90832	8.24617
	1000	13.8224	14.8115	15.7500	16.6399	17.4838	18.2806	19.0210	19.6811	20.2133
10	100	0.98110	1.10485	1.21491	1.31527	1.40961	1.50056	1.58895	1.67259	1.74426
	500	7.0268	7.51721	7.97195	8.40321	8.81973	9.22468	9.61243	9.96339	10.2354
	1000	18.4088	19.1724	19.8949	20.5881	21.2586	21.9055	22.5160	23.0598	23.4784
15	100	1.5791	1.67524	1.75813	1.83478	1.91034	1.98788	2.06789	2.14699	2.21569
	500	9.5904	9.96438	10.3063	10.6363	10.9667	11.3013	11.6325	11.9359	12.1607
	1000	22.6599	23.2264	23.7651	24.2946	24.8239	25.3513	25.8606	26.3151	26.6456
20	100	2.1508	2.21770	2.27377	2.32832	2.38744	2.45438	2.52911	2.60711	2.67655
	500	12.0353	12.2935	12.5283	12.7655	13.0202	13.2959	13.5825	13.8507	14.0411
	1000	26.7014	27.0832	27.4540	27.8376	28.2441	28.6706	29.0972	29.4805	29.7409

[a] At each temperature and salinity entry various pressures are reported.

TABLE AVI

PRANDTL NUMBER $Pr(t, s, p)$ OF WATER AT VARIOUS TEMPERATURES AND SALINITIES[a]

Temperature (°C)	Pressure (bars)	Prandtl number at salinity given								
		0‰	5‰	10‰	15‰	20‰	25‰	30‰	35‰	40‰
0	1	13.183	13.187	13.193	13.199	13.206	13.214	13.223	13.234	13.245
	100	12.878	12.874	12.851	12.848	12.884	12.911	12.930	12.948	12.965
	500	11.806	11.819	11.842	11.876	11.939	11.987	12.024	12.057	12.089
	1000	11.038	11.074	11.115	11.165	11.239	11.298	11.345	11.387	11.427
5	1	10.994	11.020	11.045	11.072	11.098	11.125	11.153	11.182	11.211
	100	10.740	10.771	10.802	10.833	10.863	10.894	10.924	10.954	10.985
	500	9.949	9.992	10.036	10.080	10.122	10.163	10.202	10.240	10.279
	1000	9.357	9.412	9.467	9.519	9.569	9.616	9.660	9.704	9.747
10	1	9.308	9.343	9.379	9.415	9.451	9.487	9.523	9.560	9.597
	100	9.114	9.152	9.190	9.229	9.267	9.304	9.342	9.379	9.146
	500	8.497	8.544	8.591	8.638	8.682	8.726	8.768	8.809	8.851
	1000	8.038	8.091	8.144	8.196	8.245	8.292	8.337	8.381	8.426
15	1	8.009	8.047	8.085	8.123	8.162	8.200	8.238	8.276	8.315
	100	7.858	7.899	7.939	7.979	8.019	8.058	8.096	8.134	8.172
	500	7.380	7.427	7.474	7.519	7.563	7.606	7.674	7.687	7.727
	1000	7.023	7.075	7.125	7.174	7.221	7.265	7.308	7.350	7.391
20	1	6.992	7.028	7.065	7.101	7.137	7.173	7.209	7.246	7.282
	100	6.878	6.916	6.955	6.933	7.030	7.066	7.102	7.138	7.173
	500	6.515	6.559	6.602	6.644	6.685	6.724	6.761	6.797	6.834
	1000	6.242	6.290	6.335	6.379	6.421	6.461	6.499	6.536	6.573

[a] Calculated using equations of Matthaus [8] for μ, Caldwell [14] for k, and C_p as calculated in this study. At each temperature and salinity entry various pressures are reported.

TABLE A VII

SCHMIDT NUMBER $Sc(t, s, p)$ OF SEAWATER ($s = 28.5‰$) AT VARIOUS TEMPERATURES AND PRESSURES[a]

Temperature (°C)	Schmidt number at pressure given							
	1 bars	100 bars	200 bars	300 bars	400 bars	500 bars	600 bars	700 bars
0	2654.9	2596.9	2542.1	2491.0	2443.3	2399.0	2357.8	2319.8
5	1926.1	1885.9	1847.9	1812.5	1779.5	1748.8	1720.4	1694.1
10	1429.1	1401.2	1374.9	1350.3	1327.3	1306.0	1286.2	1267.9
15	1084.1	1065.0	1046.9	1029.9	1014.1	999.29	985.52	972.75
20	838.99	826.20	814.01	802.54	791.76	781.64	772.16	763.29
25	658.70	650.59	642.81	635.40	628.36	621.67	615.33	609.32
30	519.92	515.40	510.97	506.67	502.40	498.40	494.44	490.58

[a] Calculated using the equations of Matthaus [8] for μ; Gebhart and Mollendorf [1] for ρ; Chen and Millero [28] for ρ; and Caldwell and Eide [26], for D.

Appendix B: Calculation Techniques

The thermodynamic definition of thermal expansion coefficient β, is

$$\beta = -\frac{1}{\rho}\frac{\partial\rho}{\partial t}\bigg|_{p,s} \tag{B.1}$$

where units of β are $1/°C$. Because of the existence of $t_m(s, p)$, $d\rho/dt$ is defined as

$$\frac{\partial\rho}{\partial t} = -(\rho_m\alpha q|t - t_m|^{q-1}) \qquad \text{for} \quad t > t_m(s, p) \tag{B.2}$$

$$\frac{\partial\rho}{\partial t} = 0 \qquad \text{for} \quad t = t_m(s, p) \tag{B.3}$$

and

$$\frac{\partial\rho}{\partial t} = \rho_m\alpha q|t - t_m|^{q-1} \qquad \text{for} \quad t < t_m(s, p) \tag{B.4}$$

The saline expansion coefficient β^* is the change in volume of seawater held at constant temperature and pressure subjected to a salinity change. The definition of the coefficient of saline expansion is

$$\beta^* = \frac{1}{\rho}\frac{\partial\rho}{\partial s}\bigg|_{p,t} \tag{B.5}$$

where the units of β^* are $1/‰$.

The change in specific heat with pressure at constant temperature and salinity levels may be calculated from the thermodynamic relation

$$\left(\frac{\partial C_p}{\partial p}\right)_{t,s} = -T\left(\frac{\partial^2 v}{\partial t^2}\right)_{p,s} \tag{B.6}$$

where v is the specific volume, from which ΔC_p can be calculated.

Combining ΔC_p with $C_p(t, s, 1 \text{ atm abs})$ (Eq. 15), $C_p(t, s, p)$ is calculated as

$$C_p(t, s, p) = C_p(t, s, p_0) + \int_{p_0}^{p}\left[T\left(\frac{\partial^2 v}{\partial t^2}\right)_{p,s} dp\right]_{t,s} \tag{B.7}$$

where the units C_p are Btu/lbm °F, and where

$$\frac{\partial v}{\partial t} = -\rho^{-2}\frac{\partial \rho}{\partial t} \tag{B.8}$$

$$\frac{\partial^2 v}{\partial t^2} = \rho^{-2}\left[\frac{2}{\rho}\frac{dp}{dt} - \frac{\partial^2 \rho}{\partial t^2}\right] \tag{B.9}$$

$$\left(\frac{\partial^2 \rho}{\partial t^2}\right) = -\rho_m \alpha q(q-1)|t - t_m|^{(q-2)} \tag{B.10}$$

and $\partial \rho/\partial t$ is defined in Eqs. (B.2)–(B.4).

For the calculation of the change of enthalpy, the relation

$$\left(\frac{\partial h}{\partial p}\right)_{t,s} = T\left(\frac{\partial s}{\partial p}\right)_{t,s} + v \tag{B.11}$$

was used.

It then follows that

$$\left(\frac{\partial h}{\partial p}\right)_{t,s} = v - T\left(\frac{\partial v}{\partial t}\right)_{p,s} \tag{B.12}$$

and

$$\Delta h = v - \left[T\int_{p_0}^{p}\left(\frac{\partial v}{\partial t}\right)_{p,s} dp\right]_{t,s} \tag{B.13}$$

where the units of Δh are (J/g).

For calculations of change of entropy, the following relation was used,

$$\left(\frac{\partial S}{\partial p}\right)_{t,s} = -\left(\frac{\partial v}{\partial t}\right)_{p,s} \tag{B.14}$$

It then follows that

$$(\Delta S)_{t,s} = \left[-\int_{p_0}^{p} \left(\frac{\partial S}{\partial p} \right)_{t,s} dp \right]_{t,s}$$

(B.15

and

$$(\Delta S)_{t,s} = \left[-\int_{p_0}^{p} \left(\frac{\partial v}{\partial t} \right)_{p,s} dp \right]_{t,s}$$

(B.16)

where the units of ΔS are (J/g °C).

NOMENCLATURE

f_{ij}	effect of pressure on density	v	specific volume, m³/kg
g_{ij}, h_{ij}	effect of salinity on density	A_0, B_0, C_0, T_0	coefficients in Eq. (6)
h	enthalpy, J/g	A_1, A_2, A_3, A_4	coefficients in Eq. (4)
k	thermal conductivity, cal/ cm °C sec	C_1, C_2, C_3, C_4	coefficients in Eq. (3)
p	pressure, (bars absolute)	C_p	specific heat, Btu/lbm °F
p'	pressure, kilobars	D	diffusivity, cm²/sec
p_0	reference pressure, bars absolute	Le	Lewis number
q	equation of state correlation parameter	P	applied pressure, bars
		Pr	Prandtl number
s	salinity, ‰	S	entropy, J/g °C
t	temperature, °C	Sc	Schmidt number
t_m	temperature at which density extremum occurs, °C	T	absolute temperature, K

Greek Symbols

α	equation of state correlation parameter	Δ	change in
β	coefficient of thermal expansion, 1/°C	ρ	density, kg/m³
		ρ_m	maximum density, kg/m³
β^*	coefficient of saline expansion, 1/‰	μ	absolute viscosity, cP
		v	kinematic viscosity, $(v \equiv \mu/\rho)$

Special Symbols

‰	parts per thousand

REFERENCES

1. B. Gebhart and J. C. Mollendorf, *Deep-Sea Res.* **24,** 831 (1977).
2. J. L. Richardson, R. J. Getz, and G. Segovia, *U.S. Off. Saline Water, Res. Dev. Prog. Rep.* **211** (1966).

3. N. E. Dorsey "Properties of Ordinary Water Substance in All the Phases." van Nostrand-Reinhold, Princeton, New Jersey, 1940.
4. J. D. Isdale, C. M. Spence, and J. S. Tudhope, *Desalination* **10**, 319 (1972).
5. S. F. Chen, R. C. Chan, S. M. Read, and L. A. Bromley, *Desalination* **13**, 37 (1973).
6. E. M. Stanley and R. C. Batten, *J. Geophys. Res.* **74**(13), 3415 (1969).
7. R. A. Horne and D. S. Johnson, *J. Geophys. Res.* **71**(22), 5275 (1966).
8. W. Matthaus, *Monatsber. Dtsch. Akad. Wiss. Berlin* **12**(11–12), 850 (1970).
9. O. Krummel, "Handbuch der Ozeanographie," Vols. I and II. J. Engelhorn, Stuttgart, 1907, 1911.
10. L. Riedel, *Chem.-Ing.-Tech.* **23**, 54 (1951).
11. D. T. Jamieson and J. S. Tudhope, *Desalination* **8**, 393 (1970).
12. V. J. Castelli and E. M. Stanley, "The Thermal Conductivity of Pure Water and Standard Sea Water . . . ," NSRDC-3566-II (unpublished). Nav. Ship Res. Dev. Cent., 1972.
13. V. J. Castelli, E. M. Stanley, and E. C. Fischer, *Deep-Sea Res.* **21**, 311 (1974).
14. D. Caldwell, *Deep-Sea Res.* **21**, 131 (1974).
15. Clack, reported in Richardson *et al.* [2], with no reference given (1924).
16. Fuoss, reported in Richardson *et al.* [2], with no reference given (1932).
17. Onsager and Fuoss, reported in Richardson *et al.* [2], with no reference given (1932).
18. Gordon, reported in Richardson *et al.* [2], with no reference given (1937).
19. R. H. Stokes, *J. Am. Chem. Soc.* **72**, 2243 (1950).
20. Vitagliano and Lyons, reported in Richardson *et al.* [2], with no reference given (1956).
21. O'Donnel and Gosting, reported in Richardson *et al.* [2], with no reference given (1956).
22. Dunlop and Gosting, reported in Richardson *et al.* [2], with no reference given (1955).
23. J. L. Richardson, P. Bergsteinsson, D. L. Peters, and R. W. Sprague, "Sea Water Mass Diffusion Coefficient Studies," Philco Aeronautic Div., Publ. No. U-3021, W. O. 2503, ONR Contract No. NONR-4061(00) (unclass.). Off. Nav. Res., Washington, D.C., 1964.
24. D. Caldwell, *Deep-Sea Res.* **20**, 1029 (1973).
25. D. Caldwell, *Deep-Sea Res.* **21**, 369 (1974).
26. D. Caldwell and S. A. Eide, *Deep-Sea Res.* **28A**, 1605 (1981).
27. R. A. Horne, "Marine Chemistry." Wiley (Interscience), New York, 1969.
28. C. T. Chen and F. J. Millero, *Deep-Sea Res.* **23**, 595 (1976).
29. F. J. Millero, C. T. Chen, A. Bradshaw, and K. Schliecher, *Deep-Sea Res.* **27A**, 255 (1980).
30. A. Poisson, C. Burnet, and J. C. Brun-Cottan, *Deep-Sea Res.* **27**, 1013 (1980).
31. F. J. Millero and A. Poisson, *Deep-Sea Res.* **28A**, 625 (1981).
32. A. Bradshaw and K. Schleicher, *Deep-Sea Res.* **17**, 691 (1970).
33. W. Wilson and D. Bradley, "Specific Volume, Thermal Expansion and Isothermal Compressibility of Sea Water," Noltr 66–103. U.S. Nav. Ordnance Lab., White Oak, Maryland, 1966.
34. N. P. Fofonoff, Physical properties of sea water. *In* "The Sea" 1 (M. N. Hill, gen. ed.), Vol. 1. Wiley (Interscience), New York, 1962.
35. M. Knudsen, "Hydrographical Tables According to the Measurings of Carl Forch, P. Jacobsen, Martin Knudsen, and S. P. L. Sorensen." G.E.C. Gad. Copenhagen; Williams Norgate, London, 1901.
36. C. Eckart, *Am. J. Sci.,* **256**, 225 (1958).
37. W. Sturges, *Deep-Sea Res.* **17**, 637 (1970).
38. W. Wilson and D. Bradley, *Deep-Sea Res.* **15**(3), 355 (1968).
39. D. Caldwell and B. E. Tucker, *Deep-Sea Res.,* **17**, 707 (1970).
40. H. Bryden, *Deep-Sea Res.* **20**, 401 (1972).

41. E. LaFond, "Processing Oceanographic Data," U.S. Navy Hydrog. Office, Publ. 614, 1951.
42. J. Crease, *Deep-Sea Res.* **9,** 209 (1962).
43. F. J. Millero and F. K. Lepple, *Mar. Chem.* **1,** 89 (1973).
44. D. P. Wang and F. J. Millero, *J. Geophys. Res.* **78,** 7122 (1973).
45. R. A. Fine, T. P. Wang, and F. J. Millero, *J. Mar. Res.* **32,** 5529 (1974).
46. R. T. Emmet and F. J. Millero, *J. Geophys. Res.* **79,** 3463 (1974).
47. R. A. Cox, M. J. McCartney, and F. Culkin, *Deep-Sea Res.* **17,** 679 (1970).
48. R. A. Cox and N. D. Smith, *Proc. Soc. London* **252,** 51 (1959).
49. L. A. Bromley, V. A. Desaussure, J. C. Clipp, and J. S. Wright, *J. Chem. Eng. Data* **12,** 202 (1962).
50. L. A. Bromley, A. E. Diamond, E. Salami, and D. G. Wilkins, *J. Chem. Eng. Data* **15,** 246 (1970).
51. L. A. Bromley, V. A. Desaussure, J. C. Clipp, and J. S. Wright, *J. Chem. Eng. Data* **12,** 202 (1967).
52. V. F. Kovalenko, "Termicheskoe Opresnenie Morskoi Vody" (Thermal distillation of sea water) (Fundamentals of the theory of marine evaporators), Rep. AEC-tr-6893. Transport Publ. House, Moscow, 1966.
53. F. J. Millero, G. Perron, and J. E. Desnoyers, *J. Geophys. Res.* **78**(21), 4499 (1973).
54. H. U. Sverdrup, M. W. Johnson, and R. H. Fleming, "The Oceans." Prentice-Hall, Englewood Cliffs, New Jersey, 1942.
55. V. Ekman, *Ann. Hydrogr. Mar. Meteorol.* **42,** 340 (1974).
56. N. P. Fofonoff and C. Frocse, "Program for Oceanographic Computations and Data Processing on the Electronic Digital Computer ALWAC-111-E. PWS-1 Programs for Properties of Sea Water," No. 27. Pac. Oceanogr. Group, Nanaimo, B.C., Canada, 1958.
57. G. S. Kell and E. Whalley, *Philos. Trans. R. Soc. London, Ser. A* **565** (1965).
58. A. M. Sirota and Z. K. Shrago, *Therm. Eng.* **18,** 114 (1968).
59. C. T. Chen, High pressure P-V-T properties of sea water. M.S. Thesis, University of Miami, Coral Gables, Florida (1974).
60. N. P. Fofonoff and R. C. Millard, Jr., Algorithms for computation of fundamental properties of sea water. UNESCO *Tech. Pap. Mar. Sci.* **44** (1983).
61. J. Kestin and J. H. Whitelaw, *J. Eng. Power, Ser. A* **82,** No. 1, 99 (1966).
62. R. J. Zaworski and J. H. Keenan, *J. Appl. Mech.* **34,** 478 (1967).
63. Engineering Sciences Data Item (ESDI), "Prandtl Number of Water Substance," No. 68011. ESDI, London, 1968.

Subject Index

A

Acoustic methods, transition boiling studies, 265–267

Aerowindow, 41

d'Alembert paradox, 90

Altered free-volume state model, 226–227

Aspect ratio, flat-plate collectors, 12–15

Average velocity, fluid flow in a restricted channel, 99

B

Balance equation, heat transfer in incompressible liquid media, 162–163

Banks
closely spaced, heat transfer, 145–147
in crossflow, heat transfer calculations, 154–156
hydraulic drag
calculation methods, 148–150
mean drag, 150–153
proposals for calculation of, 153–154
in-line, fluid flow pattern changes, 103
staggered, 100

Biological liquids, thermal conductivity, 209–211

Blockage ratio, fluid flow in a restricted channel, 97–98

Blood, heat conductivity, 210–211

Boundary layer
flow, in solar buildings, 68–72
laminar–turbulent transition in fluid flow, 91–92

separation in fluid flow past a single tube, 90

stability, in fluid flow past a single tube, 92

Bounds, thermal conductivity of suspensions, 197–200

Bridgman model, 216–217

Brownian motion, thermal conductivity of suspensions and, 206–207

Bulk density driven flow, in solar buildings, 67–68

C

Cavity receivers
central receiver systems, 35–41
flow pattern, 35
parabolic dish systems, 41–46

Central receiver systems
aerowindow, 41
cavity receivers, 35–41
convective losses from, 36–41
external receivers, 30–34
schematic diagram of external and cavity receivers, 4

Chain branching, effect on thermal conductivity of polymeric liquids, 218

Channels, restricted
average velocity, 99
blockage ratio, 97–98
flow in, 97–99
mean velocity in minimum free cross section, 99
reference velocity, 99
single-tube heat transfer in, 129–131